METHODS IN MOLECULAR BIOLOGY

Series Editor
John M. Walker
School of Life and Medical Sciences
University of Hertfordshire
Hatfield, Hertfordshire, AL10 9AB, UK

For further volumes:
http://www.springer.com/series/7651

Histones

Methods and Protocols

Edited by

Benoit Guillemette

Department of Biology, Université de Sherbrooke, Sherbrooke, QC, Canada

Luc Gaudreau

Department of Biology, Université de Sherbrooke, Sherbrooke, QC, Canada

Humana Press

Editors
Benoit Guillemette
Department of Biology
Université de Sherbrooke
Sherbrooke, QC, Canada

Luc Gaudreau
Department of Biology
Université de Sherbrooke
Sherbrooke, QC, Canada

ISSN 1064-3745 ISSN 1940-6029 (electronic)
Methods in Molecular Biology
ISBN 978-1-4939-6628-8 ISBN 978-1-4939-6630-1 (eBook)
DOI 10.1007/978-1-4939-6630-1

Library of Congress Control Number: 2016958302

Printed on acid-free paper

This Humana Press imprint is published by Springer Nature
The registered company is Springer Science+Business Media LLC
The registered company address is: 233 Spring Street, New York, NY 10013, U.S.A.

Preface

Histones are the fundamental building blocks upon which eukaryotic cells rely to compact and organize their DNA in the nucleus. The essential nature of histones for the composition of chromatin has been known for many years, but we are merely beginning to understand how histones contribute in organizing the genome in the three dimensions of the nucleus and impact in regulating gene expression, DNA replication, and repair. Histone research has greatly expanded in the last 15 years, and an increasing number of laboratories have taken interest in studying them. This book aims at providing a comprehensive list of protocols for molecular biologists, biochemists, and geneticists who wish to study the many aspects of histone biology both in vivo and in vitro: from the discovery and characterization of histone modifications to the impact of histones on chromatin structure. This book also includes protocols that allow the study of protein complexes that modify chromatin either by adding or removing post-translational modifications or by exchanging histone variants within the nucleosome.

Sherbrooke, QC, Canada

Benoit Guillemette
Luc Gaudreau

Contents

Contributors

JUAN AUSIÓ • *Department of Biochemistry and Microbiology, University of Victoria, Victoria, BC, Canada*

WAJID BHAT • *St-Patrick Research Group in Basic Oncology, Laval University Cancer Research Center, CHU de Québec Research Center- Oncology axis, Hôtel-Dieu de Québec, Quebec City, QC, Canada*

TIZIANA BONALDI • *Department of Experimental Oncology, European Institute of Oncology, Milan, Italy*

MYLÈNE BRUNELLE • *Département de biologie, Faculté des sciences, Université de Sherbrooke, Sherbrooke, QC, Canada*

MANJINDER S. CHEEMA • *Department of Biochemistry and Microbiology, University of Victoria, Victoria, BC, Canada*

JACQUES CÔTÉ • *St-Patrick Research Group in Basic Oncology, Laval University Cancer Research Center, CHU de Québec Research Center- Oncology axis, Hôtel-Dieu de Québec, Quebec City, QC, Canada*

ALESSANDRO CUOMO • *Department of Experimental Oncology, European Institute of Oncology, Milan, Italy*

ARNOB DUTTA • *Department of Cell and Molecular Biology, University of Rhode Island, Kingston, RI, USA*

BENJAMIN A. GARCIA • *Epigenetics Program, Department of Biochemistry and Biophysics, Perelman School of Medicine, University of Pennsylvania, Philadelphia, PA, USA*

LUC GAUDREAU • *Department of Biology, Université de Sherbrooke, Sherbrooke, QC, Canada*

VINCENT GÉLI • *Cancer Research Center of Marseille (CRCM), U1068 Inserm, UMR7258 CNRS, Aix Marseille University (AMU), Institut Paoli-Calmettes, Marseille, France*

NICOLAS GÉVRY • *Département de biologie, Faculté des sciences, Université de Sherbrooke, Sherbrooke, QC, Canada*

BENOIT GUILLEMETTE • *Department of Biology, Université de Sherbrooke, Sherbrooke, QC, Canada*

IAN HAMMOND-MARTEL • *Centre de Recherche de l'Hôpital Maisonneuve-Rosemont, Montréal, QC, Canada*

CATHERINE A. HAZZALIN • *Nuclear Signalling Laboratory, Department of Biochemistry, University of Oxford, Oxford, UK*

MICHAEL J. HENDZEL • *Departments of Oncology and Cell Biology, Faculty of Medicine and Dentistry, University of Alberta, Edmonton, AB, Canada*

PIERRE-ÉTIENNE JACQUES • *Département de biologie, Faculté des sciences, Université de Sherbrooke, Sherbrooke, QC, Canada; Département d'informatique, Faculté des sciences, Université de Sherbrooke, Sherbrooke, QC, Canada; Centre de recherche du Centre hospitalier universitaire de Sherbrooke, Sherbrooke, QC, Canada*

CÉLIA JERONIMO • *Institut de recherches cliniques de Montréal, Montréal, QC, Canada*

FREDERIC JOURQUIN • *Cancer Research Center of Marseille (CRCM), U1068 Inserm, UMR7258 CNRS, Aix Marseille University (AMU), Institut Paoli-Calmettes, Marseille, France*
NICOLAS LACOSTE • *St-Patrick Research Group in Basic Oncology, Laval University Cancer Research Center, CHU de Québec Research Center- Oncology axis, Hôtel-Dieu de Québec, Quebec City, QC, Canada; Institut Curie/CNRS, UMR3664, Paris, France*
LIETTE LAFLAMME • *Département de biologie, Faculté des sciences, Université de Sherbrooke, Sherbrooke, QC, Canada*
LOUIS C. MAHADEVAN • *Nuclear Signalling Laboratory, Department of Biochemistry, University of Oxford, Oxford, UK*
JEAN MBOGNING • *Department of Pharmacology and Therapeutics, McGill University, Montreal, QC, Canada*
VANESSA MENIL-PHILIPPOT • *UMR CNRS 6286 UFIP, Université de Nantes, Nantes, France*
JEAN-FRANÇOIS MILLAU • *Département de biologie, Faculté des sciences, Université de Sherbrooke, Sherbrooke, QC, Canada*
FRANÇOIS ROBERT • *Institut de recherches cliniques de Montréal, Montréal, QC, Canada; Département de médecine, Faculté de médecine, Université de Montréal, Montréal, QC, Canada*
SÉBASTIEN RODRIGUE • *Département de biologie, Faculté des sciences, Université de Sherbrooke, Sherbrooke, QC, Canada*
SIMONE SIDOLI • *Epigenetics Program, Department of Biochemistry and Biophysics, Perelman School of Medicine, University of Pennsylvania, Philadelphia, PA, USA*
MONICA SOLDI • *Department of Experimental Oncology, European Institute of Oncology, Milan, Italy*
HILMAR STRICKFADEN • *Department of Oncology and Cell Biology, Faculty of Medicine and Dentistry, University of Alberta, Edmonton, AB, Canada*
JASON C. TANNY • *Department of Pharmacology and Therapeutics, McGill University, Montreal, QC, Canada*
CHRISTOPHE THIRIET • *UMR CNRS 6286 UFIP, Université de Nantes, Nantes, France*
ALAIN VERREAULT • *Institut de Recherche en Immunologie et Cancérologie (IRIC), Université de Montréal, Montréal, QC, Canada; Département de Pathologie et Biologie Cellulaire, Université de Montréal, Montréal, QC, Canada*
JERRY L. WORKMAN • *Stowers Institute for Medical Research, Kansas City, MO, USA*
HUGO WURTELE • *Centre de Recherche de l'Hôpital Maisonneuve-Rosemont, Montréal, QC, Canada; Département de Médecine, Université de Montréal, Montréal, QC, Canada*

Chapter 1

In Vitro Assembly of Nucleosomes for Binding/Remodeling Assays

Arnob Dutta and Jerry L. Workman

Abstract

In order to study the functions and activities of chromatin remodeling enzymes in vitro, it is necessary to be able to reconstitute nucleosomes on DNA templates. In this chapter, we describe procedures for purification of histones from *E. coli*, formation of octamers, and reconstitution of nucleosomes that can be further modified by chromatin modifiers. In addition, we describe methods to purify nucleosomes from human cells. Finally, we also describe assays to test binding and remodeling activities of chromatin remodelers.

Key words Nucleosome, Chromatin remodeler, Histone, Chromatin modifying enzyme

1 Introduction

DNA in eukaryotic cells is packaged by wrapping DNA around the histone octamer to form nucleosomes, which is further assembled into higher order chromatin structure. Access to the underlying DNA sequences by RNA polymerases for efficient gene expression, requires the action of ATP dependent chromatin remodeling complexes that either slide nucleosomes along DNA or evict histones [1]. Studies in vivo, either in strains with deletions of subunits of chromatin remodeling complexes or by chromatin immunoprecipitation to access genomic occupancy of these factors, allow for study of their roles in regulating gene expression. However, to understand in detail the biochemical properties of these complexes, one must carry out in vitro assays using purified chromatin remodeling factors along with nucleosome templates assembled in vitro [2–5].

In this chapter, we first describe a method for assembly of histone octamers from recombinant histones purified from *Escherichia coli*. This allows for using histones from various organisms including *Xenopus* and human. Also this affords the user the ability to assemble nucleosomes using histone variants, as well as histones with chemically modified amino acid residues that mimic various

Benoit Guillemette and Luc Gaudreau (eds.), *Histones: Methods and Protocols*, Methods in Molecular Biology, vol. 1528,
DOI 10.1007/978-1-4939-6630-1_1, © Springer Science+Business Media New York 2017

histone-post translational modifications (PTMs) [6–8]. We next describe how these histone octamers can be reconstituted into nucleosomes by salt dialysis. Many studies have used chemically modified peptides to identify proteins/complex that recognize the PTMs [9, 10]. Though these methods have been successful in identifying binding targets, peptides alone do not provide the in vivo context of the nucleosomes, where additional interactions with histones and DNA may govern binding. Since these nucleosomes are reconstituted from histones purified from bacteria they do not carry post-translational modifications. These nucleosomes can be post-translationally modified by chromatin modifiers such as histone acetyltransferases and histone methyltransferases. This allows for testing for effects of histone modifications on actions of chromatin remodelers and also identification of new factors that recognize specific histone marks. Also, in vivo, nucleosomes may not contain single histone modifications but may be marked by several of them simultaneously [11]. We have also described a method for purification of chromatin from cells (we describe purification from human cells) that can be further assembled onto DNA templates by salt dialysis. This allows for use of nucleosomes in vitro that closely mimics conditions in vivo. Finally, we described a generic method for carrying out assays to test binding of chromatin modifying complexes to nucleosomes and remodeling of nucleosomes in vitro.

2 Materials

All buffers are prepared using ultrapure distilled and deionized water (ddH_2O). DTT, β –mercaptoethanol, and protease inhibitors must be added fresh to all buffers just before using.

2.1 Expression and Purification of Recombinant Histones

1. pET-histone expression plasmids.
2. Transformation-competent *Escherichia coli* BL21(DE3) pLysS cells.
3. 2× TY-AC media: 16% (w/v) bacto-tryptone, 10% (w/v) yeast extract, and 5% (w/v) NaCl, 100 μg/l ampicillin and 25 μg/l chloramphenicol.
4. AC agar plates: 10% (w/v) bacto-tryptone, 5% (w/v) yeast extract, 8% (w/v) NaCl, 1.5% (w/v) Agar, 100 μg/l ampicillin, and 25 μg/l chloramphenicol.
5. 1 M Isopropyl-ß-D-thiogalactopyranoside (IPTG) in water.
6. Wash buffer: 50 mM Tris–HCl (pH 7.5), 100 mM NaCl, 1 mM Na-EDTA, 1 mM Benzamidine-HCl, and 5 mM β-mercaptoethanol.
7. Branson sonifier model 450.
8. TW buffer: Wash buffer with 1% (v/v) Triton X-100.

9. Unfolding buffer: 7 M guanidinium HCl, 20 mM Tris–HCl (pH 7.5), and 10 mM DTT.
10. SAU-200: 7 M urea, 20 mM sodium acetate (pH 5.2), 200 mM NaCl, 5 mM β-mercaptoethanol, and 1 mM Na-EDTA.
11. SAU-600: 7 M urea, 20 mM sodium acetate (pH 5.2), 600 mM NaCl, 5 mM β-mercaptoethanol, and 1 mM Na-EDTA.
12. Dialysis tubing 3000 MWCO.
13. Hi-Trap SP-Sepharose column.
14. Destain solution: 40% methanol, 10% acetic acid in H_2O.
15. Coomassie blue solution: 0.1% Coomassie brilliant blue R250 in destain solution.

2.2 Reconstitution of Histone Octamers from Recombinant Histones

1. Lyophilized recombinant histones H2A, H2B, H3, and H4 purified as in Subheading 3.1.
2. Unfolding buffer: 20 mM Tris–HCl (pH 7.5), 7 M guanidinium HCl, and 10 mM DTT.
3. Dialysis tubing 3500 MWCO.
4. Refolding buffer: 20 mM Tris–HCl (pH 7.5), 2 M NaCl, 1 mM EDTA (pH 8.0), and 10 mM β-mercaptoethanol.
5. Superdex 200 column (GE Healthcare).
6. Amicon ultra centrifugal filters 30,000 MWCO (Millipore).

2.3 Purification of Short Oligonucleosomes (SON) and Long Oligonucleosome (LON)

1. Phosphate buffer saline (pH 7.4): 137 mM NaCl, 2.7 mM KCl, 10 mM Na_2HPO_4, and 1.8 mM KH_2PO_4
2. Extraction Buffer 1: 10 mM HEPES (pH 7.9), 1.5 mM $MgCl_2$, 10 mM KCl, 1 mM PMSF, 1 mM DTT, and protease inhibitor cocktail (complete mini, Roche).
3. 10% NP-40.
4. Low Salt Buffer: 20 mM HEPES (pH 7.9), 25% glycerol, 1.5 mM $MgCl_2$, 0.02 M KCl, 0.2 mM EDTA, 0.2 mM PMSF, and 0.5 mM DTT.
5. High Salt Buffer: 20 mM HEPES (pH 7.9), 25% glycerol, 1.5 mM $MgCl_2$, 1.2 M KCl, 0.2 mM EDTA, 0.2 mM PMSF, and 0.5 mM DTT.
6. Nuclear Extraction (NE) Dialysis Buffer: 20 mM HEPES (pH 7.9), 20% glycerol, 1.5 mM $MgCl_2$, 100 mM KCl, 0.2 mM EDTA, 0.2 mM PMSF, and 0.5 mM DTT.
7. Medium Salt Buffer (MSB) + 0.2% NP-40: 20 mM HEPES (pH 7.5), 5% glycerol, 0.4 M NaCl, 1 mM EDTA (pH 8.0), 0.2% NP40, 1 mM β-mercaptoethanol, 1 μM pepstatin A, and 1 μM leupeptin.

8. Medium Salt Buffer (MSB): 20 mM HEPES (pH 7.5), 5% glycerol, 0.4 M NaCl, 1 mM EDTA (pH 8.0), 1 mM B-mercaptoethanol, 1 μM pepstatin A, and 1 μM leupeptin.
9. Wheaton douncer, loose pestle.
10. Branson sonifier model 450.
11. High Salt Buffer (HSB): 20 mM HEPES (pH 7.5), 0.34 M sucrose, 0.65 M NaCl, 1 mM EDTA (pH 8.0), 1 mM B-mercaptoehanol, 0.5 mM PMSF.
12. Branson sonifier model 450.
13. Proteinase K (20 mg/ml).
14. Dialysis tubing 12–14,000 MWCO.
15. Low Salt Buffer (LSB): 20 mM HEPES (pH 7.5), 0.34 M sucrose, 100 mM NaCl, 1 mM EDTA (pH 8.0), 1 mM β -mercaptoethanol, 0.5 mM PMSF.
16. 100 mM $CaCl_2$.
17. MNase 10 units/ul (Worthington) in 10 mM Tris–HCl (pH 7.5), 50 mM NaCl, 1 mM EDTA, 50% glycerol.
18. 0.5 mM EDTA.
19. 2 M NaCl.
20. 0.6 M HB (LON column running buffer): 20 mM HEPES (pH 7.5), 10% glycerol, 0.6 M NaCl, 1 mM EDTA (pH 8.0), 1 mM β-mercaptoethanol, 0.5 mM PMSF, 5 μg/ml aprotinin, 2 μg/ml pepstatin A, and 2 μg/ml leupeptin (filter before using on Sepharose CL-6b column).
21. Sepharose CL-6b column (1.6×76 cm) (GE Healthcare).

2.4 Reconstitution of Mononucleosomes

1. 601 nucleosome positioning sequence containing plasmid (pGEM-3Z-601).
2. Cy5- and biotin-labeled primers for PCR (*see* **Note 1**).
3. *Taq* DNA polymerase.
4. 10× Taq buffer.
5. 2.5 mM dNTP mix.
6. QIAquick PCR purification kit (Qiagen).
7. 10× TAE buffer (pH 8.0): 400 mM Tris base, 200 mM acetic acid, and 10 mM EDTA.
8. 100 bp DNA ladder (NEB).
9. QIAquick gel extraction kit (Qiagen).
10. γ-^{32}P-ATP.
11. T4 Polynucleotide kinase (T4 PNK) 10 units/uL (NEB).
12. ProbeQuant G-50 mini column (GE healthcare).

13. 5 M NaCl.
14. 1 mg/ml BSA.
15. Initial Dilution Buffer: 10 mM Tris–Cl (pH 7.4), 1 mM EDTA (pH 8.0), 5 mM DTT and 0.5 mM PMSF.
16. Final Dilution Buffer: 10 mM Tris–Cl (pH 7.4), 1 mM EDTA (pH 8.0), 5 mM DTT, 0.1% NP-40, 20% Glycerol, and 200 μg/ml BSA.
17. Dynabeads M-280 Streptavidin coated magnetic beads (Invitrogen).
18. Wash Buffer H: 25 mM HEPES (pH 7.5), 300 mM KCl, 0.5 mM EDTA (pH 8.0), 0.1 mM EGTA (pH 8.0), 10% glycerol, 0.02% NP-40, 1 mM DTT, and 100 μg/ml BSA.
19. Digestion Buffer : 10 mM Tris–HCl (pH 7.9), 100 mM NaCl, 2 mM $MgCl_2$, 5 mM β-mercaptoethanol, and 100 μg/ml BSA.
20. BamHI 20 units/μl (NEB).
21. Digestion Buffer with 50% glycerol: 10 mM Tris–HCl (pH 7.9), 100 mM NaCl, 2 mM $MgCl_2$, 50% glycerol, 5 mM β-mercaptoethanol, and 100 μg/ml BSA.
22. 10× TBE buffer (pH 8.3): 890 mM Tris base, 890 mM boric acid, and 20 mM EDTA.
23. 5× HAT buffer: 250 mM Tris–HCl (pH 7.9), 25% glycerol, 250 mM KCl, 0.5 mM EDTA (pH 8.0), 50 mM sodium butyrate (add fresh), 5 mM DTT, and 5 mM PMSF.
24. 100 μM Acetyl-CoA.
25. 10 mg/ml ultrapure BSA.
26. Purified histone acetyltransferase complex (*see* **Note 2**).

2.5 EMSA and Chromatin Remodeling Assay

1. Binding and remodeling buffer: 20 mM HEPES -KOH (pH 7.9), 70 mM KCl, 10 mM $MgCl_2$, 2 mM DTT, 10% glycerol, 0.05% NP-40, 100 μg/ml BSA, and 0.5 mM PMSF.
2. Calf thymus DNA 10 mg/ml.
3. STOP buffer: 750 ng calf thymus DNA, 500 ng long oligonucleosomes (LON), 6× orange G dye in 50% glycerol.
4. 10 mM ATP.
5. 10× TBE buffer: 890 mM Tris base, 890 mM boric acid, and 20 mM EDTA.
6. 40% Acrylamide:Bis-acrylamide::37.5:1.
7. 40% Acrylamide:Bis-acrylamide::60:1.
8. Tetramethylethylenediamine (TEMED).
9. 10% w/v ammonium persulfate (APS).

3 Methods

3.1 Expression and Purification of Recombinant Histones

1. Transform *Escherichia coli* BL21 (DE3) pLysS competent cells with plasmids, each expressing H2A, H2B, H3, and H4 [12].
2. Select four colonies from each plate, for each histone clone, and inoculate into 5 ml of 2X TY-AC.
3. Grow at 37 °C in a shaking incubator until the absorbance (600 nm) is ~ 0.6.
4. Remove 0.5 ml of each culture into sterile 1.5 ml tubes. Add 0.2 ml of sterile glycerol, mix and store at −80 °C. Remove an additional 0.5 ml into separate tubes that serve as uninduced controls.
5. To the remaining 4 ml cultures add IPTG to a final concentration of 0.2 mM and grow at 37 °C in a shaking incubator for 3 h. Harvest cells by centrifugation at 5000 × *g*, 5 min.
6. Resuspend the cell pellets in 200 μl of SDS-PAGE loading buffer and boil for 5 min.
7. Centrifuge at maximum speed in a microcentrifuge for 1 min.
8. Load supernatant on a 15 % SDS-polyacrylamide gel to determine the colony with maximum expression. After the migration, stain the gel with Coomassie blue solution for 30 min and then incubate in destain solution until bands are visible.
9. For large-scale expression of histones, streak out cultures from glycerol stocks onto AC agar plate, and incubate at 37 °C overnight.
10. The next morning, inoculate into two tubes containing 4 ml 2× TY-AC medium and grow at 37 °C in a shaking incubator until the absorbance (600 nm) is ~ 0.3.
11. Inoculate each 4 ml culture into 500 ml of 2× TY-AC media and grow at 37 °C in a shaking incubator until the absorbance (600 nm) is ~ 0.6.
12. Induce each flask with IPTG to a final concentration of 0.2 mM and grow for an additional 2 h for H3 and H4 expression and 3 h for H2A and H2B expression.
13. Harvest cells by centrifugation at 5000 × *g*, 20 min at room temperature and resuspend cells in 20 ml of wash buffer. Flash freeze in liquid nitrogen and store at −80 °C.
14. Thaw cells in a 30 °C water bath with constant stirring using a glass rod. Bring the volume up to 25 ml. Sonicate on ice, in pulses of 15 s for a total of six times using a Branson Sonifier (model 450) at setting 6 and 50 % duty cycle. Keep cells on ice for 2 min between pulses.
15. Centrifuge at 26,900 × *g* in a JA-17 rotor for 20 min at 4 °C to pellet the inclusion bodies.

16. Resuspend the pellet in 25 ml TW buffer and break up the pellet as much as possible using a 10 ml pipette. Spin at 26,900 × *g* for 15 min at 4 °C. Wash the pellet an additional two times with 25 ml of TW buffer, followed by washing two times with 25 ml of wash buffer. The pellets can be stored at −80 °C at this point.
17. Add 350 μl of DMSO to each pellet and soak at room temperature for 30 min. Resuspend the pellet in 15 ml of Unfolding buffer by gentle rotation for 1 h at room temperature.
18. Centrifuge at 26,900 × *g* for 20 min in a JA-17 rotor at room temperature to remove debris. Keep the supernatant. Resuspend the pellet again in 15 ml of Unfolding buffer and repeat the spin as before. Combine the supernatants.
19. Using 3 kDa cutoff dialysis bags, dialyze the supernatant in SAU-200 buffer: two times in 1 l of buffer for 1 h each followed by dialysis overnight in 1 l of buffer.
20. Centrifuge at 26,900 × *g* for 20 min in a JA-17 rotor.
21. Load the supernatant on a Hi-Trap SP-sepharose column, pre-equilibriated in SAU-200 and elute using a gradient of SAU-200 to SAU-600 at a flow rate of 2 ml/min. Collect the flowthrough and the elution fractions. Run the peak fractions on a 15% SDS-polyacrylamide gel along with input and flowthrough controls.
22. Pool the peak fractions. Using 3 kDa cutoff dialysis bags, dialyze at 4 °C against cold water containing 5 mM 2-mercaptoethanol and 0.2 mM PMSF: twice in 1 l for 1 h each followed by overnight dialysis in 1 l.
23. After dialysis, centrifuge at 26,900 × *g* for 20 min in a JA-17 rotor at 4 °C.
24. Determine the concentration of protein by measurement of absorbance at 276 nm for each histone (*see* **Note 3**).
25. Lyophilize each histone in 2 mg aliquots and store at −80 °C.

3.2 Reconstitution of Histone Octamers from Recombinant Histones (*See* Note 4)

1. Dissolve each histone pellet in 1 ml of Unfolding buffer for 1 h at room temperature by head to head rotation in an orbital shaker.
2. Measure the absorbance at 276 nm for each histone: calculate the histone concentrations (*see* **Note 3**). Mix equimolar concentrations of each histone and adjust the concentration to 1 mg/ml using the Unfolding buffer. Set the tube at room temperature for 30 min.
3. Add histone mix to a dialysis bag of 3500 MWCO. Dialyze against cold Refolding buffer at 4 °C in the following steps: twice in 1 l for 1 h each, followed by once in 1 l overnight.
4. Centrifuge the dialyzed histone mix for 10 min at 23,000 × *g* in a SS-34 rotor for 10 min at 4 °C to remove any aggregates.

5. Load onto a Superdex 200 column pre-equilibrated in Refolding buffer. Run at a flow rate of 1 ml/min and collect 1 ml fractions.
6. Analyze the peak fractions by SDS-PAGE in a 15% gel. Check fractions for the stoichiometry of all four histones in the histone octamer.
7. Pool the octamer fractions and concentrate using an Amicon Ultra with 30,000 MWCO to ~1 ml.
8. Measure the absorbance at 276 nm and calculate the octamer concentration [A_{276} = 0.45 for (octamer) = 1 mg/ml].
9. Aliquot and store at −80 °C (*see* **Note 5**).

3.3 Preparation of Nuclei for Nuclear Extract and Chromatin

1. Centrifuge 24 l of HeLa S3 cells (0.63 × 10^6 cells/ml) at 4000 × *g* in a Beckman JLA 8.1000 rotor for 6 min at 4 °C. Cell pellets are buff to yellowish in color (*see* **Note 6**).
2. Resuspend the cell pellets in 50 ml of ice cold PBS by pipetting. Spin cells in conical bottom bottles at 650 × g in Sorvall H6000A rotor for 5 min at 4 °C. Wash cells two more times in PBS.
3. Estimate volume of cells (we use conical bottom bottles that are graduated). Add 5× cell volume of Extraction buffer I. Mix cells by pipetting and vortex cells to mix at speed set to 6–7 for 10 s. Incubate cells in ice for 5 min.
4. Add 10% NP-40 to cells, so that the final concentration of NP-40 is 1%. Mix by swirling the cells in the bottle and then vortex as above for 15 s. Incubate cells on ice for 5 min. Vortex for 5 s every minute and mix by swirling cells around. This will help to break open cells more efficiently.
5. Centrifuge at 3000× *g* for 5 min, to precipitate the nuclei. Remove supernatant completely (this is the cytoplasmic extract, store at −80 °C if you need this for some experiments (*see* **Note 7**).
6. Add 1/2 pellet volume of Low Salt Buffer and stir gently at 4 °C. Add equal volume of High Salt Buffer drop wise while gently stirring at 4 °C. Gently stir for an additional 30 min at 4 °C.
7. Centrifuge at 20,000 × *g* in a JA-17 for 30 min at 4 °C. The supernatant is the nuclear extract (NE) and the pellet is the chromatin fraction.
8. Dialyze supernatant (NE) at 4 °C using 3–5000 MWCO dialysis tubing against 50 volumes (4 l) of dialysis buffer until conductivities of NE and dialysis buffer are equal (usually takes dialysis overnight at 4 °C). Spin dialyzed NE at 20,000 × g in JA-17, for 20 min, 4 °C. Save small protein pellet to re-extract DNA later if needed. Flash freeze several 1.5 ml aliquots, labeled NE, store at −80 °C. Save 5 μl of NE for analysis later.

3.4 Short Oligonucleosomes (SON) and Long Oligonucleosome (LON) Preparation from HeLa Cells

1. Resuspend the HeLa chromatin fraction in ~2 ml of MSB for each ml of pellet volume, using a glass rod. Homogenize with a Wheaton Douncer, loose pestle, on ice. Spin at 20,000 × *g* in a JA-17 at 4 °C for 20 min. (Supernatant will be clear. Save for analysis.). Repeat this four more times.
2. Resuspend pellet by homogenization in 30 ml of MSB + 0.2% NP-40 and pellet nuclear material by centrifugation at 20,000 × *g* in a JA-17 rotor for 5 min at 4 °C. Repeat one more time.
3. Resuspend pellet by homogenization in 30 ml of MSB (without NP-40). Pellet nuclear material by centrifugation at 20,000 × *g* in JA-17 rotor for 5 min 4 °C. Repeat one more time.
4. Resuspend pellet in 3–4 volumes (50 ml total volume) in HSB and homogenize using a loose pestle and dounce for ~50 strokes.
5. Pellet nuclei for 20 min at 19,000 × *g* at 4 °C in a SA600 rotor. The pellet should look goopy (*see* **Note 8**).
6. Sonicate bulk of chromatin (cloudy white supernatant) on ice, using a micro-tip, in pulses of 15 s (50% duty) for a total of 1.75 min using a Branson Sonifier (model 450) at setting 6 and 50% duty cycle. This greatly improves the solubility of the chromatin.
7. Add 16 ml of MSB to the chromatin pellet. Sonicate as described above for the supernatant fraction.
8. At this point, analyze the quality of the chromatin by digesting 10 μl of chromatin with 10 μg proteinase K digestion at 55 °C for 2 h followed by phenol/chloroform extraction and agarose gel electrophoresis. The fragments should be approximately 4 kb in size.
9. Dialyze both supernatant and pellet fractions in separate dialysis bags 12–14,000 MWCO against 4 l LSB overnight at 4 °C. Make sure to keep the dialysis tubing loose as it will swell during dialysis (*see* **Note 9**).
10. Take 500 μl of dialyzed chromatin. Add $CaCl_2$ to final concentration of 5 mM. Warm at 37 °C to help solubilize the chromatin. Carry out a series of digestions with MNase (0 mU, 50 mU, 125 mU, 500 mU, 1 U, 2 U, and 4 U) to 10 μl of chromatin at room temperature for 10 min. Treat samples with 10 μg proteinase K followed by phenol/chloroform extraction and agarose gel electrophoresis to determine the concentration of Mnase that results in fragments of ~400 bp average length (Fig. 1a).
11. To the bulk dialyzed chromatin add 0.03 volumes of 0.1 M $CaCl_2$ and warm at 37 for 5 min. Add the appropriate concentration of MNase as determined in the last step and incubate at room temperature for 20 min. Stop the MNase digestion by adding 0.1 volumes of 0.5 mM EGTA and put on ice. Add 2 M NaCl drop wise with constant mixing to a final concentration of 0.6 M NaCl.

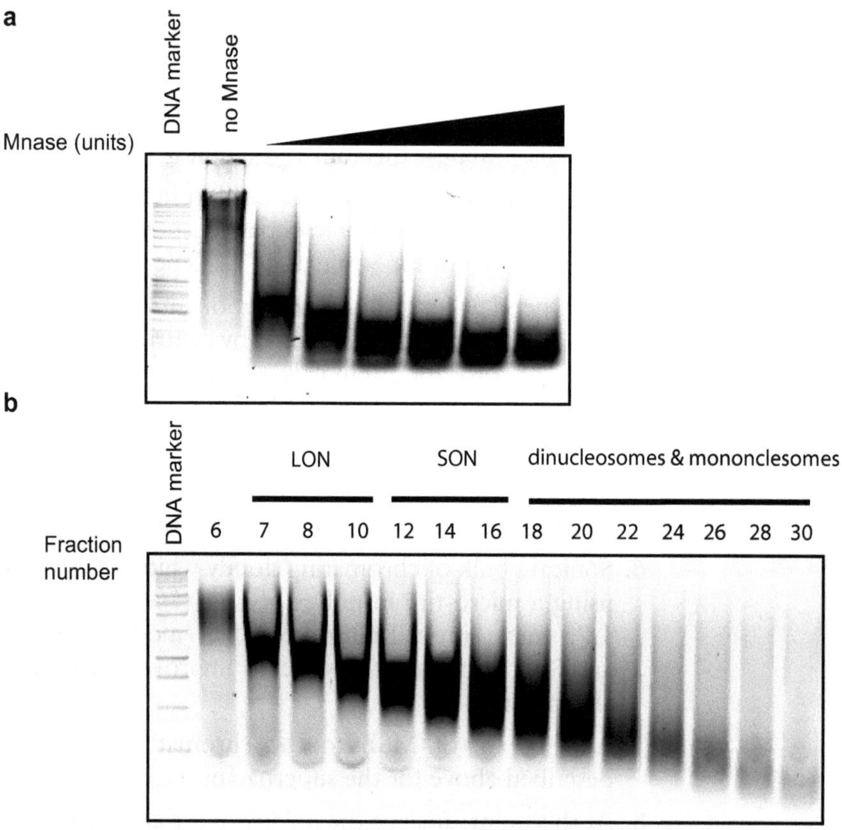

Fig. 1 (**a**) To check for appropriate concentration to MNase to digest soluble chromatin, aliquots of soluble chromatin were digested with increasing concentration of MNase at room temperature for 10 min. Samples were treated with proteinase K, followed by phenol/chloroform extraction and run on a 1.3 % agarose gel in 1X TAE buffer. (Subheading 3.2, **step 17**) (**b**) Fractions collected from running soluble MNase digested chromatin on a Sepharose CL-6b column, were treated with proteinase K, extracted by phenol/chloroform extraction and resolved on a 1.3 % agarose gel in 1× TAE buffer. Fractions 6–30 are shown. (Subheading 3.2, **steps 20–22**)

12. Centrifuge at 20,000 × *g* in a JA-17 rotor for 15 min at 4 °C to remove any insoluble material.
13. Load the digested chromatin on a Sepharose CL-6b column (1.6 × 70 cm) pre-equilibrated in 0.6 M HB buffer. Run at 0.2 ml/min and collect 1.5 ml fractions (*see* **Note 10**).
14. Analyze peak fractions by SDS-PAGE in a 15 % gel. Check to ensure all fractions have equimolar ratios of histones. Also, treat fractions with proteinase K followed by phenol/chloroform extraction and agarose gel electrophoresis (Fig. 1b).
15. Pool LON fractions (~ fractions 7–11, oligonuclesomes with 6 or more nucleosomes) and SON fractions (~ fractions 12–15, oligonuclesomes with 3–5 nucleosomes) and store at −80 °C. Mono/dinucleosomes (~fractions 21–35) are not pooled and are stored individually at 80 °C.

3.5 Preparation of Labeled DNA-Template for Mononucleosome Reconstitution

We propose two alternative methods for labeling the DNA probe containing the 601 nucleosome positioning sequence, depending on your specific needs.

3.5.1 Alternative 1: Generation of Cy5-Labeled DNA Probes Containing the 601 Nucleosome Positioning Sequence

1. Different assays use varying lengths of linker DNA on both 5′ and 3′ of the -601- positioning sequence (*see* **Note 11**).
2. PCR amplify the biotinylated Cy5-labeled 601 DNA probe: mix 10 μl of 10× Taq buffer, 8 μl of 2.5 mM dNTP mix, 2 μl each of Cy5-forward and Biotin-reverse primers from 20 μM stocks of primers, 1 μl of plasmid pGEM-3Z-601 (10 ng/μl), 76.5 μl of H_2O and 0.5 μl of Taq polymerase. PCR conditions: 30 cycles of PCR with annealing at 57 °C for 1 min followed by extension time at 72 °C for 30 s.
3. Purify PCR product using the Qiagen PCR purification kit. Run the DNA probe in a 2.0% agarose gel in 1.5× TAE Buffer. Excise the DNA band of correct size and purify using the Qiagen Gel extraction purification kit. Quantify the DNA using both a UV spectrophotometer (we use a NanoDrop (Thermo scientific)) and by running on a 2% agarose gel in 1.5× TAE Buffer along with appropriate DNA ladder.

3.5.2 Alternative 2: Generation of Radio-Labeled DNA Probes Containing the 601 Nucleosome Positioning Sequence

1. End-label the primer without the biotin tag: 1 μl of 20 μM primer, 1 μl of 10× Kinase Buffer (T4 PNK NEB), 0.5 μl of γ-^{32}P -ATP, 6.5 μl of H_2O, 1 μl of T4 PNK. Incubate at 37 °C for 60 min.
2. Heat at 70 °C for 10 min to heat inactivate the PNK.
3. Generate PCR products as in **step 2** in the previous section by adding the entire end-labeled primer (10 μl) to the reaction mix.
4. Purify the PCR product using a ProbeQuant G-50 mini-column to remove unincorporated γ-^{32}P ATP. Gel purify the PCR product using the QIAgen gel extraction kit (*see* **Note 12**).

3.6 Reconstitution of Mononucleosomes by Salt Dialysis Method and Purification of Mononucleosomes

1. The initial mix is in a volume of 25 μl. Add -601- containing DNA and histone octamers purified earlier to a molar ratio of 1:1.3. Add 5 M NaCl and 1 mg/ml BSA to a final concentration of 1 M NaCl and 0.1 mg/ml BSA. Bring the volume to 25 μl by adding Dilution buffer. Mix by pipetting and incubate at 37 °C for 20 min (*see* **Note 13**).
2. Add Dilution buffer in steps of 6.26, 10.4, 20.84, and 62.5 μl. After each addition mix by pipetting and incubate at 30 °C for 20 min. Add 125 μl of Final Dilution buffer, mix by pipetting and incubate at 37 °C for 20 min. This dilution series reduces the salt concentration from 1 M NaCl to 100 mM NaCl.
3. Run 2–3 μl of reconstituted nucleosomes on a 4.5% native gel (made from 40% acrylamide:Bis-acrylamide::37.5:1, in 0.3× TBE, containing 0.1% APS and 40 μl TEMED) to access the quality of nucleosomes.

4. Add 50 µl of streptavidin beads to a tube and wash three times with 500 µl of Final Dilution buffer containing 100 mM NaCl. Remove the supernatant. Add the reconstituted nucleosomes to the streptavidin beads and bind overnight at 4 °C by end to end rotation in an orbital mixer.
5. Wash the nucleosome bound streptavidin beads twice with Final Dilution buffer containing 100 mM NaCl and once with Wash Buffer H, followed by one wash with Digestion buffer. Each wash is for 5 min by end to end rotation in an orbital mixer. After this step either proceed to digestion (**step 6**) to release the nucleosomes from the streptavidin beads or to the next section for modification of histones.
6. Add 50 U of BamHI to 100 µl Digestion buffer and mix. Add this mix to the nucleosome bound beads and incubate at 37 °C by end to end rotation in an orbital mixer for 4 h.
7. Remove the supernatant containing the nucleosomes to a fresh tube. Wash the beads with 100 µl of Digestion buffer containing 50% glycerol. Add to the nucleosomes and mix. These nucleosomes that are now in 25% glycerol buffer can be stored at 4 °C for up to a month.
8. Run 2 µl of nucleosomes along with a titration of Cy5-labeled DNA probe generated earlier in a on a 5% native gel (made from 40% acrylamide:Bis-acrylamide::37.5:1) in 0.3× TBE to assess the quality of nucleosomes and quantify the amount of nucleosomes (Fig. 2a).

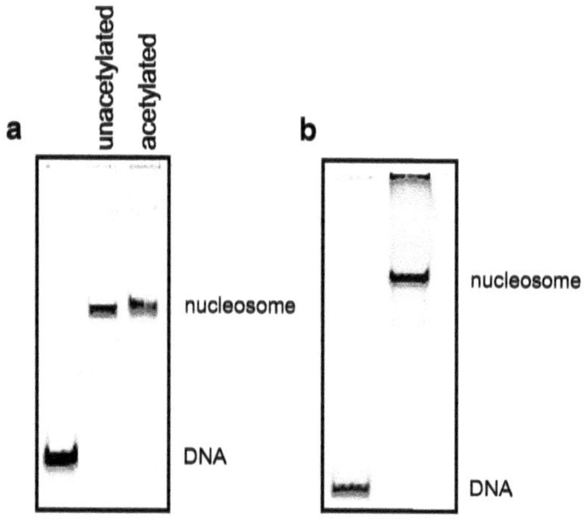

Fig. 2 Mononucleosomes reconstituted on DNA with a -601- positioning sequence (**a**) Unacetylated and acetylated nucleosomes reconstituted from octamers using recombinant *Xenopus* histones purified from *E. coli.* (**b**) Nucleosomes reconstituted from SON. (Subheading 3.3)

3.7 Modification of Nucleosomes by Histone Acetyltraferases (HAT)

The nucleosomes bound to streptavidin beads as described in the previous section (**step 5**) can be modified by other chromatin modifiers such as histone methyltransferases and kinases. Here, we have described the procedure to acetylate the nucleosomes by HATs.

1. For modification of nucleosomes following wash with Wash Buffer H, wash the nucleosome bound streptavidin beads in 500 μl 1× HAT buffer. Repeat the wash step one more time. Proceed to modify the nucleosomes. Remove the supernatant of the last wash just before addition of HAT reaction mix.
2. HAT reaction mix: add 20 μl of 5× HAT buffer, 1 μl 10 mg/ml BSA, acetyl-CoA to a final concentration of 100 μM, appropriate amount of HAT enzyme and water to final volume of 100 μl and mix.
3. Add the HAT reaction mix to the nucleosome bound streptavidin beads. Mix and incubate at 30 °C (yeast HATs) or appropriate temperature (room temperature for fly HATs and 37 °C for human HATs) for 4–5 h by end to end rotation in an orbital mixer.
4. Wash the nucleosome bound streptavidin beads three times with Wash Buffer H and once with Digestion buffer.
5. Proceed to release nucleosomes from streptavidin beads as in the previous section (**step 6**). Unacetylated and acetylated nucleosomes reconstituted from *Xenopus* octamers are shown in Fig. 2a.

3.8 Reconstitution of Mononucleosomes from HeLa SON

Oligonucleosomes purified from HeLa cells contain a mixture of histone modifications that vary from SON to LON. This can be determined by western blot analysis using histone modification antibodies. Nucleosomes can be prepared by octamer transfer from HeLa SON by the salt dialysis method described above.

1. The initial mix is in a volume of 25 μl. Add 0.5 pmol of -601-containing DNA and 4 μg of SON (1 μg/μl). Add 5 M NaCl and 1 mg/ml BSA to a final concentration of 1 M NaCl and 0.1 mg/ml BSA. Bring the volume to 25 μl by adding Dilution buffer. Mix by pipetting and incubate at 37 °C for 20 min.
2. Add Dilution buffer in steps of 6.26, 10.4, 20.84, and 62.5 μl. After each addition mix by pipetting and incubate at 30 °C for 20 min.
3. Add 125 μl Final Dilution buffer, mix by pipetting and incubate at 37 °C for 20 min. This dilution series reduces the salt concentration from 1 M NaCl to 100 mM NaCl.

Follow **steps 3–8** in Subheading 3.6. Nucleosomes reconstituted from SON are shown in Fig. 2b (*see* **Note 14**).

3.9 EMSA

1. For EMSA assays to test for binding of chromatin remodeling complex to nucleosomes set up 15 μl reactions each. Aliquot increasing amount of complex into separate tubes. Add buffer (same buffer in which the chromatin remodeling complex is resuspended) to bring up to the same volume in each tube. This ensures that salt concentrations are the same in all reaction tubes.
2. Make a master mix containing 5× Binding and remodeling buffer, nucleosomes and water to volume needed to bring reaction volume up to 15 μl. Mix well by pipetting. For N reactions make reaction mix for N + 2 reactions.
3. Add master mix to the chromatin remodeling complexes in each tube and mix by pipetting. Incubate at the appropriate temperature for 30 min. The incubation temperature depends on the species the complex is purified from.
4. Resolve the bound complexes from free nucleosomes by running on a 4% native gel (made from 40% acrylamide:bis-acrylamide 60:1) in 0.3× TBE (*see* **Note 15**).
5. When using ^{32}P- labeled probes dry the gel and expose to a phosphoimager screen. Scan on Typhoon (GE healthcare) and quantify the % of nucleosomes bound. For Cy5 labeled DNA scan directly using Typhoon (GE healthcare) (Fig. 3a).

3.10 Nucleosome Remodeling Assay

1. For nucleosome sliding assays to test for remodeling as a function of chromatin remodeling complex concentration, set up reactions as in **steps 1** and **2** of the previous section.
2. Incubate at the appropriate temperature for 10 min.
3. Add ATP to a final concentration of 250 μM to each tube and incubate for a time period as needed. Start the timer when ATP is added to the first tube.
4. To stop the reaction add 3 μl of Stop buffer to each tube, mix and incubate at room temperature for 10 min. Place on ice.
5. Resolve by native PAGE (made from 40% Arylamide:Bis-acrylamide::37.5:1) in 0.3× TBE.
6. For nucleosome sliding assays to test for remodeling as a function of time, aliquot the Stop buffer into tubes. Make a master mix containing 5× Binding and remodeling buffer, nucleosomes, and water. Add chromatin remodeling complex and mix. Incubate at an appropriate temperature for 10 min. Add ATP to a final concentration of 250 μM and start the remodeling reaction. If taking N time points, make reaction mix sufficient for N + 2 reactions. At appropriate time remove aliquots equal to one reaction mix volume (15 μl) from the remodeling reaction and add to tubes with Stop buffer. Mix and incubate at room temperature for 10 min. Place on ice.
7. Resolve unremodeled and remodeled nucleosomes by native PAGE as in **step 5** (Fig. 3b).

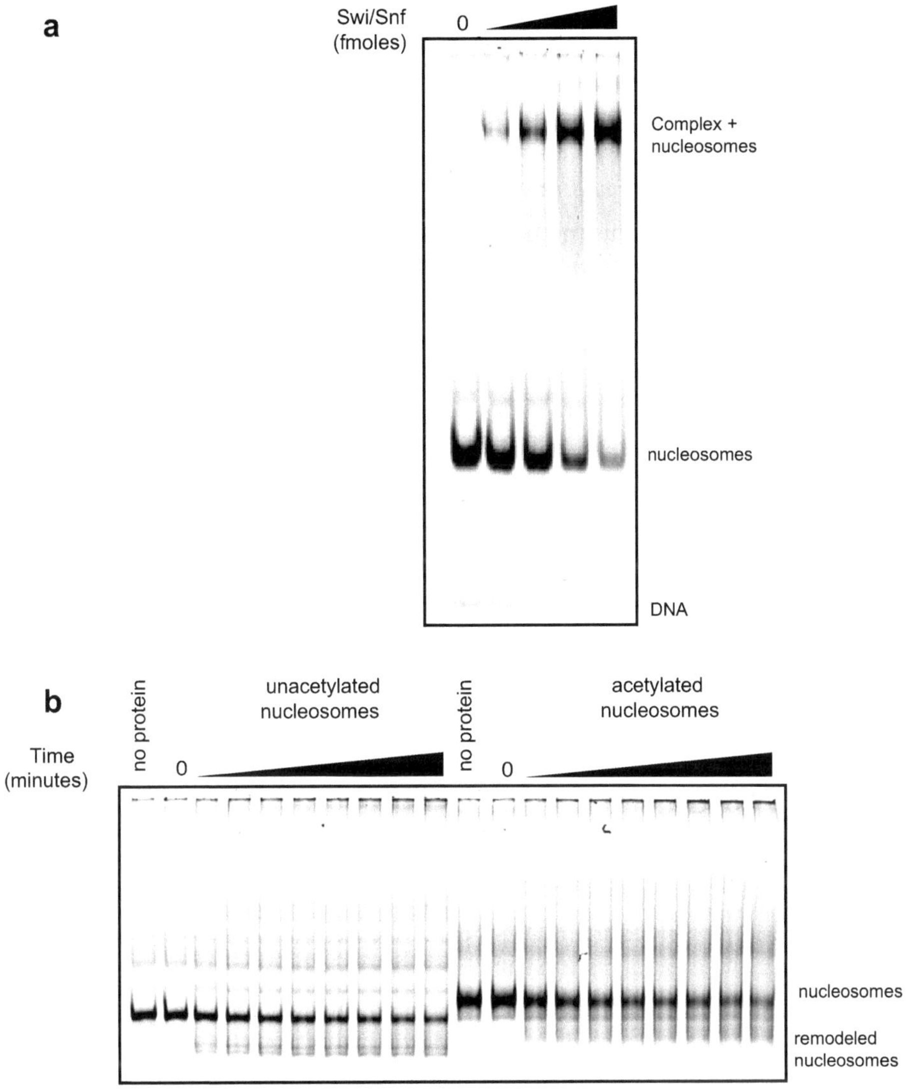

Fig. 3 (**a**) **EMSA.** 20 fmol of 5′Cy5 labeled mononucleosomes (reconstituted from octamers using *Xenopus* histones) was incubated with increasing concentration of Swi/Snf complexes (0, 20, 40, 60 and 80 fmol) and analyzed by native PAGE. Bands showing unbound mononucleosome and Swi/Snf bound mononucleosomes are labeled. (**b**) **Nucleosome remodeling assay.** Remodeling of unacetylated and acetylated nucleosomes was carried out as a function of time. In each case 40 fmol mononucleosomes reconstituted on DNA with a -601- positioning sequence with octamers containing *Xenopus* histones, were incubated with 40 fmol of Swi/Snf complex. Reactions were stopped at 0, 2.5, 7.5, 10, 12.5, 15, 20 and 30 min and analyzed by native PAGE. (Subheading 3.4)

4 Notes

1. The primers are designed based on required length of linker DNA on the 5′ and 3′ around the -601- positioning sequence.
2. We have used Gcn5 containing histone acetyltransferase complex purified from yeast using TAP-tag on Ada2 or Sgf73 subunits for

acetylation of nucleosomes. The histone acetyltransferase complexes should be tested for activity prior to use for bulk acetylation of nucleosomes.

3. Calculate the theoretical extinction coefficient (ε) for each histone using the ProtParam tool in Expasy (http://web.expasy.org/protparam/). Abs = εcl, c = concentration, l = path length.
4. This method for generating octamers works best for *Xenopus*, *Drosophila*, and human histones but not for yeast histones, for which the yields by this method are very low.
5. Make aliquots of small volume and flash freeze in liquid nitrogen. Do not freeze thaw tubes multiple times as this leads to clipping of the histone tails specifically histone H3.
6. We have successfully purified oligonucleosomes from both HeLa cells and *Drosophila melanogaster* S2 cells using the method described.
7. We get ~ 24 ml of nuclear pellet from 24 l of cells.
8. ~20 ml pellet size and ~36 ml cloudy white sup is usually obtained at this step.
9. Final volume of dialyzed chromatin will be ~35 ml and white, glue consistency without any precipitates.
10. Do not start collecting fractions until 30 ml. Broad peak comes off around 37–110 ml (fraction 7–30). We collect 35 fractions.
11. The primers are designed accordingly to amplify a -601- containing fragment from pGEM-3Z-601. The forward primer is designed to contain a 5′-Cy5 probe and the reverse primer contains a 5′-biotin followed by a linker –GCGC- and a BamHI digestion site.
12. Instead of end-labeling the primer, the probe can be body labeled by incorporating α-^{32}P –dATP in the PCR reaction mix and using unlabeled forward primer and biotinylated reverse primers. Measure the amount of PCR product by running on a 2% agarose gel in 1.5× TAE Buffer along with appropriate DNA ladder. Measure the radioactivity incorporated into the DNA probe by measuring on a scintillation counter. Calculate the counts per million (cpm) per ng of DNA probe.
13. To calculate DNA concentration from ng/μl to nmol/μl: (nmol/μl) = (ng/μl of DNA)/(659 × bp of DNA.
14. Nucleosomes can also be generated from LON by this method as well.
15. When resolving nucleosome bound complexes by native PAGE do not add any dye to the samples. Run the dye separately in an empty lane. We typically use gels that are 16 × 20 cm for separation of nucleosome bound complexes from free nucleosomes. The gels are run at 200 V for ~4–4.5 h.

Acknowledgments

We thank current and former members of the Workman laboratory for help with developing these protocols and many useful discussions. We acknowledge Malini Natarajan for critical reading of the manuscript. This work was supported by National Institutes of Health grant R01GM047867 to J.L.W. and funding from the Stowers Institute for Medical Research.

References

1. Clapier CR, Cairns BR (2009) The biology of chromatin remodeling complexes. Annu Rev Biochem 78:273–304
2. Dutta A et al (2014) Swi/Snf dynamics on stress-responsive genes is governed by competitive bromodomain interactions. Genes Dev 28(20):2314–2330
3. Chatterjee N et al (2011) Histone H3 tail acetylation modulates ATP-dependent remodeling through multiple mechanisms. Nucleic Acids Res 39(19):8378–8391
4. Gutierrez JL et al (2007) Activation domains drive nucleosome eviction by SWI/SNF. EMBO J 26(3):730–740
5. Owen-Hughes T et al (1999) Analysis of nucleosome disruption by ATP-driven chromatin remodeling complexes. Methods Mol Biol 119:319–331
6. Simon MD et al (2007) The site-specific installation of methyl-lysine analogs into recombinant histones. Cell 128(5):1003–1012
7. Yun M et al (2012) Reconstitution of modified chromatin templates for in vitro functional assays. Methods Mol Biol 833:237–253
8. Dyer PN et al (2004) Reconstitution of nucleosome core particles from recombinant histones and DNA. Methods Enzymol 375:23–44
9. Neumann H et al (2009) A method for genetically installing site-specific acetylation in recombinant histones defines the effects of H3 K56 acetylation. Mol Cell 36(1):153–163
10. Shogren-Knaak M et al (2006) Histone H4-K16 acetylation controls chromatin structure and protein interactions. Science 311(5762):844–847
11. Kouzarides T (2007) Chromatin modifications and their function. Cell 128(4):693–705
12. Luger K et al (1999) Expression and Purification of Recombinant Histones and Nucleosome Reconstitution. Methods Mol Bio 119:1–16

Chapter 2

An Assay for Measuring Histone Variant Exchange within Nucleosomes In Vitro

Liette Laflamme, Benoit Guillemette, and Luc Gaudreau

Abstract

The incorporation of histone variants into specific chromatin regions is a mechanism by which cells can regulate many important biological processes. One such example is H2A.Z, a highly conserved variant of H2A that is incorporated in genomic regulatory regions and contributes to control gene expression. H2A.Z variant exchange involves the removal of H2A-H2B dimers from a preassembled nucleosome and their replacement with H2A.Z-H2B dimers. A specific family of chromatin remodeling complexes, homologous to the yeast Swr1 complex, have been shown to be capable of this histone exchange activity both in vivo and in vitro. Here, we describe an assay to measure the histone H2A.Z exchange activity of recombinant human p400 on immobilized mononucleosomes in vitro. The assay can be adapted to other histone exchange complexes/catalytic subunits purified from any species.

Key words Chromatin, Histone variant, Exchange, Nucleosome, p400, Swr1, H2A.Z

1 Introduction

The four canonical histones, H3, H4, H2A, and H2B, are the basic, fundamental proteins that serve as building blocks for chromatin. These core histones assemble together in an octamer (with two copies of each histone) around which DNA is wrapped to form the nucleosome, the fundamental unit of chromatin [1]. Cells not only rely on histones to compact chromatin, but also to organize the nucleus and to help regulate all the functions pertaining to DNA such as transcription, replication, and repair. Many mechanisms are used by the cell to control the access to DNA via the histones, such as post-translational modifications, ATP-dependent chromatin remodeling, and histone variants. Noncanonical histone variants can be incorporated in the nucleosome to confer special functions at specific genomic regions. H2A.Z is a highly conserved variant of H2A which is enriched in gene regulatory regions and possesses specific functions in gene expression control and genomic stability [2]. The incorporation of H2A.Z (like most noncanonical variants) is not

Benoit Guillemette and Luc Gaudreau (eds.), *Histones: Methods and Protocols*, Methods in Molecular Biology, vol. 1528,
DOI 10.1007/978-1-4939-6630-1_2, © Springer Science+Business Media New York 2017

dependant on the cell cycle and involves specific ATP-dependant chromatin remodeling complexes. The archetype of this family of complexes is the yeast SWR1, a multisubunit complex conserved across evolution. SWR1 targets nucleosomes in specific gene regulatory regions, removes an H2A-H2B dimer, and replaces it with an H2A.Z-H2B dimer [3]. There are two known human homologues of Swr1: p400 and SRCAP that have been shown to possess H2A.Z exchange activity [4, 5]. Although both proteins are found in large complexes in vivo, purified p400 is capable of H2A.Z/H2B dimer incorporation in nucleosomes in vitro [4].

We present here a series of protocols to perform histone variant exchange assays on immobilized nucleosomes using affinity-purified recombinant p400. The first four protocols serve to prepare the reagents necessary for the fifth protocol, the actual histone exchange assay. We start with the purification of Flag-tagged recombinant p400 from baculovirus infected Sf9 insect cells. We then isolate core histones from HeLa nuclear extracts, followed by reconstitution of mononucleosomes on the sea urchin 5S rDNA template, a well-characterized nucleosome positioning sequence. Mononucleosomes carry a biotin tag on the 5′ ends and are immobilized with streptavidin magnetic beads. We also describe the purification of recombinant human Flag-H2A.Z and H2B from *E. coli*, and their assembly into dimers. Finally, we describe a protocol for histone exchange in vitro using immunoblotting to reveal the exchange activity. With minor adjustments, these protocols can be adapted to study other chromatin remodeling complexes.

2 Materials

2.1 Preparation of Recombinant p400 from SF9 Cells

1. Baculovirus containing the Flag-tagged human p400 cDNA.
2. Sf9 cells.
3. TNM-FH insect medium.
4. 150 mm plastic petri dishes.
5. 500 ml spinner flask.
6. 50 ml conical tubes.
7. 15 ml Dounce homogenizer type B pestle.
8. 250 ml polycarbonate or polypropylene centrifuge bottles.
9. 50 ml polycarbonate or polypropylene round bottom centrifuge tubes with cap.
10. Anti-Flag M2-agarose resin.
11. PBS: phosphate buffered saline, 137 mM NaCl, 2.7 mM KCl, 4.3 mM Na_2HPO_4, 1.4 mM KH_2PO_4, pH7.3.
12. Lysis buffer: 20 mM Tris–HCl pH 7.5, 500 mM NaCl, 1.5 mM $MgCl_2$, 0.2 mM EDTA, 20% glycerol. Added fresh before using: 2 mM DTT, 1 mM PMSF, protease inhibitor cocktail.

13. Dilution buffer: 20 mM Tris–HCl pH 7.5, 1.5 mM $MgCl_2$, 0.2 mM EDTA, 10% glycerol, 0.5% Igepal. Added fresh before using: 2 mM DTT, 1 mM PMSF, protease inhibitor cocktail.
14. Wash buffer: 20 mM Tris–HCl pH 7.5, 150 mM NaCl, 1.5 mM $MgCl_2$, 0.2 mM EDTA, 10% glycerol, 0.2% Igepal. Added fresh before using: 2 mM DTT, 1 mM PMSF.
15. Low-salt elution buffer: 20 mM Tris–HCl pH 7.5, 100 mM NaCl, 0.2 mM EDTA, 15% glycerol. Added fresh before using: 1 mM PMSF, 0.2 mg/ml Flag peptide, 0.5 mg/ml human insulin.
16. High-salt elution buffer: 20 mM Tris–HCl pH 7.5, 500 mM NaCl, 0.2 mM EDTA, 15% glycerol. Added fresh before using: 1 mM PMSF, 0.2 mg/ml Flag peptide, 0.5 mg/ml human insulin.
17. 6× protein loading buffer: 60 mM Tris pH 6.8, 47% v/v glycerol, 12% SDS, 0.06% w/v bromophenol blue.
18. Protein prestained molecular weight marker.
19. SDS-PAGE apparatus.
20. Western blot wet transfer apparatus.
21. PVDF membranes.
22. SDS-PAGE 6% resolving gel: 6% acrylamide:bis-acrylamide 29:1, 375 mM Tris–HCl pH 8.8, 0.1% SDS, 0.1% ammonium persulfate (APS), 0.1% TEMED.
23. SDS-PAGE 4% stacking gel: 4% acrylamide:bis-acrylamide 29:1, 125 mM Tris–HCl pH 6.8, 0.1% SDS, 0.1% ammonium persulfate (APS), 0.1% TEMED.
24. Running buffer: 25 mM Tris, 192 mM glycine, 0.1% SDS.
25. Wet Transfer Buffer: 25 mM Tris–HCl pH 8.3, 192 mM glycine, 10% methanol, 0.01% SDS.
26. Blotting paper.
27. TBS: Tris buffered saline, 50 mM Tris–HCl pH 7.5, 150 mM NaCl.
28. 5% milk-TBS: 5% fat free powdered milk in TBS.
29. Monoclonal mouse anti-Flag (M2) antibody.
30. Anti-mouse HRP-conjugated antibody.
31. Chemiluminescence detection kit (ECL prime (GE Healthcare) or equivalent).

2.2 Purification of Human Core Histones from HeLa Cells

1. HeLa cells.
2. DMEM: Dulbecco's minimum essential medium.
3. FBS: fetal bovine serum.
4. 6 l spinner flask.

5. PBS: *See* Subheading 2.1.
6. Buffer A: 20 mM HEPES-KOH pH 7.5, 0.25 M sucrose, 3 mM $MgCl_2$, 0.2% Igepal. Added fresh before using: 3 mM β-mercaptoethanol, 1 mM PMSF, protease inhibitor cocktail (Roche).
7. Dounce homogenizer with type B pestle.
8. Buffer B: 20 mM HEPES-KOH pH 7.5, 3 mM $MgCl_2$, 0.2 mM EGTA. Added fresh before using: 3 mM β-mercaptoethanol, 1 mM PMSF, protease inhibitor cocktail (Roche).
9. Buffer C: 20 mM HEPES-KOH pH 7.5, 3 mM $MgCl_2$, 0.2 mM EGTA, 0.6 M KCl, 10% glycerol. Added fresh before using: 3 mM β-mercaptoethanol, 1 mM PMSF, protease inhibitor cocktail (Roche).
10. Bio-Gel HTP hydroxyapatite (Bio-Rad).
11. HAP buffer: 50 mM sodium phosphate pH 6.8, 0.6 M NaCl, Added fresh before using: 1 mM β-mercaptoethanol, 0.5 mM PMSF.
12. HAP buffer + NaCl: 50 mM sodium phosphate pH 6.8, 2.5 M NaCl, Added fresh before using: 1 mM β-mercaptoethanol, 0.5 mM PMSF.
13. FPLC system.
14. Chromatography column (2.5 × 20 cm).
15. Amicon Ultra-15 Centrifugal Filter, 3 kDa MWCO (EMD Millipore, or equivalent).

2.3 *In Vitro* Assembly of Biotin-Labelled Mononucleosomes

1. pIC208-5S plasmid: Yeast centromeric plasmid containing the sea urchin 5S rDNA nucleosome positioning sequence.
2. EcoRI-HF restriction enzyme 100,000 U/ml (New-England Biolabs).
3. Cutsmart buffer 10× (New-England Biolabs).
4. Agarose.
5. 6× OG DNA loading buffer: 20 mM Tris–HCl ph 8.0, 15% Ficoll, 0.9% Orange G.
6. TBE 1×: 89 mM Tris base, 89 mM Boric acid, 2 mM EDTA.
7. Qiaquick Gel Extraction Kit (Qiagen).
8. 0.4 mM biotin-14-dATP.
9. dBTP mix: 1 mM dGTP, 1 mM dCTP, 1 mM dTTP.
10. 1M $MgCl_2$.
11. 1 M DTT.
12. DNA Polymerase I, Large Klenow Fragment, 50,000 U/ml.
13. NEBuffer 2: 50 mM NaCl, 10 mM Tris–HCl pH 7.9, 10 mM $MgCl_2$, 1 mM DTT.

14. 500 mM EDTA pH 8.0.
15. 3 M sodium acetate pH 5.2.
16. 100% ethanol, −20 °C.
17. 20 μg/μl glycogen.
18. 5 M NaCl.
19. 10 mM Tris–HCl pH 7.6.
20. 50% glycerol.
21. 5% polyacrylamide DNA gel: 5% acrylamide:bis-acrylamide (29:1), 0.5× TBE, 0.1% ammonium persulfate (APS), 0.1% TEMED.
22. SYBR Green I (10,000×).
23. 50 mM Tris–HCl pH 7.5.
24. Streptavidin-coupled magnetic beads.
25. 1 mg/ml BSA (Bovine Serum Albumin) ultrapure.
26. Wash buffer 100: 10 mM Tris–HCl pH 7.5, 0.1 M NaCl.
27. Wash buffer 600: 10 mM Tris–HCl pH 7.5, 0.6 M NaCl.

2.4 Preparation of Recombinant Flag-H2A.Z/H2B Dimers

1. pET expression plasmids containing human Flag-H2A.Z (N-terminal tag) and H2B encoding genes.
2. *E. coli* BL21 codon + RIL strain.
3. LB-agar plates.
4. 2× TY-AC: 16% (w/v) bacto-tryptone, 10% (w/v) yeast extract, and 5% (w/v) NaCl.
5. 1000× ampicillin: 50 mg/ml in H_2O.
6. 1000× chloramphenicol: 34 mg/ml chloramphenicol in methanol.
7. 500 mM IPTG (Isopropyl β-D-1-thiogalactopyranoside).
8. 30 mM Rifampicin: 24.7 mg/ml Rifampicin in methanol.
9. Wash buffer: 50 mM Tris–HCl (pH 7.5), 100 mM NaCl, 1 mM Na-EDTA, 1 mM Benzamidine-HCl and 5 mM β-mercaptoethanol.
10. Lysozyme: lyophilized powder, ≥40,000 units/mg of protein.
11. TW buffer: Wash buffer with 1% (v/v) Triton X-100.
12. Unfolding buffer: 7 M guanidinium HCl, 20 mM Tris–HCl (pH 7.5), and 10 mM DTT.
13. SAU-200: 7 M urea, 200 mM NaCl, 20 mM sodium acetate (pH 5.2), 5 mM β-mercaptoethanol, and 1 mM EDTA (*see* **Note 1**).
14. SAU-600: 7 M urea, 600 mM NaCl, 20 mM sodium acetate (pH 5.2), 5 mM β-mercaptoethanol, and 1 mM EDTA (*see* **Note 1**).

15. 3000 MWCO dialysis tubing.
16. SP-Sepharose containing FPLC column.
17. Q-Sepharose containing FPLC column.
18. 1.6×20 cm FPLC column.
19. Refolding buffer: 20 mM Tris–HCl pH 7.5, 2 M NaCl, 1 mM EDTA (pH 8.0), 10 mM β-mercaptoethanol.
20. Superdex 200 column (GE Healthcare).
21. Amicon Ultra-15 Centrifugal Filter, 10 kDa MWCO (EMD Millipore, or equivalent).

2.5 Histone Variant Exchange Assay and Western Blot

1. Exchange buffer: 25 mM HEPES-KOH pH 7.6, 0.37 mM EDTA, 0.35 mM EGTA, 5 mM $MgCl_2$, 70 mM KCl, 10% glycerol, 0.02% Igepal, 1 mM DTT, 0.1 mg/ml BSA.
2. Exchange wash buffer: Exchange buffer containing 400 mM KCl.
3. Exchange dilution buffer: Exchange buffer containing 70 mM KCl.
4. 1 mM ATP.
5. 50 U/ml apyrase.
6. 6x protein loading buffer: *See* Subheading 2.1.
7. 15% SDS-PAGE resolving gel: 15% acrylamide:bis-acrylamide 29:1, 375 mM Tris–HCl pH 8.8, 0.1% SDS, 0.1% ammonium persulfate (APS), 0.1% TEMED.
8. 6% SDS-PAGE stacking gel: 6% acrylamide:bis-acrylamide 29:1, 125 mM Tris–HCl pH 6.8, 0.1% SDS, 0.1% ammonium persulfate (APS), 0.1% TEMED.
9. Semi-dry transfer buffer: 29.9 mM Tris, 24.5 mM glycine, 0.024% SDS, 5% methanol.
10. TBS: *See* Subheading 2.1.
11. 5% milk-TBS: *See* Subheading 2.1.
12. Anti-H3 C-terminus Polyclonal rabbit antibody (ab1791, Abcam).
13. Anti-rabbit HRP-conjugated antibody.

3 Methods

3.1 Preparation of Recombinant p400 from Sf9 Cells

Recombinant human Flag-p400 is expressed from Sf9 insect cells infected with a bacoluvirus containing the construct [4]. Initial steps must be performed in order to amplify the baculovirus to high titers (**steps 1–7**). In parallel, Sf9 insect cells are grown up to inoculate a spinner flask, into which the amplified baculovirus is finally added. In the later steps, cells are harvested and lysed, the protein is purified using Anti-Flag M2-agarose resin and eluted with Flag competitor peptide.

1. Seed 2×10^7 Sf9 cells on two 150 mm plates.
2. Allow the cells to attach for 15 min.
3. Aspirate medium and add fresh TNM-FH medium.
4. Add between 100 μl and 1 ml of low titer recombinant baculovirus stock to the plate. If you know the virus titer of your stock solution, use a multiplicity of infection (MOI) smaller than 1.
5. Incubate the cells at 27 °C for 5 days. Check for signs of infection 2 days after the start of infection (*see* **Note 2**).
6. Harvest the supernatant from the plate and transfer to a sterilized 40 ml round bottom centrifuge tubes.
7. Remove cellular debris by centrifugation at 10,000 × *g*, 10 min at room temperature. Keep the supernatant (approximately 100 ml) in sterilized 50 ml conical tubes (*see* **Notes 3** and **4**).
8. Grow enough Sf9 cells to inoculate 400 ml of TNM-FH medium in a 500 ml spinner flask and grow to a density of 1×10^6 cells/ml on a magnetic platform at 27 °C.
9. Add 100 ml of amplified baculovirus to the spinner flask, for a total volume of 500 ml.
10. Incubate at 27 °C for 36 h on a magnetic platform.
11. Collect the cells by transferring the contents of the spinner flask in two 250 ml centrifuge bottles, centrifuge at the speed of 1000 × *g* for 10 min at 4 °C.
12. Carefully decant the supernatant and resuspend the cells in a small volume of PBS (approximately 10 ml) by gently pipetting the PBS "up and down" over the cell pellet with a 10 ml pipette until it is fully suspended.
13. Transfer each cell suspension in a 50 ml conical tube and add PBS up to 50 ml.
14. Centrifuge at 1000 × *g* for 5 min at 4 °C.
15. Resuspend each pellet in 15 ml of Lysis buffer, rest on ice for 15 min.
16. Transfer the lysate to a 15 ml homogenizer and grind with 25 strokes using the "B" (or tight) pestle.
17. Incubate on ice for 15 min.
18. Transfer the lysate to a 50 ml round bottom centrifuge tube and centrifuge at 27,000 × *g* in a Beckman JA-20 rotor (or equivalent) for 20 min at 4 °C.
19. Transfer the supernatant to two 50 ml conical tubes. Keep a small sample (e.g., 15 μl) aside as a total extract control for SDS-PAGE analysis.

20. To each lysate, add 15 ml of Dilution buffer and 150 μl of M2-Anti-Flag-agarose resin (*see* **Note 5**).
21. Incubate on a rotating platform for 2–3 h at 4 °C.
22. All washes and elution steps are done using centrifugation and aspiration with a pipette.
23. Centrifuge at 4000 × *g*, 5 min, 4 °C. Carefully remove supernatant with a pipette (a little bit of supernatant can be left to avoid losing resin). Keep a small sample, e.g., 15 μl, aside as a flow through control for SDS-PAGE analysis.
24. Wash the resin: add 10 ml of Wash buffer; gently resuspend by inversion and flicking; incubate on a rotating platform 5 min at 4 °C; centrifuge at 4000 × *g* 5 min at 4 °C; carefully remove supernatant with a pipette (a little bit of supernatant can be left to avoid losing resin).
25. Repeat **step 24** three more times for a total of four washes.
26. After the final wash, resuspend the resin of each tube in 300 μl of Wash buffer and transfer the suspension to a fresh 1.5 ml tube (a total of two tubes). Add another 300 μl of Wash buffer to the old 15 ml tube to harvest as many resin beads as possible (rinse the sides of the tube), and pool together with the first 300 μl in the 1.5 ml tube.
27. Centrifuge at 5000 × *g* 5 min in a microcentrifuge at 4 °C.
28. Carefully remove as much supernatant as possible by using a fine-point tip.
29. Elution: add 100 μl of Low Salt Elution buffer; resuspend and incubate on a rotating platform for 10 min at 4 °C; centrifuge at 5000 × *g* 5 min in a microcentrifuge at 4 °C; remove and keep supernatant in a fresh 1.5 ml tube. Pool together the eluates from both tubes.
30. Repeat the elution step, but with 100 μl of High Salt Elution buffer per tube. Aliquots (5–15 μl) should be kept at −80 °C.
31. Analyze the enrichment of Flag-p400 obtained in your experiment by SDS-PAGE and Western blot using an appropriate anti-p400 or anti-Flag antibody.
32. Prepare a standard SDS-PAGE setup with a 6% resolving gel and a 4% stacking gel.
33. Load 5 μl of the Low Salt and High Salt eluates, along with 15 μl of total extract and flow through samples kept at the earlier steps. Run at 150 V for at least 60 min. Monitor the migration of proteins in real time by using a protein prestained molecular marker. Make sure the Flag-p400 protein does not exit the gel (the observed migration is around 350 kDa).
34. Transfer the proteins from the gel to a PVDF membrane using a standard wet transfer apparatus at 4 °C in wet transfer buffer for 4 h at 60 V.

35. Pre-block the membrane in 5% milk-TBS for 1 h at room temperature.
36. Hybridize the membrane with 1:1000 dilution of anti-Flag M2 monoclonal antibody in 5% milk-TBS overnight at 4 °C with gentle shaking. Make sure the solution covers the membrane completely and that it does not evaporate.
37. Remove the antibody solution and place the membrane in a petri dish.
38. Add enough TBS to cover the membrane, incubate at room temperature with shaking for 5 min and remove TBS.
39. Repeat **step 38** three more times (for a total of four washes).
40. Hybridize the membrane with 1:10 000 dilution of anti-mouse HRP-conjugated antibody in 5% milk-TBS for 1 h at room temperature with shaking. Make sure the antibody solution covers the membrane completely and that it does not evaporate.
41. Remove the antibody solution and place the membrane in a petri dish.
42. Add enough TBS to cover the membrane, incubate at room temperature with shaking for 5 min, and remove TBS.
43. Repeat **step 42** three more times (for a total of four washes).
44. Reveal the location of HRP-conjugated antibodies by incubating the membrane with chemiluminescence solution (follow the manufacturer's guidelines) and exposing on film. Expect a strong enrichment of Flag-p400 in the elution samples.

3.2 Purification of Human Core Histones from HeLa cells

1. Grow HeLa cells in 3 l of DMEM + 10% FBS in a spinner flask, up to a density of $0.5–2 \times 10^5$ cells/ml.
2. Harvest the cells in 500 ml centrifuge bottles. Centrifuge at $3000 \times g$ for 5 min, room temperature.
3. Carefully decant supernatant and add 200 ml of cold PBS to each bottle to resuspend the cell pellet. Centrifuge at $3000 \times g$ for 5 min, 4 °C. All subsequent steps should be performed on ice or in a cold room.
4. Carefully decant the supernatant and add approximately 10 ml of cold PBS to each bottle to resuspend the pellets. Split all the suspensions pellets into three 50 ml conical tubes.
5. Centrifuge at $3000 \times g$ for 5 min at 4 °C.
6. Carefully decant the supernatant and weigh the cell pellet. A total of 2–6 g of cells should be obtained.
7. Resuspend the pellets in 20 volumes of Buffer A (20 ml per gram of cell pellet).
8. Lyse the cells with 20 strokes of Dounce homogenizer with a type B pestle. Monitor cell lysis under the light microscope: greater than 90% of lysed cells should be obtained.

9. Transfer the lysates to 50 ml conical tubes and centrifuge at 3000 × *g* for 15 min at 4 °C.
10. Carefully decant the supernatant. The pellet contains unwashed nuclei.
11. Wash the nuclei by carefully resuspending the pellet in 50 ml of Buffer A. Centrifuge at 3000 × *g* for 15 min at 4 °C. Carefully decant the supernatant.
12. Repeat the previous step once more for a total of two washes with Buffer A.
13. Resuspend the pellets in 50 ml of Buffer B with 2–3 strokes of the Dounce homogenizer. Centrifuge at 3000 × *g* for 15 min at 4 °C. Carefully decant the supernatant.
14. Resuspend the pellets in three pellet volumes of Buffer B and pool together in a single 50 ml conical tube.
15. Add dropwise one total pellet volume of Buffer C while stirring gently. Incubate 10 min at 4 °C with gentle stirring.
16. Transfer the suspension to 1.5 ml centrifuge tubes (split the volume equally among the tubes) and centrifuge a max speed 30 min at 4 °C in a microcentrifuge.
17. Carefully aspirate supernatants with a pipette. The pellets contain isolated nuclei. They can be frozen in dry ice and stored indefinitely at −80 °C or processed immediately for core histone purification.
18. Estimate the amount of nuclear material by measuring the amount of DNA in a pellet (*see* **Note 6**).
19. Pool together enough nuclear pellets to add up to the equivalent of 10 mg of DNA. Add 30 ml of HAP buffer and mix gently for 10 min at 4 °C.
20. Add 20 g of Bio-Gel HTP hydroxyapatite and mix gently for 10 min at 4 °C.
21. Pour the suspension in a 2.5 × 20 cm chromatography column and set up the column on an FPLC system. Prepare at least 1000 ml of HAP buffer as buffer A and 500 ml of HAP buffer + NaCl as buffer B to perform a step elution.
22. Wash the column with ten column volumes of HAP buffer at a flow rate of 1 ml/min. This should take roughly 10 h (600 ml).
23. Elute the histones by applying HAP buffer + NaCl (without gradient) and by collecting 8 ml fractions. Measure the absorbance at 230 or 280 nm and pool together the peak fractions.
24. Concentrate the samples to a concentration of at least 2 mg/ml in an Amicon Ultra-15 Centrifugal Filter with a 3 kDa MWCO. If you reach a concentration higher than 2 mg/ml, adjust the concentration to 2 mg/ml with HAP + NaCl buffer.
25. Divide into 20 μl aliquots in 1.5 ml tubes, snap freeze on dry ice, and store at −80 °C.

3.3 In Vitro Assembly of Biotin-Labeled Mononucleosomes

1. First prepare the Sea urchin 5S rDNA fragment that serves as a nucleosome positioning sequence by incubating the pIC208-5S plasmid with EcoRI endonuclease. Prepare the following reaction in a 1.5 ml tube: 20 μl of Cutsmart buffer 10×, 5 μl of EcoRI-HF 100,000 U/ml, 100 μg of pIC208-5S DNA, complete to 200 μl with H_2O (*see* **Note 7**).
2. Incubate the restriction reaction at 37 °C for at least 3 h up to overnight.
3. Add 40 μl of 6× OG DNA loading buffer and mix well.
4. Prepare a 2 % agarose gel in TBE with at least 10 wells (>30 μl in volume).
5. Split the reaction into 10 wells (24 μl per well) of the 2 % agarose gel and run at 100 V for 60 min.
6. Cut the ~200 bp DNA bands and purify by using the Qiaquick gel extraction kit by following the manufacturer's instructions. Make sure to take into account the maximal agarose amount for a single column. Measure the concentration of the purified DNA with a UV spectrophotometer (Nanodrop or equivalent) (*see* **Note 8**).
7. To label the EcoRI-digested 5S rDNA fragments with biotin-14-dATP, prepare the following reaction in a 1.5 ml tube: 7.5 μl NEBuffer 2 (10X), 10 μg of 5S rDNA EcoRI fragment, 2 μl of biotin-14-dATP, 2 μl of dBTP mix, 5 mM $MgCl_2$, 1.5 mM DTT, 1 μl DNA polI (Klenow), complete the volume to 75 μl with H_2O.
8. Incubate 30 min at room temperature.
9. Add 1.5 μl of 500 mM EDTA and incubate at 70 °C.
10. Precipitate the DNA by adding the following: 1 μl of glycogen, 7.5 μl of 3 M sodium acetate pH 5.2, and 200 μl of 100 % ethanol. Mix well and incubate at −20 °C for 20 min.
11. Centrifuge at maximum speed in a microcentrifuge for 10 min. Remove supernatant and add 500 μl of 70 % ethanol.
12. Remove supernatant completely and let the DNA pellet air dry for 5–10 min.
13. Suspend the pellet in 20 μl of H_2O, leave at room temperature for 10 min. Measure the concentration of the purified DNA with a UV spectrophotometer (Nanodrop or equivalent) (*see* **Note 9**).
14. To assemble mononucleosomes on the biotin-labeled-rDNA by salt dilution, prepare the following reaction in a 1.5 ml tube: 400 ng (3.1×10^{-12} mol) of biotin-labeled rDNA, 8 μg (74.1×10^{-12} mol) of HeLa core histones and 2 μl of 5 M NaCl (for a final concentration of 2 M NaCl), complete the volume to 10 μl with H_2O. This represents a 25:1 ratio of histone octamer to DNA molecule ratio. Mix well and incubate at room temperature for 20 min.

15. Successively add the following volumes of 10 mM Tris–HCl pH 7.6 and mix with a pipette: 10, 5, 5, 70, and 100 μl. In between each volume, incubate 20 min at room temperature (*see* **Note 10**).
16. Keep the assembly reaction at 4 °C. Proceed to the following step for quality analysis of mononucleosome assembly. Proceed to **step 23** for immobilization of the mononucleosomes to streptavidin-coated beads.
17. Analysis of the assembly of mononucleosomes is done by running the sample on a native polyacrylamide gel, in parallel with nonassembled DNA. First prepare a 5% polyacrylamide DNA gel using a standard mini gel apparatus (0.75 mm thick).
18. Pre-run the gel in 0.5× TBE at 150 V for 30 min.
19. Prepare the samples for loading: take 20 μl of the assembly reaction, add 2 μl of 50% glycerol. As a negative control, take 40 ng of nonassembled biotin-labeled-rDNA, add 2 μl of 50% glycerol, and complete the volume to 20 μl with H_2O. Also load a lane containing a standard DNA molecular weight marker with OG loading dye, simply to monitor gel migration.
20. Load the samples on the gel and run in TBE at 150 V until the OG dye reaches the bottom of the gel.
21. Remove the gel from the apparatus and carefully place in a container containing a solution of 1× SYBR Green I in 50 mM Tris–HCl pH 7.5.
22. Incubate at room temperature for 10 min. The gel can then be directly analyzed by using a UV transilluminator (*see* **Note 11**).
23. Pipette 60 μl of Streptavidin-coupled magnetic beads in a clean 1.5 ml tube. Add 1 ml of Wash buffer 100, mix and place on the magnet. Remove supernatant and add 1 ml of Wash buffer 100. Repeat this Step twice more for a total of three washes.
24. After the final wash, resuspend the beads in 60 μl of Wash buffer 100.
25. To the remaining 180 μl of assembled mononucleosomes from **step 16**, add 60 μl of prewashed Streptavidin-coupled magnetic beads. Incubate at 4 °C for several hours (overnight) on a rotating platform.
26. Place the sample on the magnet, leave for ~1 min to let the beads settle. Remove the supernatant and add 500 μl of Wash buffer 100. Incubate on a rotating platform for 10 min at 4 °C.
27. Repeat the previous step, but with Wash buffer 600 twice, and finally with Wash buffer 100.
28. After the final wash, resuspend the beads in 400 μl of Wash buffer 100. This represents the immobilized mononucleosome templates at a concentration of~ 1 μg/ml. Store at 4 °C.

3.4 Preparation of Recombinant Flag-H2A.Z/H2B Dimers (See Note 12)

1. Transform the pET expression plasmids containing Flag-H2A.Z and H2B in competent *E. coli* BL21 codon+RIL strain. Plate the transformation on LB-agar containing ampicillin and chloramphenicol (*see* **Note 13**).
2. With a single colony, inoculate 5 ml of 2x-TY-AC with 1× ampicillin and 1× chloramphenicol. Incubate with shaking at 37 °C until the density (OD at 600 nm) reaches 0.6 (*see* **Note 14**).
3. Transfer the 5 ml culture into 1000 ml of 2× TY-AC with 1× ampicillin and 1× chloramphenicol and incubate at 37 °C with shaking until the density (OD at 600 nm) reaches 0.6.
4. Add 0.5 mM IPTG and incubate for 2 h at 37 °C with shaking.
5. Add 150 μM Rifampicin and incubate for an additional 2 h as above.
6. Harvest the cells by centrifugation at 5000 × *g*, 20 min at 4 °C in two 500 ml centrifuge bottles.
7. Discard supernatant and resuspend the cells in 17 ml of wash buffer. Flash freeze in liquid nitrogen and store at −80 °C.
8. Thaw the cell suspension in a 37 °C water bath with constant stirring. Bring the volume up to 25 ml with wash buffer and add 5 mg of Lysozyme.
9. Sonicate on ice, in pulses of 15 s for a total of 5 times at 50% amplitude. Keep cells on ice for >2 min between pulses.
10. Centrifuge at 16,000 × *g* for 20 min at 4 °C. Discard the supernatant (*see* **Note 15**).
11. Resuspend the pellet in 25 ml of TW buffer. Use a 10 ml pipette to mince the pellet as much as possible with up/down motions.
12. Centrifuge at 16,000 × *g* for 20 min at 4 °C and discard the supernatant.
13. Repeat the previous step for an additional time with 25 ml of TW buffer, followed by washing two times with 25 ml of wash buffer. The pellets can be stored at −80 °C at this point.
14. To the thawed pellet, add 350 μl of DMSO, mix with the pipette tip and incubate at room temperature for 30 min.
15. Add 15 ml of unfolding buffer, incubate with gentle rotation for 1 h at room temperature.
16. Centrifuge at 16,000 × *g* for 20 min at room temperature and keep the supernatant aside.
17. Resuspend the pellet, this time in 5 ml of unfolding buffer and Centrifuge at 16,000 × *g* for 20 min at room temperature. Combine the supernatants with the one of **step 16**.
18. Dialyze the supernatant in a 3 kDa MWCO bag with four steps, each for 1 h in 1 l of SAU-200 buffer (*see* **Note 16**).

19. Centrifuge at 16,000 × *g* for 20 min at room temperature.
20. Load the supernatant on the FPLC system with tandem Sepharose Q and Sepharese SP columns, pre-equilibrated in SAU-200. Wash with three column volumes (CV) of SAU-200.
21. Before starting the elution step, remove the Sepharose Q column.
22. Elute using a gradient of SAU-200 to SAU-600 over 7 CV at a flow rate of 2 ml/min. Run the fractions surrounding and within the peak on a 15% SDS-polyacrylamide gel, analyse with an in-gel protein stain (e.g., Coomassie blue).
23. Pool the peak fractions. Dialyze the supernatant at 4 °C in a 3 kDa MWCO bag with three steps of 1 l of H2O with 5 mM 2-mercaptoethanol and 0.2 mM PMSF: twice for 1 h and once overnight.
24. After dialysis, centrifuge at 16,000 × *g* for 20 min.
25. Determine the concentration of the histone solution by measuring the absorbance at 276 nm (*see* **Note 17**).
26. Lyophilize each histone in 2 mg aliquots and store at −80 °C.
27. Dissolve a 2 mg of lyophilized histone Flag-H2A.Z and H2B aliquots in 1 ml of unfolding buffer for 1 h at room temperature with gentle rotation.
28. Calculate the concentration of each histone (*see* **Note 17**). Mix equimolar amounts of each histone and adjust the concentration to 1 mg/ml total using unfolding buffer.
29. Incubate at room temperature for 30 min with gentle shaking.
30. Dialyze the mix at 4 °C in a 3 kDa MWCO bag with 4 steps of 2 l of refolding buffer with three 1 h steps, followed by an overnight step (*see* **Note 18**).
31. Centrifuge the dialyzed histone mix for at 16,000 × *g* for 20 min at 4 °C to remove aggregates.
32. Concentrate the sample to a volume of approximately 1 ml with an Amicon Ultra-15 Centrifugal Filter, 10 kDa MWCO.
33. Centrifuge the sample at maximum speed in a microcentrifuge for 10 min at 4 °C to remove any debris.
34. Load the sample onto a Superdex 200 column, pre-equilibrated in refolding buffer. Run at a flow rate of 0.5 ml/min and collect 0.5 ml fractions. Flag-H2A.Z/H2B dimers should elute at around 74 ml.
35. Run the fractions surrounding and within the peaks on a 15% SDS-polyacrylamide gel, analyze with an in-gel protein stain (e.g., Coomassie blue). Check fractions for the stoichiometry of both histones in the dimer.

36. Pool the dimer fractions and concentrate using an Amicon Ultra-15 Centrifugal Filter, 10 kDa MWCO to ~ 1 ml.
37. Calculate the concentration of the dimer (*see* **Note 19**).
38. Aliquot in 50 μl samples and store at −80 °C.

3.5 Histone Variant Exchange Assay

1. For each histone exchange reaction, transfer the equivalent of 30 ng of immobilized mononucleosome templates on streptavidin-coupled dynabeads from Subheading 3.3 (hereby referred to as *nucleosome-beads*) to a 1.5 ml tube. Multiply the volume of nucleosome-beads according to the number of reactions. We typically prepare eight reactions (*see* **Note 20**).
2. Place the tube on the magnet and let the nucleosome-beads settle for 1 min on ice. Remove the supernatant and add 500 μl of exchange buffer. Repeat this wash step twice more for a total of three washes.
3. After the final wash, resuspend the nucleosome-beads in 46 μl of exchange buffer per reaction. Split the nucleosome-beads by pipetting 46 μl into separate 1.5 ml tubes per reaction.
4. Add the purified Flag-p400 (from Subheading 3.1) in varying amounts in separate tubes. We typically add 0 μl, 1 μl, 2 μl, and 4 μl of Flag-p400 and adjust all volumes by adding 4 μl, 3 μl, 2 μl, and 0 μl of exchange buffer respectively.
5. Add 1 μl of 50 U/ml apyrase (total of 50 mU per reaction) to the control reactions (without ATP). To the exchange reactions, add 1 μl of exchange buffer.
6. Incubate for 15 min at 30 °C on a rotating platform.
7. To the reaction, add 50 μl of exchange buffer containing 300 ng of Flag-H2A.Z/H2B dimer (from Subheading 3.4) (*see* **Note 21**).
8. Add 1 μl of 1 mM ATP to the exchange reactions. To control reactions, add 1 μl of exchange buffer.
9. Incubate 60 min at room temperature on a rotating platform.
10. To stop the reactions, place the tubes on the magnet for 1 min and remove the supernatant. Add 500 μl of exchange wash buffer. Mix at room temperature for 5 min.
11. Wash again with 500 μl of exchange dilution buffer.
12. After the wash, centrifuge the tubes at 1000 × *g* and place on the magnet. Carefully remove all the remaining liquid.
13. To elute the nucleosome-beads, add 25 μl of H_2O followed by 5 μl of 6x protein loading buffer.
14. Vortex and incubate at 100 °C for 5 min.
15. Centrifuge at maximum speed in a microcentrifuge for 2 min. Place the tubes on the magnet and transfer the supernatant to

fresh 1.5 ml tubes. Reactions can be analyzed immediately by Western blot or kept at –20 °C for subsequent analysis.

16. For Western blot analysis, first prepare a 15% SDS-PAGE resolving gel with a 6% SDS-PAGE resolving gel in a 0.75 mm-thick mini-gel montage.
17. Load half of the samples (15 μl) on the gel. Run at 200 V until the dye reaches the bottom of the stacking gel.
18. Dismount the gel and soak in semi-dry transfer buffer for 10 min.
19. Prepare a PVDF membrane by first soaking it in 100% methanol for 1 min, followed by 2 min in H_2O and 5 min in semi-dry transfer buffer.
20. Place the gel over the membrane, in between three slices of blotting paper soaked in semi-dry transfer buffer on each side.
21. Place the "sandwich" in a semi-dry transfer apparatus. Make sure the membrane points toward the anode. Transfer at 1 mA/cm^2 of membrane for 2 h.
22. Pre-block the membrane in 5% milk-TBS for 1 h at room temperature.
23. Hybridize the membrane with 1:1000 dilution of anti-Flag M2 monoclonal antibody in 5% milk-TBS over-night at 4 °C with gentle shaking. Make sure the solution covers the membrane completely and that it does not evaporate.
24. Remove the antibody solution and place the membrane in a petri dish.
25. Add enough TBS to cover the membrane, incubate at room temperature with shaking for 5 min and remove TBS.
26. Repeat **step 25** three more times (for a total of four washes).
27. Hybridize the membrane with 1:10 000 dilution of anti-mouse HRP-conjugated antibody in 5% milk-TBS for 1 h at room temperature with shaking. Make sure the antibody solution covers the membrane completely and that it does not evaporate.
28. Remove the antibody solution and place the membrane in a petri dish.
29. Add enough TBS to cover the membrane, incubate at room temperature with shaking for 5 min, and remove TBS.
30. Repeat **step 29** three more times (for a total of four washes).
31. Reveal the location of HRP-conjugated antibodies by incubating the membrane with chemiluminescence solution (follow the manufacturer's guidelines) and exposing on film in a dark room. Alternatively, any suitable imaging system is capable of supporting detection around wavelengths of 428 nm.

32. The following steps serve as a loading control by immunoblotting nonmodified H3. Wash the membrane three times for 5 min each with TBS.
33. Hybridize the membrane with a 1:10 000 dilution of anti-histone H3 (Abcam ab1791) antibody in 5% milk-TBS overnight at 4 °C with gentle shaking.
34. Follow the **steps 24–31**, except incubate with 1:10 000 dilution of anti-rabbit HRP-conjugated (instead of anti-mouse-HRP) antibody in 5% milk-TBS (*see* **Note 22**).

4 Notes

1. SAU-200 and 600 buffers must be prepared fresh on the day and used within 24 h in order to avoid protein damage with isocyanate. Incubate urea solutions with a mixed bed ion exchange resin such as Amberlite MB-20 to eliminate protein-damaging ions.
2. Early signs of infection include cessation of cell growth, enlarged cell and nuclei diameter and reduced refractivity in contrast phase microscopy. Later signs include decrease in cell density due to lysis. Cells may contain large vacuoles and granular nuclei.
3. Store the virus supernatant in a sterile tube at 4 °C for up to 6 months. For longer storage periods, virus supernatant should be frozen in aliquots at –80 °C.
4. This protocol typically gives a titer around 2×10^8/ml, which is then sufficient for good expression of Flag-p400 in our hands. Optionally, you can determine the virus titer of your amplification solution using either the end-point dilution assay or the plaque assay. Amplification can be repeated two or three times to attain a higher viral titer.
5. Prewash M2-agarose resin in a 15 ml conical tube by centrifugation at $4000 \times g$, 5 min, 4 °C. Remove supernatant and add 10 ml of Dilution buffer. Repeat this wash twice more, for a total of three washes.
6. Take one pellet and estimate the volume of nuclei with a pipette (it should be typically between 50 and 100 μl). Resuspend the pellet in 100 volumes of a 2 M NaCl solution, mix well, and measure the concentration of DNA in the solution on a UV spectrophotometer at 260 nm (or a Nanodrop). The solution might need to be diluted (1:5 or 1:10) in order to obtain an absorbance reading within 0.1 and 0.5 on the spectrophotometer. The amount of DNA can be calculated with the following formula (do not forget to multiply the concentration by the dilution factor if applicable):

$C = A/(\varepsilon l)$

where:

C: the concentration of the nucleic acid, in μg/ml.

A: Absorbance (OD_{260nm}).

l: width of the cuvette which used to hold the solution, in cm, usually is 1 cm.

ε: Extinction coefficient of double stranded DNA: 1 OD_{260} Unit = 50 ug/ml.

7. Make sure to have a high quality large-scale prep of pIC208-5S plasmid, (such as a Maxi-prep) with at least 100 μl at a concentration of 1 μg/μl.
8. A minimal amount of 10 μg of EcoRI-digested rDNA should be obtained. If you do not have enough, repeat the pIC208-5S restriction digestion and purification steps.
9. You should get a concentration around 300 ng/μl of biotin-labeled rDNA, this is sufficient for at least a dozen nucleosome assembly reactions.
10. When taking into account the concentration of NaCl in the volume of HeLa histones, the initial concentration is 2 M. This step-wise salt dilution process is to ensure that the histones do not simply aggregate with the DNA. The histone octamer dissociates from DNA at roughly 0.8 M NaCl. This is why two "small" dilution steps are required (adding 5 μl in the second and third steps) when the concentration of NaCl goes from 1 M to 0.8 M to 0.667 M respectively.
11. You should expect an upward shift of the band in the assembled mononucleosome sample. Over 90% of the DNA should be shifted.
12. For this assay, we use an allele of the human H2A.Z-encoding gene carrying a single Flag-tag on the N-terminal end.
13. *E. coli* BL21 codon + RIL carries a chloramphenicol-resistant plasmid expressing tRNA for rare codons in *E. coli*: arg (AGA, AGG), ile (AUA), and leu (CUA).
14. Before selecting a colony for the large-scale histone preparation, it is important to test many colonies on a small-scale. Pick several colonies and inoculate 5 ml of 2x-TY-AC with 1× ampicillin and 1× chloramphenicol and grow at 37 °C until the OD 600 nm reaches 0.6. Add 0.5 mM IPTG and let grow at 37 °C for 2 h (also keep a culture with noninduced cells as a control). Resuspend the pellet in 1× SDS-loading buffer, boil for 5 min, and load on a 15% SDS-PAgel. Verify the expression (compared to the un-induced control) by staining with any appropriate in-gel protein stain (e.g., Coomassie blue).

15. Recombinant histones are expressed in *E. coli* in insoluble inclusion bodies, so the pellet is kept at this point. The following steps serve to wash the inclusion bodies, after which histones are about 80% pure.
16. During the dialysis step, prepare the FPLC columns containing Sepharose Q and Sepharose SP. For each column, add 30 ml of resin to the HR 16/20 column and let settle. The bed volume should be about half (i.e., 15 ml). Set up the packed columns in tandem on the FPLC, starting with the Q followed by the SP. Equilibrate the columns with SAU-200 buffer.
17. Use the formula presented in Note 6, with the following Molar Extinction coefficients:

    Flag-H2A.Z: $\varepsilon = 1$ OD_{276} Unit = 3.25 mg/ml.

    H2B: $\varepsilon = 1$ OD_{276} Unit = 1.9 mg/ml.
18. During this step, prepare a Superdex-200 column on the FPLC and equilibrate it with refolding buffer.
19. Use the formula presented in Note 6, with the following Molar Extinction coefficient for the Flag-H2A.Z-H2B dimer: $\varepsilon = 1$ OD_{276} Unit = 2.4 mg/ml
20. Plan ahead to prepare enough mononucleosome-containing beads to perform: the exchange reaction with varying amounts of FLAG-p400 (including none) and controls containing apyrase without ATP. Apyrase catalyses the hydrolysis of ATP into AMP and inorganic phosphate. The addition of this enzyme is to ensure that any ATP carried over from the preparation steps is eliminated in the control reaction.
21. Dilute the Flag-H2A.Z/H2B dimer sample to 0.6 μg/μl by adding exchange buffer.
22. Nonmodified H3 serves as a crucial loading control. Any specific antibody against other histones (except H2A or H2A.Z) could be suitable as a loading control.

References

1. Luger K (2003) Structure and dynamic behavior of nucleosomes. Curr Opin Genet Dev 13:127–135
2. Billon P, Cote J (1819) Precise deposition of histone H2A.Z in chromatin for genome expression and maintenance. Biochim Biophys Acta 1819:290–302
3. Mizuguchi G, Shen X, Landry J, Wu WH, Sen S, Wu C (2004) ATP-driven exchange of histone H2AZ variant catalyzed by SWR1 chromatin remodeling complex. Science 303:343–348
4. Gevry N, Chan HM, Laflamme L, Livingston DM, Gaudreau L (2007) p21 transcription is regulated by differential localization of histone H2A.Z. Genes Dev 21:1869–1881
5. Ruhl DD, Jin J, Cai Y, Swanson S, Florens L, Washburn MP, Conaway RC, Conaway JW, Chrivia JC (2006) Purification of a human SRCAP complex that remodels chromatin by incorporating the histone variant H2A.Z into nucleosomes. Biochemistry 45:5671–5677

Chapter 3

Purification of Yeast Native Reagents for the Analysis of Chromatin Function-I: Nucleosomes for Reconstitution and Manipulation of Histone Marks

Nicolas Lacoste, Wajid Bhat, and Jacques Côté

Abstract

Purification of native biological material provides powerful tools for the functional analysis of enzymes and proteins in chromatin. In particular, histone proteins harbor numerous post-translational modifications, which may differ between species, tissues, and growth conditions and are lacking on recombinant histones. Moreover, the physiological substrate of most enzymes that modify histones is chromatin and the majority of these enzymes need to be part of a multiprotein assembly to be able to act on chromatin. For the yeast *Saccharomyces cerevisiae* different chromatin purification protocols are available but often result in poor yields or rely on genetic manipulation. We present a simple purification protocol that can yield up to 150 μg of pure native chromatin per liter of yeast culture. The purified material can be obtained from mutant cells lacking specific histone modifications and can be used in in vitro chromatin assembly for biochemical studies. Based on the extremely high degree of conservation throughout eukaryotes, this modifiable native chromatin can be used in studies with factors from other organisms including humans.

Key words Chromatin, Histones, Acetylation, Methylation, Purification, Yeast

1 Introduction

Purification of biologically relevant material is of great interest to chromatin investigators who want to decipher the precise functions of enzymes. In some cases prokaryotic and eukaryotic expression systems facilitate this process as proteins can be produced easily and in copious amounts. However, the two major drawbacks are the lack of or dissimilar patterns of post-translational modifications and the absence of other associated protein partners. Thus, working with native reagents approaches more closely cellular conditions while still achieving excellent levels of purity.

In the chromatin field the advantage of using native purified components is three-fold. First, most enzymes that target histones for modification actually act on histones in the context of chromatin in vivo, which has been shown in several instances to change their

Benoit Guillemette and Luc Gaudreau (eds.), *Histones: Methods and Protocols*, Methods in Molecular Biology, vol. 1528, DOI 10.1007/978-1-4939-6630-1_3, © Springer Science+Business Media New York 2017

specificity for a histone tail and/or residue [1]. Second, histones contain a large number of conserved and highly regulated post-translational modifications that can modulate the action of chromatin modifying and remodeling activities [2–4]. And thirdly, typically, enzymes that modify histones need to be associated with cofactors in stable multisubunit complexes to be able to act on chromatin, while it is often not the case on free histones [1]. In addition, a few specific histone-modifying enzymes have been shown to be only active on chromatin substrates, being unable to target free histones [1, 5, 6]. Over the years, laboratories working in the field of chromatin discovered multiple ways to bypass the need of purifying native chromatin. Calf thymus could be a great source of histones that are available commercially (Sigma, Worthington Biochemical Corporation), but only as free histones, not chromatin. It is also possible to produce all four core histones as recombinant forms in *E. coli*, to purify and reconstitute them into nucleosomes [7, 8]. The major problem with these is the absence of native modifications on the histone tails, while it is well known that there are critical cross-talks between histone modifications as well as reader modules for these marks within a large number of nuclear factors implicated in all DNA-based processes [4, 9–11]. Indeed, many chromatin modifying/remodeling enzymes themselves contain protein domains that recognize specific histone marks [10, 12]. To circumvent this problem research groups have developed technologies using chemical ligation, methyl-lysine analogs, and bacterial genetic manipulation in order to introduce homogenous specific modification on recombinant histones [13–16]. While these are powerful approaches, they are in some instance expensive with limited versatility or yield.

An old method to purify histones is to take advantage of their highly basic nature by performing acid extraction of nuclei. This simple method produces highly purified histones but suffers lower recovery of acetylated or phosphorylated isoforms and poor solubility/stoichiometry that impairs their use in nucleosome reconstitution. Furthermore, if you do not use particular sequences of DNA to reconstitute chromatin with free/recombinant histones, nucleosomes would not be regularly spaced along the DNA. To produce regularly spaced nucleosomes many labs use cellular extracts [17, 18]. However, this technique produces a mix of chromatin and other proteins that can be a significant problem during subsequent manipulations. A protocol to purify native chromatin from human cells is available [19, 20] and its use in nucleosome reconstitution by the transfer method has been widely used. Purification of native chromatin is also advantageous when analyzing regulation of histone modifications in vivo. Typically, data in the literature present western analysis of whole cell extracts (RIPA or acid extracted). These do not distinguish newly synthesized/free histones from the ones assembled in chromatin, while it has been known for years that they carry different specific modifications.

Yeast has been a model of choice to study chromatin function over the past 20 years because of its powerful genetic, biochemical, and proteomic/genomic techniques. While histone modifications and chromatin remodelers have been intensely studied in yeast, the use of purified native chromatin from these cells has been very limited. Yeast nucleosomes are more opened and less stable than human nucleosomes [21] and heterochromatin is virtually absent in yeast chromatin compared to human chromatin. Even if histones are some of the most conserved proteins in eukaryotes, differences still exist particularly in the N-terminal domain where the majority of the post-translational modifications are found [22]. For example, histone H3 lysine 9 and histone H4 lysine 20 are not methylated in *S. cerevisiae.* It is therefore strange to realize that most reports on yeast chromatin modifiers/remodelers used heterologous histones/chromatin for in vitro assays. Thus, the enzymatic characterization of yeast activities requires the use of their physiological substrate, i.e., purified yeast chromatin. For example, the yeast NuA4 acetyltransferase complex was initially thought to target mostly histone H4 but also histone H2A to a lesser extent [23]. More recent work using yeast chromatin as substrate clearly indicates that both tails are equally targeted by the complex [24].

Different protocols for purifying chromatin from yeast cells are available but most of them target plasmidic chromatin also known as minichromosome [25, 26]. Yields of purification are low and not suitable for the use as substrate in enzymatic assays. Another way to purify yeast chromatin uses epitope-tagged histones [27, 28], but it can create a problem in the enzymatic assays since the tags are located where the major post-translational modifications occur (N- or C-terminal tails). Furthermore, in order to obtain chromatin from different mutant backgrounds each yeast strain needs to be tagged.

Importantly, due to the compact yeast genome with limited intergenic sequences, yeast chromatin contains high physiological levels of specific modifications linked to transcription like H3K4/36/79me, which can be completely removed through genetic manipulation of the enzymes responsible for them. This proves to be very useful when studying the function of reader modules within mammalian protein complexes (e.g., [29]).

For all those reasons we develop and describe here an efficient protocol to purify large amounts of native chromatin from any yeast strain.

2 Materials

1. Sensitizing Buffer: 100 mM Tris–HCl pH 9.4, 10 mM DTT.
2. Spheroplasting buffer: 1.2 M sorbitol, 20 mM Hepes-NaOH pH 7.5.
3. 30 mg/ml Lyticase in sodium phosphate pH 7, 50% glycerol.

4. Wash Buffer: 1.2 M sorbitol, 20 mM Pipes pH 6.4, 1 mM $MgCl_2$.
5. Nuclear Isolation Buffer (NIB): 250 mM sucrose, 60 mM KCl, 1 mM PMSF, 14 mM NaCl, 0.8 % Triton-X-100, 5 mM $MgCl_2$, 1 mM $CaCl_2$, 15 mM MES.
6. Buffer A: 10 mM Tris–HCl pH 8, 0.5 % NP-40, 75 mM NaCl, 5 mM β-glycerophosphate, 5 mM Na-butyrate, 1 mM PMSF.
7. B400: 10 mM Tris–HCl pH 8, 400 mM NaCl, 5 mM β-glycerophosphate, 5 mM Na-butyrate, 1 mM PMSF.
8. B600: 10 mM Tris–HCl pH 8, 600 mM NaCl, 5 mM β-glycerophosphate, 5 mM Na-butyrate, 1 mM PMSF.
9. Medium Salt Buffer (MSB): 20 mM Hepes-NaOH pH 8, 400 mM NaCl, 1 mM EDTA, 5% glycerol, 1 mM PMSF, 1 mM β-mercaptoethanol, 2 μg/ml pepstatin, 2 μg/ml leupeptin, 5 μg/ml aprotinin, 5 mM Na-butyrate, 5 mM β-glycerophosphate, 0.5 mM spermidine, 0.15 mM spermine.
10. High Salt Buffer (HSB): 20 mM Hepes-NaOH pH 8, 650 mM NaCl, 1 mM EDTA, 340 mM Sucrose, 1 mM PMSF, 1 mM β-mercaptoethanol, 2 g/ml pepstatin A, 2 μg/ml leupeptin, 5 μg/ml aprotinin, 5 mM Na-butyrate, 5 mM β-glycerophosphate, 0.5 mM spermidine, 0.15 mM spermine.
11. Low Salt Buffer (LSB): 20 mM Hepes-NaOH pH 8, 100 mM NaCl, 1 mM EDTA, 1 mM PMSF, 1 mM β-mercaptoethanol, 2 μg/ml pepstatin A, 2 μg/ml leupeptin, 5 μg/ml aprotinin, 5 mM Na-butyrate, 5 mM β-glycerophosphate, 0.5 mM spermidine, 0.15 mM spermine.
12. Dialysis membrane 6000–8000 Da cut off.
13. Micrococcal Nuclease (Sigma).
14. 1 M $CaCl_2$.
15. 0.5 M EGTA (pH 8).
16. TE: 10 mM Tris–HCl pH8, 1 mM EDTA pH 8.
17. 10% SDS.
18. Centrifuge with SLA3000/GSA (Sorvall) or JA10 rotor (Beckman) and SS34/JA20 rotor.
19. Benchtop Centrifuge.
20. Ultra-centrifuge with 70 Ti rotor (Beckman).
21. Centriprep 30 (Amicon).
22. FPLC with superose 6 column.
23. Dialysis Buffer: 20 mM Hepes-NaOH pH 7.5, 1 mM EDTA, 1 mM PMSF, 1 mM β-mercaptoethanol, 2 μg/ml pepstatin A, 2 μg/ml leupeptin, 5 μg/ml aprotinin, 2 mM Na-butyrate, 5 mM β-glycerophosphate, 0.5 mM spermidine, 0.15 mM spermine.

3 Description of the Method

We developed an efficient protocol for the purification of yeast native chromatin based on a protocol used for human cells [19]. The following protocol starts with 1 l of yeast cells grown to late log phase (OD_{600} between 2 and 2.5 representing approximately 10^{11} cells) and the recovery can be up to 150 μg of native chromatin. The protocol can be scaled up or scaled down as needed.

3.1 Preparation of Spheroplasts

1. The preparation of nuclei is an important step in the purification of chromatin (*see* **Note 1**). All steps are carried out at 4 °C unless otherwise indicated.
2. Harvest cells from 1 l of culture (at OD_{600} between 2 and 2.5) by centrifugation in SLA3000/GSA (Sorvall) or JA10 rotor (Beckman) at 5000 × *g* for 5 min.
3. Wash the cell pellet with 250 ml of cold H_2O and spin for 5 min at 5000g.
4. Resuspend cells in 25 ml of sensitizing buffer and keep on a wheel or a rocking plate for 15 min at 30 °C.
5. Pellet cells by centrifugation (5 min, 5000 × *g*) in a SS34 (Sorvall) or a JA20 (Beckman) rotor.
6. Wash pellet with 30 ml of spheroplasting buffer and then resuspend in 40 ml of spheroplasting buffer supplemented with 7.5 mg of Lyticase (in a 50 ml conical tube).
7. Incubate at 30 °C on a wheel or a rocking plate for approximately 1–1.5 h (*see* **Note 2**) and stop the reaction by spinning cells for 5 min at 5000g.

3.2 Isolation of Nuclei

1. Wash the pellet (spheroplasts) in 40 ml of wash buffer followed by centrifugation (5 min, 5000 × *g*).
2. Resuspend pellet in 25 ml of nuclear isolation buffer and incubate for 20 min on ice (*see* **Note 3**).
3. Pellet nuclear material by centrifugation for 5 min at 5000 × *g* in a SS34/JA20 rotor. The cloudy supernatant composed of cytoplasmic components is discarded.
4. Wash the white nuclear pellet two times for 5 min on ice in 25 ml of NIB to remove remaining intact cells and debris (yellow). At this point the pellet should be entirely white with a clear supernatant.
5. Wash the nuclear pellet three times with 25 ml of buffer A, each one for 15, 10, and 5 min on ice, respectively (*see* **Note 4**).
6. Wash nuclear pellet three times with 25 ml of buffer B400 for 10 min (5 min for the 2nd and 3rd washes) and centrifuge (5 min, 5000 × *g*, SS34/JA20 rotor) (*see* **Note 5**).

7. Resuspend nuclei in 8 ml of B400 buffer and transfer to a 15 ml conical tube.
8. Pellet nuclei by centrifugation for 5 min at 3000 × *g* in a bench-top centrifuge. About 2 ml of nuclei is recovered per liter of culture.
9. Eliminate supernatant, flash freeze nuclei in liquid nitrogen, and store at −80°C. Nuclei at this step can also be used for purification of core histones, *see* Chapter 5 by Jourquin and Geli in this book.

3.3 Native Chromatin Purification

1. The following chromatin purification protocol must be completed within 1 day to prevent extensive degradation of histone H3 (*see* **Note 11**).
2. All buffers used in this protocol are supplemented with sodium butyrate to prevent histone deacetylation, β-glycerophosphate to prevent dephosphorylation, a protease inhibitor cocktail and two chromatin stabilizer molecules (spermine and spermidine).
3. Thaw nuclei on ice, resuspend in 40 ml of medium salt buffer (MSB) and transfer to a 40 ml dounce homogenizer.
4. Disrupt nuclei aggregates by giving five gentle strokes with pestle B.
5. Transfer the solution to a polycarbonate SS34 rotor tube and centrifuge 20 min at 11,500 rpm (16,400 × *g*) at 4 °C.
6. Remove supernatant and resuspend nuclear pellet in 6 ml of high salt buffer (HSB).
7. Transfer to a dounce homogenizer (7 ml) and give 400 strokes with pestle B, avoiding bubbles as much as possible. This step is the most crucial with the final yield largely depending on the amount of chromatin extracted from the nuclei (*see* **Note 6**).
8. After the physical extraction with the homogenizer, centrifuge the chromatin suspension at 11,500 rpm (16,400 × *g*) for 30 min in a SS34 rotor and recover the supernatant.
9. Dialyze the chromatin solution twice for 1–2 h at 4°C against 750 ml of low salt buffer (LSB).
10. Add $CaCl_2$ to a final concentration of 3 mM.
11. The next step is to digest chromatin with MNase to obtain chromatin fragments ranging from mononucleosome to deca-nucleosome length.
12. Do a test digestion to determine the amount of MNase to be used. Take six aliquots of 50 μl for test digestion and digest with 1 μl of a series of MNase dilutions (1 U/μl–0.1 mU/μl) (*see* **Note 7**).
13. Scale up the digestion for 8 ml chromatin solution (pre-warmed at 37 °C).

14. Stop the reaction by adding 1/10 volume of EGTA (0.5 M pH 8).
15. Increase the NaCl concentration to 600 mM (by slowly adding 2 M NaCl while mixing).
16. Centrifuge the chromatin solution at 45,000 rpm (200,000 × *g*) in a 70 Ti rotor (Beckman Ultracentrifuge) at 4 °C for 45 min to eliminate insoluble material.
17. Concentrate the solution to 2 ml with a Centriprep 30 (Amicon).
18. Centrifuge for 20 min 4 °C at 20,000 × *g* to eliminate insoluble material and load onto a superose 6 column pre-equilibrated in HSB without sucrose.
19. Run the column at 0.2 ml/min and collect 500 μl fractions. Fractions can be left in the collector overnight at 4 °C to avoid a freeze/thaw cycle.
20. Analyze the fractions for both DNA length and purity (*see* **Note 8**) (Fig. 1a, b).

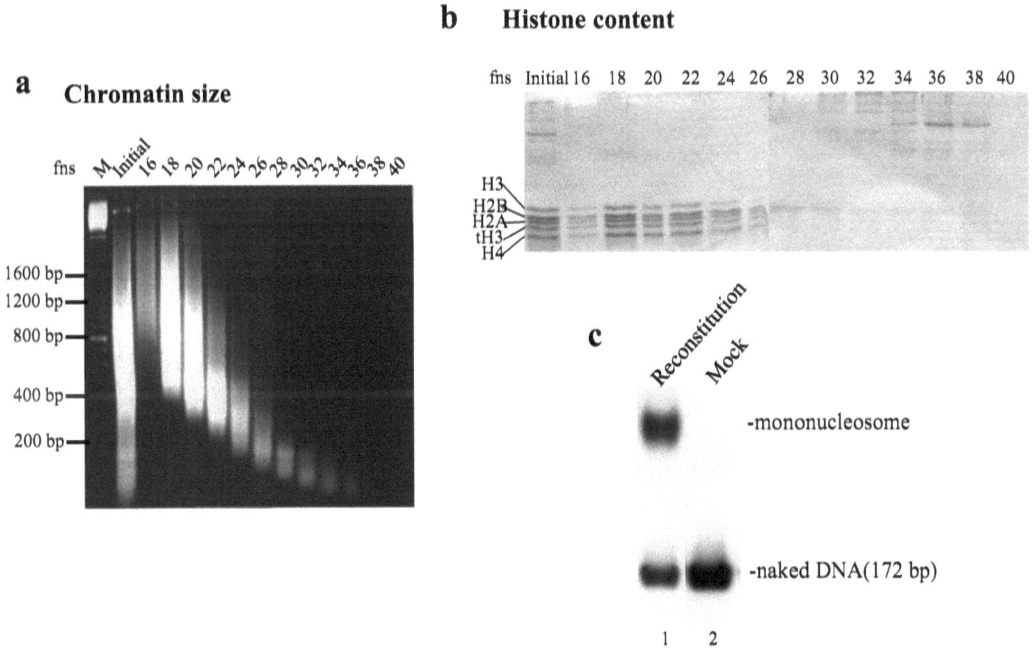

Fig. 1 Purification of yeast native chromatin. (**a**) Native yeast chromatin was purified from BY4741 yeast and aliquots from the gel filtration fractions were deproteinized and loaded on a 2 % agarose gel to analyze chromatin fragment lengths. (**b**) The same fractions were analyzed for histone content and purity on an 18 % SDS-PAGE followed by coomassie staining. The extra band represents a truncated portion of histone H3 (aa22-135, labeled tH3). (**c**) Nucleosome reconstitution with octamer transfer. A 172 bp end-labeled 5S DNA fragment (2.5 ng) was reconstituted by octamer transfer with 2.5 μg of yeast chromatin as described in [19]. Reconstitution was run in a 4 % acrylamide, 0.5× TBE, 1.5 mm gel at 200 V for 2 h, dried and exposed to film

21. Pool the fractions together taking into consideration quantity, purity and length of chromatin (*see* **Note 9**).
22. Concentrate the pooled fractions to 600 μl with a Centriprep 30 (Amicon) and dialyze twice for 1 h against 750 ml of dialysis buffer.
23. The purified chromatin is then aliquoted to reduce the number of freeze/thaw cycles (*see* **Note 11**).
24. This native chromatin can then be used for the analysis of your choice (*see* **Notes 10, 12** and **13**).

4 Notes

1. The preparation of nuclei is an important step in the purification of chromatin, because the cytosolic fraction is removed and the nuclei undergo successive washes to extract most of the nuclear proteins except those strongly associated with DNA. To prepare intact nuclei, the cell wall is digested using a lytic enzyme. In the presence of an osmotic stabilizer, such as sorbitol, yeast cells remain intact after their cell wall has been removed and are referred to as spheroplasts.
2. To follow the progress of spheroplast formation, a 40 μl aliquot of cells is taken at $t=0$ and further time points, diluted in 1% SDS, mixed and left for 2 min at which point the A600 is measured. When the A600 has dropped to 10% of the $t=0$ value, the reaction is stopped by spinning cells for 5 min at 5000g. The same time can be generally used for any yeast strain with the same lot of Lyticase.
3. At this step, it is important to resuspend the clumps as much as possible using a pipette or spatula. The cloudy supernatant composed of cytoplasmic components is discarded. The white nuclear pellet is washed twice.
4. The detergent produces holes in the nuclear membrane whereby soluble nuclear components are extracted. β-glycerophosphate prevents protein dephosphorylation and sodium butyrate inhibits deacetylation.
5. The more extensive washes in medium salt buffer B400 eliminate non-histone, chromatin-associated proteins without disrupting chromatin structure.
6. This step is the most crucial with the final yield largely depending on the amount of chromatin extracted from the nuclei. This milder method has proven the most effective with the least amount of histone degradation observed. The high salt concentration allows chromatin decondensation by disrupting linker histone interaction and soluble chromatin fragments are produced by mechanical shearing. Sonication can also be used

at this step but has a negative impact on H3 N-terminal tail stability. Extraction efficiency can be evaluated by doing a fast spin, and measuring OD_{260} of the supernatant after dilution in 2 M NaCl (concentration should be at least 1 mg/ml).

7. It is recommended to use MNase test dilutions on an aliquot of chromatin solution before doing the large-scale digestion since chromatin from different genetic backgrounds (e.g., certain mutants) could be digested differently. After dialysis, the $CaCl_2$ concentration is raised to 3 mM (required for MNase) and 6 aliquots of 50 μl are taken. To each aliquot 1 μl of a series of MNase dilutions (1 U/μl–0.1 mU/μl) is added and incubated 5 min at 37 °C. Reactions are stopped by adding 1/10 volume of EGTA (500 mM pH 8) to chelate calcium and 50 μl TE, 5 μl of SDS 10%, and 5 μl of 10 mg/ml proteinase K are added and incubated 1 h at 50 °C. After a phenol-chloroform extraction followed by an ethanol precipitation, the DNA pellet is resuspended in 50 μl of TE and 1 μl of RNAse A (10 mg/ml) is added. After incubation at 37 °C for 30 min, 10–20 μl are analyzed on a 2% agarose gel. The concentration of MNase that generates a smear averaging between 200 and 2000 bp is optimal.

8. Starting from the void volume, 15 μl of even numbered fractions are mixed with 15 μl of TE, 1.5 μl of SDS 10%, and 1.5 μl of proteinase K (10 mg/ml) and incubated 1 h at 50 °C. Samples are brought up to 200 μl with water, followed by phenol-chloroform extraction and ethanol precipitation. To analyze the length of the chromatin, 20–40% of the resuspension is loaded onto a 2% TAE agarose gel (Fig. 1a). 10 μl of the same fractions are analyzed by 18% SDS-PAGE in Tris/glycine buffer (Tris 50 mM, glycine 196 mM, SDS 0.1%) run for 2 h in a mini protean II apparatus (Biorad) and stained with Coomassie (Fig. 1b).

9. Fractions are pooled taking into consideration quantity, purity, and length of chromatin. We do not take fractions with DNA smaller than 200 bp since these fragments are not likely to harbor intact nucleosomes. A typical pool includes fraction 16–25 or 26. Alternatively, different lengths of chromatin can be grouped by pooling long oligonucleosomes (LON, >1000 bp) and short oligonucleosomes (SON, 200–1000 bp).

10. This yeast native chromatin can be used as substrate for various biochemical assays, whether it is enzymatic activities (e.g., HAT assay, HMT assay, e.g., Fig. 2a, *see* Chapter 4) or in vitro binding experiments. One useful experiment with native chromatin is to perform histone octamer transfer onto a radiolabeled DNA probe (*see* detailed protocol in [19, 20]). This is also possible with yeast chromatin as shown on the mononucleosome gel shift assay presented in Fig. 1c. These reconstitutions and biochemical assays with chromatin from any growth

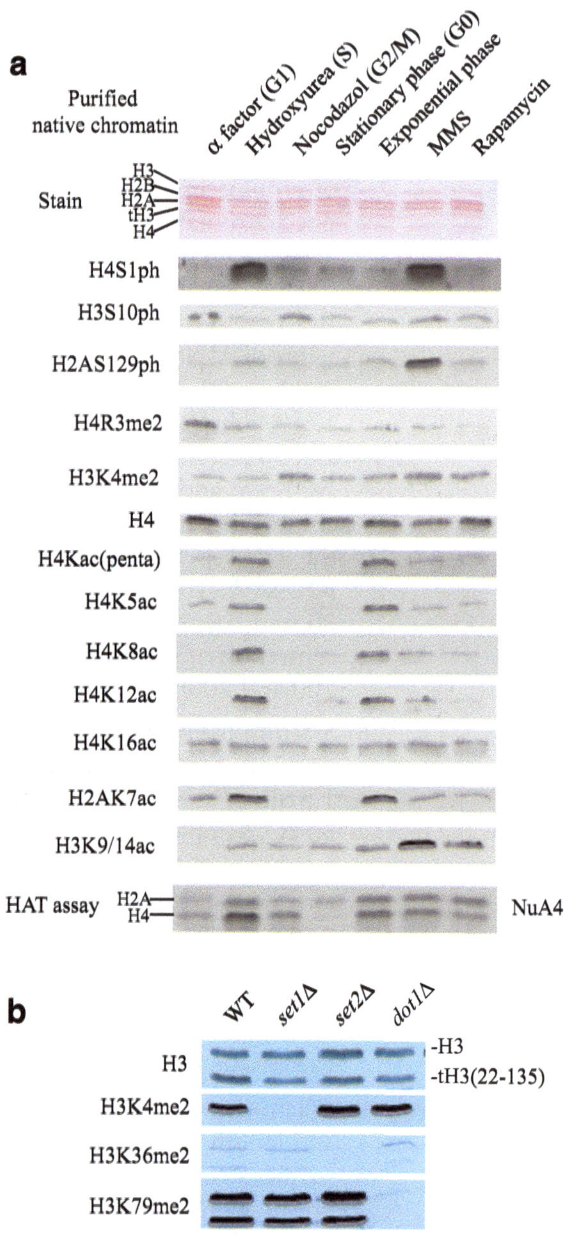

Fig. 2 Analysis of yeast native chromatin purified from cells grown in different conditions or from specific histone modifiers mutants. (**a**) Native chromatin was purified from yeast cells blocked in different phases of the cell cycle or under stress (MMS: genotoxic; rapamycin: nutrient deprivation) and analyzed by western blot for several histone post-translational modifications (modified from [31]) and histone acetyltransferase assay with the NuA4 complex (last panel; with 3H-acetyl-CoA after gel migration and fluorography, (*see* Chapter 4). (**b**) Native chromatin was purified from yeast cells mutated for the different H3 methyltransferases (Set1, Set2, and Dot1) and analyzed by western blot for the corresponding histone modifications

conditions or mutant background (Fig. 2a, b) can be performed to study altered interactions with factors in binding assays, chromatin remodeling with SWI/SNF family ATPases, in vitro transcription assays, and native chromatin immunoprecipitation analysis (N-ChIP).

11. Speed is important while performing this protocol since part of histone H3 N-terminal tail has a high propensity to get lost during the process, the cleavage occurring after alanine 21 [5]. Repeated freeze/thawing also increases the loss of histone H3 tail. Histone H3 partial truncation is the major drawback of the protocol presented here (*see* Fig. 1b. lanes 16–26, also Fig. 2b) but the purified chromatin is still suitable for most studies. Speed remains the key and preparations with time cut to minimum by non-stop work from start to finish led to chromatin with no detectable/minimal H3 truncation (*see* examples in [24, 30]).
12. The concentration of the purified chromatin ranges from 0.25 to 0.5 mg/ml, which is enough for most enzymatic analysis, but some experiments require higher concentrations. In these cases, purification can be performed as described but starting with nuclei from 2 to 3 l of yeast culture to obtain a more concentrated but less pure chromatin preparation.
13. Using this protocol, it is possible to recover up to 150 μg of pure chromatin from one liter of yeast culture. It represents about 300 reactions using any of the enzymatic assays described above. Nevertheless, the protocol presents some drawback. The method uses a lot of protease inhibitors to prevent the loss of H3 N-terminus, especially when doing dialysis, and we do not know what could happen if these were omitted from the buffer. The optimal MNase quantity is determined for each purification since digestion levels seem to vary from one experiment to another. Furthermore, when purifying chromatin from strains with altered chromatin structure, it is essential to perform the MNase titration assay.

Acknowledgments

We are grateful to Rhea Utley for significant corrections to this manuscript. This work was supported by a grant from the Canadian Institutes of Health Research (CIHR; MOP-14308). N.L. was supported by a Canadian Government Foreign Affairs/International Trade studentship. J.C. holds the Canada Research Chair in Chromatin Biology and Molecular Epigenetics.

References

1. Lalonde ME, Cheng X, Cote J (2014) Histone target selection within chromatin: an exemplary case of teamwork. Genes Dev 28:1029–1041. doi:10.1101/gad.236331.113
2. Huang H, Sabari BR, Garcia BA et al (2014) SnapShot: histone modifications. Cell 159: 458–458.e1
3. Kouzarides T (2007) Chromatin modifications and their function. Cell 128:693–705
4. Zentner GE, Henikoff S (2013) Regulation of nucleosome dynamics by histone modifications. Nat Struct Mol Biol 20:259–266. doi:10.1038/nsmb.2470
5. Lacoste N, Utley RT, Hunter JM et al (2002) Disruptor of telomeric silencing-1 is a chromatin-specific histone H3 methyltransferase. J Biol Chem 277:30421–30424. doi:10.1074/jbc.C200366200
6. Strahl BD, Grant PA, Briggs SD et al (2002) Set2 is a nucleosomal histone H3-selective methyltransferase that mediates transcriptional repression. Mol Cell Biol 22:1298–1306
7. Tanaka Y, Tawaramoto-Sasanuma M, Kawaguchi S et al (2004) Expression and purification of recombinant human histones. Methods 33:3–11. doi:10.1016/j.ymeth.2003.10.024
8. Dyer PN, Edayathumangalam RS, White CL et al (2004) Reconstitution of nucleosome core particles from recombinant histones and DNA. Methods Enzymol 375:23–44
9. Fischle W, Wang Y, Allis CD (2003) Histone and chromatin cross-talk. Curr Opin Cell Biol 15:172–183
10. Musselman CA, Lalonde ME, Cote J et al (2012) Perceiving the epigenetic landscape through histone readers. Nat Struct Mol Biol 19:1218–1227. doi:10.1038/nsmb.2436
11. Suganuma T, Workman JL (2011) Signals and combinatorial functions of histone modifications. Annu Rev Biochem 80:473–499. doi:10.1146/annurev-biochem-061809-175347
12. Yun M, Wu J, Workman JL et al (2011) Readers of histone modifications. Cell Res 21:564–578. doi:10.1038/cr.2011.42
13. He S, Bauman D, Davis JS et al (2003) Facile synthesis of site-specifically acetylated and methylated histone proteins: reagents for evaluation of the histone code hypothesis. Proc Natl Acad Sci 100:12033–12038
14. Simon MD, Shokat KM (2012) A method to site-specifically incorporate methyl-lysine analogues into recombinant proteins. Methods Enzymol 512:57–69. doi:10.1016/B978-0-12-391940-3.00003-2
15. Neumann H, Hancock SM, Buning R et al (2009) A method for genetically installing site-specific acetylation in recombinant histones defines the effects of H3 K56 acetylation. Mol Cell 36:153–163. doi:10.1016/j.molcel.2009.07.027
16. Shogren-Knaak MA, Peterson CL (2004) Creating designer histones by native chemical ligation. Methods Enzymol 375:62–76
17. Becker PB, Tsukiyama T, Wu C (1994) Chromatin assembly extracts from Drosophila embryos. Methods Cell Biol 44:207–223
18. Wippo CJ, Korber P (2012) In vitro reconstitution of in vivo-like nucleosome positioning on yeast DNA. Methods Mol Biol 833:271–287. doi:10.1007/978-1-61779-477-3_17
19. Côté J, Utley RT, Workman JL (1995) Basic analysis of transcription factor binding to nucleosomes. Methods Mol Genet 6:108–128
20. Utley RT, Owen-Hughes TA, Juan LJ et al (1996) In vitro analysis of transcription factor binding to nucleosomes and nucleosome disruption/displacement. Methods Enzymol 274:276–291
21. White CL, Suto RK, Luger K (2001) Structure of the yeast nucleosome core particle reveals fundamental changes in internucleosome interactions. EMBO J 20:5207–5218. doi:10.1093/emboj/20.18.5207
22. Malik HS, Henikoff S (2003) Phylogenomics of the nucleosome. Nat Struct Biol 10:882–891. doi:10.1038/nsb996
23. Allard S, Utley RT, Savard J et al (1999) NuA4, an essential transcription adaptor/histone H4 acetyltransferase complex containing Esa1p and the ATM-related cofactor Tra1p. EMBO J 18:5108–5119. doi:10.1093/emboj/18.18.5108
24. Altaf M, Auger A, Monnet-Saksouk J et al (2010) NuA4-dependent acetylation of nucleosomal histones H4 and H2A directly stimulates incorporation of H2A. Z by the SWR1 complex. J Biol Chem 285:15966–15977
25. Kim Y, Shen CH, Clark DJ (2004) Purification and nucleosome mapping analysis of native yeast plasmid chromatin. Methods 33:59–67. doi:10.1016/j.ymeth.2003.10.021
26. Unnikrishnan A, Akiyoshi B, Biggins S et al (2012) An efficient purification system for native minichromosome from Saccharomyces cerevisiae. Methods Mol Biol 833:115–123. doi:10.1007/978-1-61779-477-3_8
27. Lorch Y, Kornberg RD (1994) Isolation of the yeast histone octamer. Proc Natl Acad Sci U S A 91:11032–11034

28. Griesenbeck J, Boeger H, Strattan JS et al (2003) Affinity purification of specific chromatin segments from chromosomal loci in yeast. Mol Cell Biol 23:9275–9282
29. Saksouk N, Avvakumov N, Champagne KS et al (2009) HBO1 HAT complexes target chromatin throughout gene coding regions via multiple PHD finger interactions with histone H3 tail. Mol Cell 33:257–265. doi:10.1016/j.molcel.2009.01.007
30. Altaf M, Utley RT, Lacoste N et al (2007) Interplay of chromatin modifiers on a short basic patch of histone H4 tail defines the boundary of telomeric heterochromatin. Mol Cell 28:1002–1014, doi: S1097-2765(07)00827-1 [pii]
31. Utley RT, Lacoste N, Jobin-Robitaille O et al (2005) Regulation of NuA4 histone acetyltransferase activity in transcription and DNA repair by phosphorylation of histone H4. Mol Cell Biol 25:8179–8190

Chapter 4

Purification of Yeast Native Reagents for the Analysis of Chromatin Function-II: Multiprotein Complexes and Biochemical Assays

Nicolas Lacoste*, Wajid Bhat*, and Jacques Côté

Abstract

Post-translational modifications of histones play essential roles in regulating chromatin structure and function. These are tightly regulated in vivo and there is an intricate cross-talk between different marks as they are recognized by specific reader modules present in a large number of nuclear factors. In order to precisely dissect these processes in vitro native reagents like purified chromatin and histone modifying/remodeling enzymes are required to more accurately reproduce physiological conditions. The vast majority of these enzymes need to be part of stable multiprotein complexes with cofactors enabling them to act on chromatin substrates and/or read specific histone marks. In the accompanying chapter, we have described the protocol for purification of native chromatin from yeast cells (Chapter 3). Here, we present the methods to obtain highly purified native chromatin modifying complexes from *Saccharomyces cerevisiae*, based on Tandem Affinity Purification (TAP). We also present possible applications and useful functional assays that can be performed using these yeast native reagents.

Key words Yeast, Chromatin, Histones, TAP, HAT, HMT, Purification

1 Introduction

The genome of eukaryotic cells is packaged into a highly complex and dynamic structure, chromatin. Modulation of the chromatin structure is a key aspect of gene regulation, as well as DNA repair/recombination/replication. Many factors act on chromatin to regulate its dynamic/structure but also its function as a signaling platform, such as ATP-dependent remodeling complexes, histone chaperones, histone variants, and chromatin-modifying complexes. Post-translational modifications (PTMs) mostly occur on the N-terminal tails of histones. They play a major role in chromatin structure and function by either influencing its higher-order structural organization or creating a binding platform for different

*These authors contributed equally to this work.

Benoit Guillemette and Luc Gaudreau (eds.), *Histones: Methods and Protocols*, Methods in Molecular Biology, vol. 1528,
DOI 10.1007/978-1-4939-6630-1_4, © Springer Science+Business Media New York 2017

proteins implicated in diverse biological functions. The majority of chromatin modifying/remodeling enzymes need to be part of multiprotein complexes to be able to properly act on chromatin [1–3]. A vast number of studies have used tandem affinity purification (TAP) protocols to purify and characterize protein complexes (reviewed in [4]).

The efficient tandem affinity purification (TAP) protocol was originally described by the Seraphin lab [5, 6] and we soon adapted it to fit our needs. The insertion of the TAP-tag at the C-terminus of protein encoding genes was described in detail in the original publication. We describe the use of a slightly modified tandem-affinity purification (TAP) protocol to obtain native chromatin modifying complexes as well as specific variations allowing purification and characterization of mutant complexes that would normally be lethal for the cell. TAP protocols are carried out with cell culture volumes ranging from 250 ml to 6 l depending on the final experimental goal, e.g., analytical enzymatic assays or detection of low stoichiometry post-translational modifications.

Finally, we will discuss some applications achievable with purified native reagents such as enzymatic assays useful in the functional analysis of chromatin modifications. Four major types of modifications occur largely on the N and C terminal of histone tails, i.e., acetylation, methylation, phosphorylation, and ubiquitination [7]. The last modification is the most complicated to reproduce in vitro because three different enzymes are required to transfer ubiquitin onto the C terminal tail of H2A or H2B. This assay will not be presented here, as we will focus on acetylation, methylation, and phosphorylation.

2 Materials

2.1 TAP Purification from Large Culture (1–6 l)

1. YPD: 1% yeast extract, 2% peptone, 2% D-glucose. Filter to sterilize.
2. Wash Buffer 1: 10 mM Tris–HCl pH 8, 350 mM NaCl.
3. Extraction Buffer 1: 10 mM Tris–HCL pH 8, 350 mM NaCl, 10% glycerol, 0.1% NP-40, 1 mM PMSF, 0.5 mM DTT, 2 μg/ml pepstatin A, 2 μg/ml leupeptin, 5 μg/ml aprotinin, 5 mM β-glycerophosphate, 5 mM Na-butyrate.
4. TEV cleavage Buffer: 10 mM Tris–HCl pH 8, 150 mM NaCl, 0.1% NP-40, 0.5 mM EDTA, 1 mM DTT.
5. Calmo Binding Buffer: 10 mM Tris–HCl pH 8, 150 mM NaCl, 1 mM Mg-Acetate, 2 mM CaCl2, 1 mM imidazole, 10 mM β-mercaptoethanol, 0.1% NP-40.
6. Calmo Elution Buffer: 10 mM Tris–HCl pH 8, 150 mM NaCl/50 mM KCl, 1 mM Mg-Acetate, 10 mM EGTA, 1 mM

Imidazole, 10 mM β-mercaptoethanol, 0.1% NP-40, 1 mM PMSF, 2 μg/ml Leupeptin, 2 μg/ml Pepstatin, 5% glycerol.

7. 1 M $CaCl_2$.
8. Sepharose CL-6B (Sigma).
9. TEV protease (Invitrogen).
10. IgG-sepharose (GE Healthcare Life Sciences).
11. Protein A-sepharose.
12. Calmodulin affinity resin (Stratagene).
13. Centrifuge with JA10, JA14, and JA20 rotor (or equivalent).
14. Bead beater with 80 ml chamber (Biospec products).
15. Glass beads (0.5 mm).
16. Ultracentrifuge with 70Ti rotor.
17. Benchtop centrifuge with refrigeration.

2.2 TAP Purification from 250 ml Culture

1. Dynabeads M-270 Epoxy (Life Technologies).
2. Dimethylformamide (DMF).
3. IgG from rabbit serum (~10 μg/μl).
4. Buffer A: 0.1 M sodium phosphate buffer (pH 7.4).
5. Buffer B: 3 M ammonium sulfate dissolved in 0.1 M sodium phosphate buffer (pH 7.4).
6. Buffer C: 0.1 M citrate (pH 3.1).
7. Buffer D1: Phosphate buffered saline (PBS pH 7.4): 137 mM NaCl, 2.7 mM KCl, 10 mM Na_2HPO_4, 1.8 mM NaH_2PO_4.
8. Buffer D2: 1% Triton X-100 in PBS pH 7.4 (Buffer D1).
9. Wash Buffer 2: 20 mM HEPES-NaOH pH 7.4, 300 mM NaCl, 0.1% NP-40, 2 mM $MgCl_2$, 5% glycerol, 1 mM DTT.
10. Extraction Buffer 2: 20 mM HEPES-NaOH pH 7.4, 300 mM NaCl, 0.1% NP-40, 2 mM $MgCl_2$, 5% glycerol, 1 mM DTT, 1 mM PMSF, 2 μg/ml Leupeptin, 2 μg/ml Pepstatin, 5 μg/ml Aprotinin.
11. Magnetic rack for 1.5 ml test tubes.
12. Disposable mini-columns (Biorad).

2.3 Enzymatic Assays

1. 5× HAT buffer: 250 mM Tris–HCl pH 8, 25% glycerol, 0.5 mM EDTA pH 8, 5 mM DTT, 5 mM PMSF.
2. [^{3}H] Acetyl-CoA (4.7 Ci/mmol).
3. EN^{3}HANCE (PerkinElmer).
4. Gel dryer.
5. Scintillation Counter.
6. 5× HMT buffer: 100 mM Tris–HCl pH 8, 25% glycerol, 0.5 mM EDTA pH 8, 5 mM DTT, 1 mM PMSF.

7. SAM (S-Adenosyl-L-[methyl-^{3}H] methionine) (PerkinElmer).
8. 5× kinase buffer: 125 mM Tris–HCl, pH 8, 50 mM $MgCl_2$, 5 mM DTT, 250 μM cold ATP.
9. [γ-^{32}P] ATP (6000 Ci/mmol).
10. 1 M KCl.
11. Tris glycine buffer: 25 mM Tris, 192 mM glycine, 0.1 % SDS, pH 8.3.
12. Carbonate buffer: 50 mM carbonate NaHCO3-NaCO3 pH 9.2.
13. 4× Laemmli sample buffer: 0.25 M Tris–HCl pH 6.8, 8 % SDS, 40 % glycerol, 8 % β-mercaptoethanol, 0.02 % bromophenol blue.
14. Transfer Buffer: 20 % methanol 20 mM Tris, 25 mM glycine, 0.375 %SDS.
15. Coomassie staining/destaining solutions: 40 % methanol, 7 % acetic acid with or without 0.1 % (w/v) Coomassie Brilliant Blue R-250.

3 Methods

3.1 Large-Scale Tandem Affinity Purification (TAP) Protocol for 6 l of Yeast Culture

1. *In the following protocol all steps are done at 4 °C unless otherwise mentioned.*
2. Grow TAP-tagged yeast strains in 6 l of YPD at 30 °C to an OD_{600} between 1.5 and 2.
3. Harvested cells by centrifugation at 5000 ×*g* for 5 min.
4. Wash cell pellet with 600 ml of cold wash buffer 1.
5. Pellet cells by centrifugation as above and resuspend in one volume of extraction buffer 1 (*see* **Note 1**).
6. Disrupt cells using the 80-ml chamber of a bead beater (Biospec products) filled up with 40 ml of glass beads (0.5 mm) and the yeast preparation. Complete volume with extraction buffer 1 if necessary.
7. Disruption is performed in a cold room with ice packed around the chamber using 15-s pulses each followed by a pause of 45 s. This is repeated until 70–80 % of cell disruption is achieved (*see* **Note 2**).
8. After cell lysis, recover the supernatant using a Pasteur pipette.
9. Wash beads with 10 ml of extraction buffer 1 and pool with supernatant.
10. Clear cell lysate by centrifugation for 20 min at 20,000 ×*g* in a SS34/JA20 rotor.

11. Ultracentrifuge for 45 min at 200,000 × *g* (45,000 rpm in a 70 Ti rotor) to eliminate cellular debris and ribosomes.
12. Filter the supernatant through gauze to remove lipids.
13. Preclear cell lysate with 600 μl sepharose CL-6B (pre-equilibrated with the extraction buffer 1) for 45 min on a wheel at 4 °C.
14. After preclear, add supernatant to 600 μl of IgG-sepharose (pre-equilibrated with the extraction buffer 1) to undergo the first affinity step on a wheel for 2 h at 4 °C.
15. At the end of the incubation, pour the slurry in a 10-ml column and let the supernatant pass through the column by gravity flow (stop before the resin becomes dry).
16. Wash the beads with 15 ml of extraction buffer 1 followed by a wash with 5 ml of TEV cleavage buffer to prepare the resin for the next step.
17. Resuspend the IgG beads containing the bound material in 600 μl of TEV cleavage buffer supplemented with 300 units of TEV protease.
18. Close the column with a cap and incubate for 2 h on a wheel at 16 °C.
19. Recover the flow through by gravity flow and wash the resin with 400 μl of TEV cleavage buffer and combine it with the first TEV elution.
20. Add 30 μl of Protein A-sepharose (50% slurry pre-equilibrated with Calmo binding buffer) to remove IgG molecules that have leaked from the beads during the TEV elution.
21. Incubate for 15 min at 4 °C on a wheel.
22. Spin at 3000 rpm for 1 min at 4 °C and transfer the supernatant in a new 15 ml conical tube.
23. Add 3 μl of 1 M $CaCl_2$ per ml of TEV elution to overcome chelation by the EDTA present in the buffer.
24. Add three volumes of calmodulin binding buffer to the sample and mix with 600 μl of pre-equilibrated calmodulin affinity resin.
25. Incubate overnight at 4 °C on a wheel (*see* **Note 3**).
26. Next morning, pour the slurry into a disposable 10 ml column and allow the supernatant to flow through by gravity.
27. Wash the resin with 15 ml of calmodulin binding buffer.
28. Elute proteins with the calmodulin elution buffer. Collect 6 fractions of 600 μl each (*see* **Note 4**).
29. Fractions are analyzed by SDS-PAGE and used for enzymatic assays (Subheadings 3.3 and 3.4).
30. To increase the purity and yield of TAP purification see helpful tips in **Notes 5** and **6**.

3.2 Small-Scale Tandem Affinity Purification (TAP) Protocol for 250 ml of Yeast Culture

This method takes advantage of rabbit IgG-coupled Dynabeads M-270 Epoxy with very low nonspecific binding that can allow fast one-step purification. It is based on the method described by the Baetz lab [8] but modified in order to obtain biologically active material in solution (*see* **Note 5**). In the following sections, we present protocols for coupling of IgG to Dynabeads and protein purification.

3.2.1 Coupling IgG to Dynabeads

1. The coupling protocol has been adapted from the product information sheet for the Dynabeads M-270 Epoxy (Invitrogen).
2. Resuspend Dynabeads in DMF according to manufacturers protocol, vortex for 2 min and store at 4 °C.
3. Take 300 μl of suspension (Dynabeads in DMF) and dispense into a 1.5 ml tube.
4. Isolate beads from DMF on a magnet (wait about 30 s until beads have all stuck to the side of tube and discard the supernatant).
5. Wash beads three times with Buffer A. Each time, resuspend the beads by flicking the tube until all the beads are floating in solution. Then place the tube back onto the magnet and wait for the beads to completely stick to the side of the tube before discarding the supernatant.
6. Resuspend beads in 345 μl of Buffer B.
7. Add 15 μl of rabbit IgG (~10 μg/μl) and vortex.
8. Add 180 μl of Buffer C and mix well.
9. Incubate at 37 °C for 24 h with end-over-end rotation.
10. After 24 h, wash beads four times with 1 ml of Buffer D1 (1 min for each wash).
11. Wash once with Buffer D2 for 10 min (end-over-end rotation) followed by two 1-min washes with Buffer D1.
12. Wash with Buffer C four times (10 min each wash, end-over-end rotation).
13. Wash with Buffer D1 two times (10 min each wash, end-over-end rotation) followed by two additional washes for 1 min.
14. Resuspend in 600 μl of Buffer D and keep at 4 °C.
15. Beads are ready to use, can be kept for 2–3 weeks, and are equilibrated with extraction buffer just before adding to cell lysate (**step 29**).

3.2.2 Purification of TAP-Tagged Proteins Using IgG-Dynabeads

16. *All the steps in this protocol are carried out at 4 °C unless otherwise mentioned.*
17. Grow TAP-tagged yeast strain in 250 ml of YPD at 30 °C to an OD_{600} between 2 and 2.5 (Always use freshly streaked yeast).

18. Pellet cells by centrifugation at 5000 × *g* for 5 min and remove all the supernatant.
19. Wash pellet with 10 ml wash buffer 2, transfer into 50 ml conical tube, and centrifuge at 5000 × *g* for 5 min.
20. Resuspend pellet in 1 ml of extraction buffer 2 and transfer to 1.5 ml tubes, not more than 750 μl per tube and pellet by centrifugation at 5000 × *g* for 3 min (usually ends up with 2–3 tubes per sample at this point).
21. Flash freeze and keep at −80 °C or proceed to next step.
22. Thaw sample on ice with an equal volume of chilled extraction buffer.
23. Add 600 μl of acid washed glass beads.
24. Disrupt cells by vortexing at 4 °C (1 min on and 1 min off three to six times).
25. Stop the lysis when disruptions reach 50–70% (*see* **Note 1**).
26. Recover the lysate by piercing a hole in the bottom with a 25G needle and spinning into a new clean tube.
27. Centrifuge the lysate at 14,000 RPM, 20–30 min, 4 °C, in a microfuge.
28. Put the supernatant in a 2 ml tube.
29. To the supernatant, add Dynabeads-IgG prepared as described in the previous section (from **step 15**). Dynabeads-IgG should be pre-equilibrated with extraction buffer 2 before adding to supernatant (by washing three times with 1 ml of extraction buffer).
30. Incubate for 3–4 h at 4 °C on a wheel.
31. Using magnetic rack take out the supernatant.
32. Wash resin four times with 1 ml of extraction buffer 2.
33. Wash resin two times with 1 ml of TEV buffer.
34. Add 500 μl of TEV cleavage buffer and 13U of TEV protease (Invitrogen, 10 U/μL).
35. Close the tube and place it on a wheel at 16 °C for 2 h.
36. Recover the supernatant in 1.5 ml tube.
37. Proceed to next step or use the supernatant for assays (*see* **Note 6**).
38. Wash the resin with 500 μl of Calmo binding buffer.
39. Pool the two elutions.
40. Add 10 μl of Protein A-sepharose (50% slurry pre-equilibrated with Calmo binding buffer) to remove IgG molecules that have leaked from the beads during the TEV elution.
41. Incubate for 15 min at 4 °C on a wheel.
42. Spin at 3000 rpm for 1 min at 4 °C and transfer the supernatant in a new 15 ml conical tube.

43. Proceed to next step or use the supernatant for assays (*see* **Note 6**).
44. To the supernatant add 1.5 μl of 1 M $CaCl_2$ and three volumes of Calmo binding buffer.
45. Add 60 μl of Calmodulin resin (120 μl of 50% slurry, pre-equilibrated with Calmo binding buffer).
46. Incubate over night on a wheel at 4 °C.
47. Pour the sample in a disposable mini column (Biorad) and allow the flow through to pass by gravity flow.
48. Wash resin with 2 ml of Calmo binding buffer.
49. Wash resin with 2 ml Calmo elution Buffer without EGTA.
50. Close the column at the bottom and add 80 μl of Calmo elution Buffer.
51. Close the column at the top and place it on a wheel at 4 °C for 1 h (*see* **Note 6**).
52. Collect the first fraction by gravity (fraction 1).
53. Repeat **steps 50, 51,** and **52** two more times (for second and third fractions).

3.3 Applications of the TAP Protocol

Different applications of the TAP protocol are presented in Figs 1, 2, and 3 (*see* also **Note 7**). The first analyzes the presence of specific subunits in protein complexes purified from cells in different growth conditions or mutant backgrounds (Fig. 1, *see* **Note 8**). The second offers a method to study the activity of chromatin modifying complexes bearing mutations in subunits involved in histone mark recognition (Fig. 2, *see* **Note 9**). And the third offers the possibility of looking at complex architecture and in vitro activities when essential proteins are mutated (Fig. 3, *see* **Note 10**).

3.4 Enzymatic Assays Using Yeast Native Reagents

Three examples of enzymatic assays to study chromatin modifications are presented. These assays are straightforward and can be used with purified native chromatin and any TAP-purified activities.

3.4.1 Histone Acetyltransferase (HAT) Assay

This assay measures the transfer of radiolabeled acetyl groups from a donor molecule, Acetyl-Coenzyme A, on the Nε of lysines present on histone N terminal tails [9].

1. Use 0.5 μg of native chromatin or core histones with up to 3 μl of HAT complex in a 15 μl reaction volume. Larger volumes of HAT complexes are usually detrimental to the reaction linearity.
2. Mix chromatin or core histones, the HAT fraction, 0.125 μCi of [^{3}H] Acetyl-CoA (4.7 Ci/mmol), 3 μl of 5× HAT buffer, and 1 mM Na-butyrate.
3. KCl/NaCl total concentration should be adjusted to 50 mM taking into account all reaction components. Total salt concentration can be higher but we found that 100 mM inhibits most HAT activities.

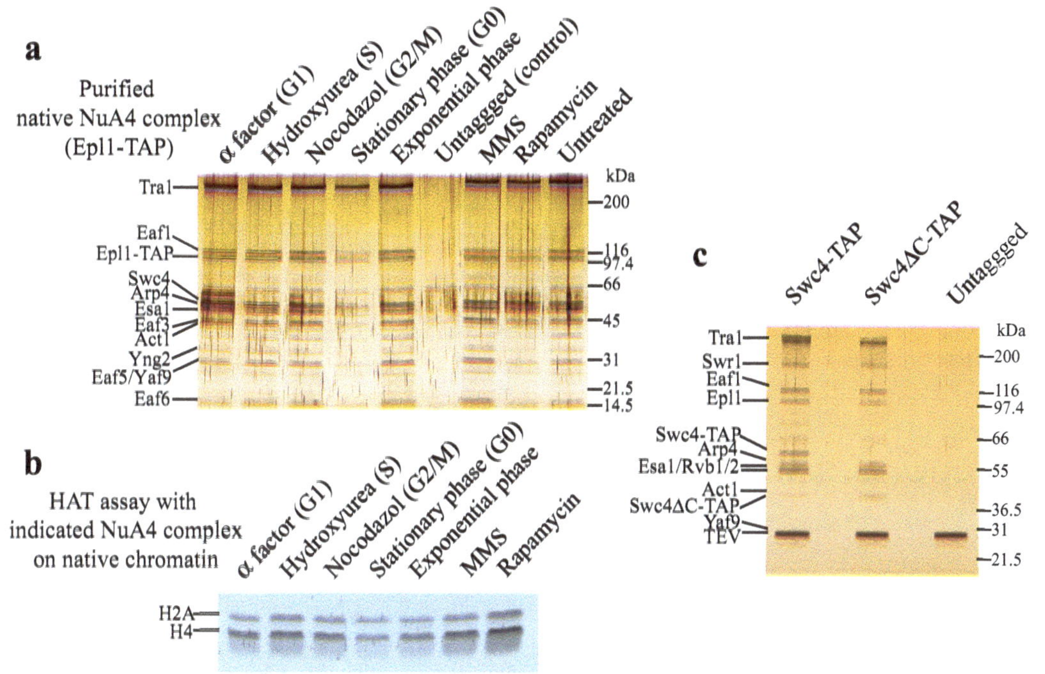

Fig. 1 Using the TAP approach to purify native chromatin modifying complexes from yeast cells in different growth conditions or genetic backgrounds. (**a**) Purification of the NuA4 HAT complex from cells in different phases of the cell cycle or stress conditions. TAP purification using the classic protocol (Subheading 3.1) was performed using Epl1-TAP expressing cells blocked in G1 (α factor), S (hydroxyurea), G2/M (nocodazole), G0 (stationary phase). Two other cultures were treated with methylmethane sulfonate (MMS) or rapamycin, drugs creating DNA damage or blocking the TOR pathway, respectively. Silver stained gel is presented with NuA4 subunits identified on the left. (**b**) Purified complexes shown in (**a**) were tested in HAT assays with 0.5 μg of native chromatin and specific histone acetylation was visualized after running on SDS-PAGE and fluorography. (**c**) Efficiency of the short TAP protocol (Subheading 3.2) using IgG-magnetic beads to analyze wild type and mutant protein complexes. IgG bound Dynabeads were used to purify wild type and truncated TAP-tagged Swc4 from small cultures through one step TEV elution. Swc4 is a subunit of the NuA4 HAT complex and the SWR1 chromatin remodeling complex [16]. Silver stain gel of the TEV elutions is shown and identified associated proteins are labeled on the *left*

4. Incubate the reaction mix at 30 °C for 30–60 min. Two alternate protocols are suggested which allows the detection of total HAT activity regardless of substrate specificity (**Step 5**) or the detection of histone specific HAT activity (**step 9**).
5. For liquid HAT assay, at the end of the incubation spot each reaction onto p81 phosphocellulose membranes.
6. Air dry the membranes and wash three times for 5 min in 50 mM carbonate ($NaHCO_3$-$NaCO_3$; pH 9.2) buffer. The highly basic histones will stay charged at this pH and remain bound.
7. Rinse the membranes briefly with acetone, air dry and place in scintillation vials.

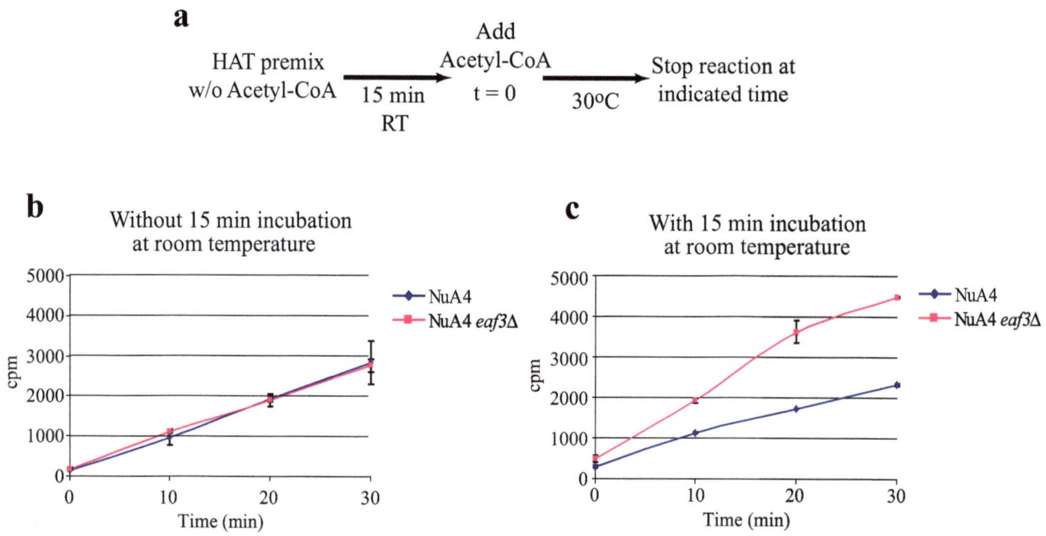

Fig. 2 In vitro analysis of subunit function in chromatin modifying complexes. Influence of preincubation in a HAT assay using the NuA4 complex lacking one of its subunits, chromodomain-containing Eaf3. (**a**) Schematic representation of the pre-incubation time course HAT assay. (**b**) Time course HAT assay performed on native chromatin as described in the protocol without preincubation with NuA4 TAP-purified from wild type cells or *eaf3*Δ mutant cells. (**c**) Time course HAT assay similar to (**b**) except that components are preincubated without acetyl-CoA at room temperature for 15 min before the enzymatic assay

8. Cover with scintillation fluid and measure counts in scintillation counter for 30 min (*see* **Note 11**).
9. Alternatively, histone tail specificity can be analyzed by submitting the HAT reaction to SDS-PAGE followed by fluorography as described below.
10. Stop reactions from **Step 4** by adding 5 μl of 4× Laemmli sample buffer and boil for 5 min.
11. Load samples on an 18% SDS-polyacrylamide gel.
12. Migrate for 2 h at 160 V in Tris glycine buffer.
13. After coomassie staining, destain until histones are readily visible (to verify equivalent histone loading between the reactions).
14. Incubate gel in EN^{3}HANCE for 30 min and then wash in H_2O for 30 min.
15. Dry the gel for 90 min at 60 °C and place against a film for 2 or 3 days (*see* **Note 9**).

3.4.2 Histone Methyltransferase (HMT) Assay

1. This assay measures the transfer of one or multiple radiolabeled methyl groups from S-adenosylmethionine (SAM) on the Nε of lysines to produce mono-, di-, or tri-methylated residues or on the terminal (or ω) guanidino nitrogen atoms of arginines to give mono- or di-methylated residues on histones, the latter being possibly symmetric or asymmetric (for reviews *see* [10, 11])

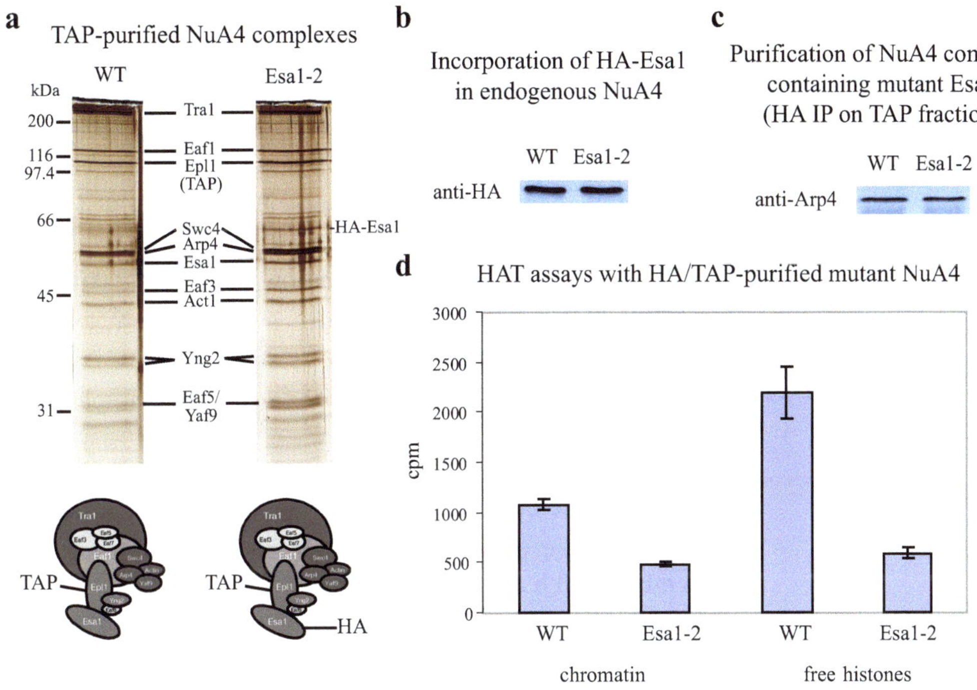

Fig. 3 Enzymatic analysis of a chromatin modifying complex containing a lethal mutation. The case of NuA4 containing the HAT enzyme Esa1 mutated in its chromodomain. (**a**) Tap purification of NuA4 containing two forms of Esa1, one endogenous and one HA tagged and mutated (Esa1-2) or not (WT) in its chromodomain. (**b**) Western blot analysis for the presence of HA-Esa1 in the Tap purification. (**c**) After HA immunoprecipitation of the TAP purified complexes, recovered NuA4complexes are assayed by western blot on the beads. (**d**) Liquid HAT assays with the WT and mutant purified complexes bound to αHA beads using free histones or chromatin as substrate

2. This method is performed similarly to the HAT assay, with a slight difference in buffer composition.
3. Mix 0.5 μg of native chromatin or core histones, histone methyltransferase, 3 μl of 5× HMT buffer, and 0.5 μCi of SAM (S-Adenosyl-L-[methyl-^{3}H] methionine in a 15 μl reaction.
4. Adjust KCl/NaCl total concentration to 50 mM taking into account all reaction components.
5. Incubate at 30 °C for 30–45 min.
6. HMT assays are monitored by liquid scintillation counting or gel/fluorography as described for HAT assays (Subheading 3.4.1, **steps 5–15**).

3.4.3 Kinase Assay

1. Histone function is also modulated by phosphorylation [12]. Most kinases require magnesium as a cofactor, which presents a caveat for chromatin study since magnesium increases compaction of chromatin thus potentially inhibiting the ability of enzymatic complexes to phosphorylate their target sites.

Generally, concentrations between 5 and 10 mM are suitable for kinase assays but one should aim for the lower end when using chromatin as substrate [13].

2. Reactions are carried out with 0.5–2 μg of histones or nucleosomes.
3. Mix histones, kinase, 3 μl of 5× Kinase Buffer, and 1 μCi [γ-^{32}P] ATP in 15 μl reaction.
4. Adjust KCl/NaCl total concentration to 80 mM taking into account all reaction components.
5. Incubate for 30 min at 30 °C.
6. Stop the reaction by adding 5 μl of 4× Laemmli sample buffer and boil for 5 min.
7. Load samples on an 18 % SDS-PAGE and run at 160 V for 2 h in Tris glycine buffer.
8. At the end of the run, transfer proteins to a nitrocellulose membrane (3 h at 4 °C, 75 V in transfer buffer).
9. Air dry the membrane and expose to film (*see* **Note 12**).

4 Notes

1. The salt concentration is increased from 150 mM in the original protocol to 350 mM in order to increase the extraction efficiency and the stringency of binding to the first resin. Phosphatase and deacetylase inhibitors can also be added.
2. To follow the level of disruption, after every two cycles, an aliquot is observed under the light microscope (400× magnification). Disruption level above 80 % should be avoided since it leads to protein degradation and higher ratio of contaminants at the end of purification.
3. The increase of incubation time with calmodulin beads from 2 h to overnight greatly improves the purification yield without significant protein degradation.
4. Increasing the EGTA concentration from 2 mM (original protocol) to 10 mM augments the efficiency of elution from the beads, though it may cause inhibition of specific enzymatic activities.
5. Using Dynabeads-IgG drastically reduces the contaminant proteins (inert and surface area). The yield and purity after TEV cleavage is sufficient to be used in biochemical assays. We have successfully and repeatedly used TEV cleaved fractions for mass spectrometric analysis and biochemical assays.
6. To get rid of small contaminants like ribosomal proteins, it is possible to concentrate the purified fractions with Centricon

device with a 30 kDa cutoff (Amicon, Millipore). To improve the efficiency of elution from the calmodulin resin salt concentration can be increased. The purification yield can also be improved by increasing the incubation time of the calmodulin elution buffer (1 or 2 h instead of just brevity flow). Depending on the batch of IgG-sepharose beads used, some immunoglobulins could leak out during TEV elution. In this situation, a preclearing step with small amount of protein A-sepharose is added before loading the sample on the calmodulin beads. In our laboratory, TAP protocols are carried out with cell culture volumes ranging from 250 ml to 12 l where larger culture volumes generally give lower ratios of contaminant proteins.

7. If tagged proteins are present in multiple distinct complexes, the purified sample can be further fractionated on gel filtration or ion exchange columns. The TEV elution or the calmodulin affinity step can be followed by a gel filtration column (for example Superose 6, GE Healthcare Life Sciences) to separate distinct protein complexes based on their sizes (*see* [14]). This is particularly important for structural/imaging studies by (Cryo-)Electron Microscopy [15].
8. Other applications for the analysis of a multisubunit complex architecture and regulation are presented in Fig. 1. Cells can be treated with different stresses/drugs or blocked in different phases of the cell cycle before performing TAP purification and assays (Fig. 1a, b). Architecture can be analyzed by deleting/mutating one subunit-encoding gene at a time in a strain expressing another subunit that is TAP-tagged [14, 16]. After TAP purification the resulting complexes can be compared enzymatic activities and protein content (Fig. 1c). In doing a comprehensive study of all nonessential subunits in each TAP-tagged background, specific determinants of complex assembly and direct protein-protein interactions can be identified, as well as functional subcomplexes and enzymatic cofactors. This type of analysis is very powerful to characterize native stable protein interactions and can also be used with truncated and mutated proteins (e.g., on specific protein domains or amino acids) to refine the architectural and functional model of the studied complex.
9. Another application of this approach is to address the altered affinities of HAT complexes for chromatin in different mutant conditions (e.g., mutant chromatin missing specific modifications or mutant complexes missing specific domains). In this kind of study, it is often important to perform preincubation of the HAT activity with chromatin substrate before adding Acetyl-CoA for the HAT assay. Usually, HAT assay over a time course allows detection of more subtle differences. For example, as shown in Fig. 2, the effect of deleting Eaf3, the

chromodomain-containing subunit of the NuA4 HAT complex, is only seen when a 15-min preincubation is performed before doing the time course HAT assay (compare Fig. 2b, c).

10. A useful application of the TAP protocol is also to study essential proteins/lethal mutations. While similar to the applications above, this one allows purification of complexes carrying mutations of essential proteins. It then becomes possible to biochemically characterize the function of essential proteins within their native environment. The first step is to clone the essential protein of interest in a low copy vector, preferably under control of its own promoter. It is particularly important not to overexpress the protein in order to avoid disrupting the stoichiometry of endogenous complexes. The protein must also be tagged in order to immunopurify the mutant complexes at the end of the TAP protocol. Tags like Flag, GST, or HA are preferable since they allow native elution from the corresponding resin. Once a gene of interest is cloned into the vector, any desired mutation/truncation can be obtained by standard mutagenesis protocols. After transformation in yeast cells, protein levels are verified by western. Continuous growth selection is required to retain the plasmid since the endogenous wild type gene is present in the cells (hence the plasmid is not required for survival). The TAP protocol is then carried out exactly as described above (Fig. 3a). As shown in Fig. 3a, b, The TAP purified complexes contain either the endogenous target protein or the tagged-one expressed from the plasmid. The complexes containing only the mutated protein can now be obtained by immunopurification using the tag (Fig. 3c). This technique allows characterization of the effect of a lethal mutation on the chromatin modifying activity of a complex (e.g., Fig. 3d) or its subunit structure.
11. Reactions are normally linear in the range of 400–8000 cpm. Background is usually between 75 and 125 cpm for chromatin and 200–350 cpm for free histones. Small peptides corresponding to the N-terminal domain of histones can also be used in this assay (300 ng per reaction).
12. Since kinases are very common enzymes, they are also typical contaminants in many purifications. For example, yeast native chromatin contains some histone H3 kinase activity that would obviously hinder any kinase study with this substrate. As such, well-controlled experiments are required to rule out this type of problem. Gels can also be directly exposed to films without transfer, but signals tend to be sharper because of less background on membranes.

Acknowledgments

We are grateful to Rhea Utley for significant corrections to this manuscript. Leslie Mitchel and Kristin Baetz for their original short TAP protocol on magnetic beads. This work was supported by a grant from the Canadian Institutes of Health Research (CIHR; MOP-14308). N.L. was supported by a Canadian Government Foreign Affairs/International Trade studentship. J.C. holds the Canada Research Chair in Chromatin Biology and Molecular Epigenetics.

References

1. Lee KK, Workman JL (2007) Histone acetyltransferase complexes: one size doesn't fit all. Nat Rev Mol Cell Biol 8:284–295
2. Carrozza MJ, Utley RT, Workman JL et al (2003) The diverse functions of histone acetyltransferase complexes. Trends Genet 19:321–329
3. Lalonde ME, Cheng X, Cote J (2014) Histone target selection within chromatin: an exemplary case of teamwork. Genes Dev 28:1029–1041. doi:10.1101/gad.236331.113
4. Xu X, Song Y, Li Y et al (2010) The tandem affinity purification method: an efficient system for protein complex purification and protein interaction identification. Protein Expr Purif 72:149–156
5. Puig O, Caspary F, Rigaut G et al (2001) The tandem affinity purification (TAP) method: a general procedure of protein complex purification. Methods 24:218–229. doi:10.1006/meth.2001.1183
6. Rigaut G, Shevchenko A, Rutz B et al (1999) A generic protein purification method for protein complex characterization and proteome exploration. Nat Biotechnol 17:1030–1032. doi:10.1038/13732
7. Zentner GE, Henikoff S (2013) Regulation of nucleosome dynamics by histone modifications. Nat Struct Mol Biol 20:259–266. doi:10.1038/nsmb.2470
8. Mitchell L, Lambert JP, Gerdes M et al (2008) Functional dissection of the NuA4 histone acetyltransferase reveals its role as a genetic hub and that Eaf1 is essential for complex integrity. Mol Cell Biol 28:2244–2256. doi:10.1128/MCB.01653-07
9. Steunou A-L, Rossetto D, Cote J (2014) Regulating Chromatin by Histone Acetylation. In: Workman JL, Abmayr SM (eds) Fundamentals of chromatin. Springer, New York, pp 147–212. doi:10.1007/978-1-4614-8624-4_4
10. Greer EL, Shi Y (2012) Histone methylation: a dynamic mark in health, disease and inheritance. Nat Rev Genet 13:343–357. doi:10.1038/nrg3173
11. Bedford MT, Clarke SG (2009) Protein arginine methylation in mammals: who, what, and why. Mol Cell 33:1–13. doi:10.1016/j.molcel.2008.12.013
12. Rossetto D, Avvakumov N, Cote J (2012) Histone phosphorylation: a chromatin modification involved in diverse nuclear events. Epigenetics 7:1098–1108. doi:10.4161/epi.21975
13. Utley RT, Lacoste N, Jobin-Robitaille O et al (2005) Regulation of NuA4 histone acetyltransferase activity in transcription and DNA repair by phosphorylation of histone H4. Mol Cell Biol 25:8179–8190
14. Rossetto D, Cramet M, Wang AY et al (2014) Eaf5/7/3 form a functionally independent NuA4 submodule linked to RNA polymerase II-coupled nucleosome recycling. EMBO J 33:1397–1415. doi:10.15252/embj.201386433
15. Chittuluru JR, Chaban Y, Monnet-Saksouk J et al (2011) Structure and nucleosome interaction of the yeast NuA4 and Piccolo-NuA4 histone acetyltransferase complexes. Nat Struct Mol Biol 18:1196–1203. doi:10.1038/nsmb.2128
16. Auger A, Galarneau L, Altaf M et al (2008) Eaf1 is the platform for NuA4 molecular assembly that evolutionarily links chromatin acetylation to ATP-dependent exchange of histone H2A variants. Mol Cell Biol 28:2257–2270. doi:10.1128/MCB.01755-07

Chapter 5

Histone Purification from *Saccharomyces cerevisiae*

Frederic Jourquin and Vincent Géli

Abstract

The nucleosome structure consists of a histone octamer made by a tetramer of H3-H4 histones and two dimers of H2A-H2B. Nucleosomes undergo extensive posttranslational modifications that regulate nucleosome interactions, position, and stability.

We describe a protocol allowing the robust purification of histones from the yeast *Saccharomyces cerevisiae*. This method appears to be suitable to quantitatively analyze specific posttranslational histone modifications.

Key words Histones, Chromatin, Purification, Budding yeast

1 Introduction

This protocol to purify yeast histones by acid extraction of isolated chromatin is adapted from Dehe et al. [1] and Poveda et al. [2]. The protocol is written for an amount of wet yeast cells of 100 mg. It can be adapted for larger amount of wet cells.

2 Materials

All solutions are prepared using ultrapure water (prepared from deionized water to reach the 18.2 MΩ cm at 25 °C).

1. YPD: 1 % Yeast extract, 2 % dextrose, 2 % peptone.
2. Buffer 1 (extraction buffer pH = 7.5): 1 M Sorbitol, 50 mM Tris–HCl pH = 7.5, 5 mM $MgCl_2$ (keep at −20 °C).
3. Buffer 2 (extraction buffer pH = 6): 1 M Sorbitol, 50 mM MES pH = 6, 5 mM $MgCl_2$ (buffered with KOH 2 M and keep at −20 °C).
4. Buffer 3 (lysis buffer K): 50 mM MES pH = 6, 75 mM KCl, 0.5 mM $CaCl_2$, 0.1 % NP 40 (keep at −20 °C).

Benoit Guillemette and Luc Gaudreau (eds.), *Histones: Methods and Protocols*, Methods in Molecular Biology, vol. 1528, DOI 10.1007/978-1-4939-6630-1_5, © Springer Science+Business Media New York 2017

5. Buffer 4 (lysis buffer HS): 10 mM MES pH = 6, 430 mM NaCl (keep at 4 °C).
6. Protease inhibitors (PI): Add one tablet of PI complete mini EDTA-free (Roche) in 20 ml of a solution at 20 μM of 3–4 dichloroisocoumarin serine protease inhibitor and 1 mM of PMSF.
7. Zymolyase (R) 20 T from *Arthrobacter luteus* (50 mg/ml in 0.1 M sorbitol).
8. 2-Mercaptoethanol (14.3 M solution).
9. IGEPAL® CA-630 detergent (Nonidet P-40 substitute).
10. HCl (37% solution corresponding to 12.1 M).
11. Acetone in solution.
12. Acidified acetone, adjusted at 120 mM HCl.
13. Loading buffer (2×): 100 mM Tris–HCl pH 7.5, 20% glycerol, 4% SDS, 2 M urea, 200 mM DTT, bromophenol blue.
14. Coomassie staining: 50% Ethanol, 7% acid acetic, 0.5 g/L Coomassie brilliant blue R250. Complete to 1 L with distilled water. Store at room temperature.
15. Coomassie destaining: 20% Ethanol; 7% acid acetic. Store at room temperature.

3 Methods

3.1 Cell Sample Preparation

1. Inoculate 100 ml YPD with a wild-type strain (W303) (*see* **Note 1**) to an OD_{600} below 0.1. Grow the culture at 30 °C.
2. Harvest the cells at $OD_{600} = 0.8$ and centrifuge the culture for 5 min at 2000 × *g*.
3. Wash the cell pellet twice in water.
4. Wash the cell pellet with buffer 1.
5. Weight the cellular pellet.
6. Prepare samples of 100 mg of wet cells in 2 ml Eppendorf tubes.
7. Wash the cell pellet with buffer 1.
8. Discard the supernatant (you can stop the protocol at this step by freezing the pellets in liquid nitrogen and storage at −80 °C).

3.2 Spheroplast Preparation

All centrifugations are done in a microcentrifuge.

1. Cell pellets of 100 mg are resuspended in 1 ml of buffer 1. Add 5.4 μl of 2-mercaptoethanol (14.3 M). Incubate in ice during 10 min.

2. Centrifuge at 4 °C for 5 min at 900×*g* and discard the supernatant.
3. Resuspend the pellet in 1 ml of buffer 1 (without 2-mercaptoethanol). Add 60 µl of zymolyase 20T (*see* **Note 2**).
4. Incubate with gentle agitation for 30 min at 35 °C.
5. Add directly in the tube 700 µl of buffer 2 containing the mixture of protease inhibitors (20 µl of PI).
6. Centrifuge for 5 min at 4 °C at 3500 rpm and discard the supernatant (to avoid any residual traces of zymolyase) (*see* **Note 3**).

3.3 Extraction

All the following steps are to be performed on ice, unless otherwise indicated.

1. Resuspend the pellet in 700 µl of buffer 3 + PI (20 µl). Incubate for 5 min on ice.
2. Centrifuge for 5 min at 13,000 rpm at 4 °C and discard the supernatant.
3. Resuspend the pellet in 700 µl of buffer 4. Add IGEPAL® CA-630 detergent to a final concentration of 0.5% and PI (20 µl) (*see* **Note 4**). Let in ice for 5 min.
4. Centrifuge at 4 °C for 5 min at 13,000 rpm and discard the supernatant.
5. Resuspend the pellet in 700 µl of buffer 4 + PI (20 µl) and let on ice for 5 min.
6. Centrifuge at 4 °C for 5 min at 13,000 rpm, and discard the supernatant.
7. Add 120 µl of 0.25 M HCl.
8. Place a small magnetic barrel in the tube and agitate for 2 h (minimum) at 4 °C (a rotor wheel can also be used).
9. Centrifuge at 4 °C for 15 min at 13,000 rpm.
10. Transfer the supernatant in a 15 ml Falcon tube.
11. Add 8 volumes of acetone for 1 volume of supernatant.
12. Mix by inversion and place the tube overnight at −20 °C.
13. Centrifuge at room temperature for 10 min at 3500 rpm and discard the supernatant.
14. Resuspend the pellet with 400 µl of acidified acetone and transfer the resuspended pellet in a 1.5 ml tube.
15. Centrifuge at RT for 5 min at 10,000 rpm. You should see a visible white pellet.
16. Discard the supernatant, add 1 ml of acetone, and carefully slowly vortex.

17. Centrifuge at RT for 5 min at 10,000 rpm and discard the supernatant.
18. Dry the pellet with a speed-vac (no more of 20 min) or let it dry at 37 °C (*see* **Note 5**).
19. Resuspend the pellet with 10 μl of distilled water.
20. Add 5 μl of loading buffer (2×) to 5 μl of purified histones. The remaining histone preparation can be kept at −20 °C.
21. Denature for 5 min at 100 °C.
22. Load 2 μl on a 16% SDS-PAGE (*see* **Note 6**) and run at 100 V for 3 h.

3.4 Coomassie Staining

1. Stain the acrylamide gel in the Coomassie staining solution for 30 min.
2. Destain for 1 h in the Coomassie destaining solution (*see* Fig. 1).

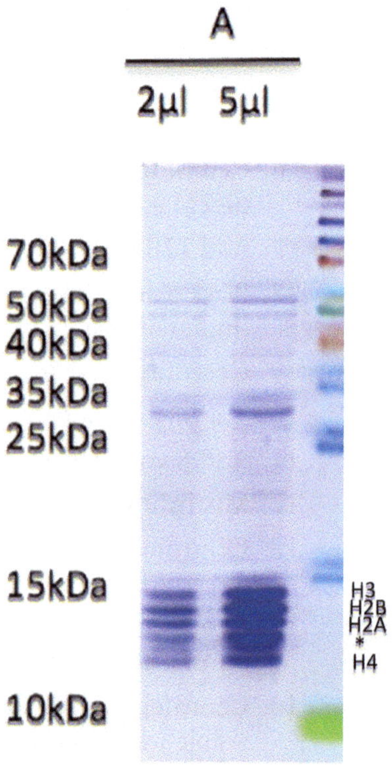

Fig. 1 Purification of histones visualized in a 16% SDS-PAGE stained with Coomassie blue. The four histones are indicated. *Asterisk* represents a degradation band

4 Notes

1. W303 *MAT*a {*leu2-3,112 trp1-1 can1-100 ura3-1 ade2-1 his3-11,15*}—Most *S. cerevisiae* strains can be used for this protocol.
2. Do not add an excess of zymolyase, as it may result in the clipping of the N-terminal tail of H3 [3].
3. IGEPAL® CA-630 is a nonionic, non-denaturing detergent. Caution: It is a severe eye irritant.
4. (Optional step to discard all traces of residual zymolyase) Add 2 ml of buffer 2 + PI without resuspension of the pellet and centrifuge at 4 °C for 5 min at 3500 rpm. Discard the supernatant.
5. Do not dry the pellet excessively.
6. Wear a mask when weighing acrylamide. Unpolymerized acrylamide is a neurotoxin and care should be exercised to avoid skin contact.

Acknowledgments

We thank Vicente Tordera and Merce Pamblanco (University of Valencia) who showed us how to purify histones from *S. cerevisiae.*

References

1. Dehé PM, Pamblanco M, Luciano P, Lebrun R, Moinier D, Sendra R, Verreault A, Tordera V, Géli V (2005) Histone H3 lysine 4 monomethylation does not require ubiquitination of histone H2B. J Mol Biol 353:477–484
2. Poveda A, Pamblanco M, Tafrov S, Tordera V, Sternglanz R, Sendra R (2004) Hif1 is a component of yeast histone acetyltransferase B, a complex mainly localized in the nucleus. J Biol Chem 279:16033–16043
3. Santos-Rosa H, Kirmizis A, Nelson C, Bartke T, Saksouk N, Cote J, Kouzarides T (2009) Histone H3 tail clipping regulates gene expression. Nat Struct Mol Biol 16:17–22

Chapter 6

Analytical Ultracentrifuge Analysis of Nucleosomes Assembled from Recombinant, Acid-Extracted, HPLC-Purified Histones

Manjinder S. Cheema and Juan Ausió

Abstract

The accumulating discoveries of new posttranslational modifications (PTMs) and the increasing relevance of histone variants within the frame of epigenetics demand the availability of methods for a rapid and efficient nucleosome reconstitution to analyze their structural and functional implications. Here we describe a method suitable for this purpose, starting from bacterially expressed histones, solubilized by acid and purified by reversed-phase high-performance liquid chromatography. This method allows the preparation of micrograms to milligram amounts of in vitro-assembled nucleosomes. Finally, we demonstrate the efficiency of this method for the structural analysis of nucleosomes in the analytical ultracentrifuge.

Key words Nucleosome, Recombinant histones, Acid extraction, Fast protein liquid chromatography (FPLC), Reversed-phase high-performance liquid chromatography (HPLC), Analytical ultracentrifuge (AUC)

1 Introduction

Nucleosomes are the basic structural subunit of chromatin. The chromatin fiber consists of a string of nucleosome core particles connected by linker DNA. Linker histones of the histone H1 family bind to these linker DNA regions, while core histones, as their name indicates, form the core of the nucleosome core particle. The latter consists of approximately 146–147 bp of DNA wrapped in 1 and ¾ left-handed coils about an octameric protein core consisting of an H3–H4 tetramer which is flanked by two H2A-H2B dimers [1].

Histones can be posttranslationally modified, and they also include several highly replication-independent histone variants, known as replacement variants [2]. The combination of the histone PTMs and/or their variants imparts on the nucleosomes individual structural and functional characteristics that have important epigenetic implications [3–5]. Fifteen chemically different PTMs have been described to date [6], and they can affect different

Benoit Guillemette and Luc Gaudreau (eds.), *Histones: Methods and Protocols*, Methods in Molecular Biology, vol. 1528,
DOI 10.1007/978-1-4939-6630-1_6, © Springer Science+Business Media New York 2017

well-defined amino acid sites in each of the histones. New relevant PTMs which affect different regions of the histones are being discovered on a regular basis [7].

An interesting feature of the nucleosome core particle is its ability to be reconstituted from its constitutive components [8]. Even poly-nucleosome arrays containing histone H1 that resemble the native chromatin fiber can be reconstituted [9]. In the past, chicken erythrocyte histones were routinely used for these reconstitutions [10], as they are easily amenable and contain relatively low amounts of PTMs [11]. However, the extensive structural homogeneity required for crystallization purposes [12] imposed the need for the preparation of highly homogeneous histone fractions obtained from bacterial expression lacking PTMs or any histone variant components [13].

The histone reconstitution method described here is based on three premises: (1) preferential solubility of histones in dilute acids such as 0.5 N HCl; (2) powerful and rapid resolution of the reversed-phase high-performance liquid chromatography (HPLC) [14]; and (3) the ability to refold the relatively simple histone fold [15] domain from its denatured state.

The procedure for in vitro reconstitution of nucleosomes presented in this chapter describes in detail the expression and purification of human recombinant histones required for the proper assembly of the nucleosome core particle. The chromatin complexes reconstituted in this way exhibit the same structural and hydrodynamic properties determined by analytical ultracentrifugation as their native counterparts. This protocol is very useful in assessing the structural/functional implications not only of individual histone variants [such as H2A,X [16], H2A.Z [17]], but also of their combinations [i.e.: H2A.Z/H3.3]—as well as in assessing the role of histone PTMs [such as H4K16ac [18]]. In this latter instance, histones modified by PTMs (methylated, acetylated) at distinctive amino acid sites can be rather easily synthesized from recombinant histones [19, 20].

The analytical ultracentrifuge (AUC) analysis provided at the end of the chapter is used here only as a "proof of principle" for validation purposes. However, the nucleosome reconstitution protocol described can be scaled up or down, and can be modified to suit the demands of any other types of structural or functional analyses.

2 Materials

Prepare all solutions using double-distilled water (ddH_2O) (prepared by purifying deionized water to attain a sensitivity of 18 MΩ cm at 25 °C). Prepare and store all buffers at room temperature unless otherwise indicated. Follow waste disposal rules and regulations to properly dispose of reagents and buffers. Each human recombinant histone (H2A, H2AZ.1, H2AZ.2, H2B, H3.1, H3.3, and H4) in this protocol is prepared from a 2 L culture.

2.1 Preparation of Human Recombinant Histones

1. *E. coli* BL 21 (DE3) pLyS competent cells.
2. Four wide-bottom 2 L Fernbach flasks.
3. 50 mg/ml Ampicillin stock (filter sterilize using 0.45 μm filter).
4. Agar plates: 1% Tryptone w/v, 0.5% yeast extract w/v, 0.5% NaCl w/v, 1.5% agar w/v, and add 50 μg/ml ampicillin just before pouring the plates (the media should not be hot; otherwise it will degrade the antibiotic).
5. Liquid broth (LB): 1% Tryptone w/v, 0.5% yeast extract w/v, 0.5% NaCl w/v. Add 50 μg/ml ampicillin to the media when it is not hot.
6. Spectrophotometer, NanoDrop.
7. 1 M Isopropyl β-D-1-thiogalactopyranoside (IPTG) stock.
8. 1 M Tris–HCl pH 7.5: Add 121.14 g of Tris base in 800 ml of distilled water and adjust pH with 12 N HCl while stirring. Bring to 1 L with ddH_2O; autoclave and store at room temperature.
9. 0.5 M Tris-ethylenediaminetetraacetic acid (EDTA) pH 8.0: Add 186.12 g EDTA in 800 ml ddH_2O and adjust pH to 8.0 with NaOH pellets. Bring to 1 L with ddH_2O; autoclave and store at room temperature.
10. Expression plasmid vectors pET 22b, pET 11A (EMD Millipore), and pTYB1 (New England Biolabs) containing cDNAs; store at −20 °C.
11. 37 °C Shaking incubator.
12. 30 ml and 250 ml polycarbonate centrifuge tubes, Corex tubes 30 ml (Corning) (*see* **Note 1**).
13. Lyophilizer.
14. Liquid nitrogen.
15. −80 °C Freezer, −20 °C freezer, and 4 °C fridge.
16. 15% Sodium dodecyl sulfate (SDS)-polyacrylamide gel electrophoresis (PAGE): The separating (15% polyacrylamide), bottom part of the gel is prepared from a stock solution of 30% by weight of acrylamide to 0.8% by weight of N,N′-bis-methylene acrylamide in 0.375 M Tris–HCl pH 8.8 and 0.1% SDS, and gels were polymerized with 0.045% by volume of tetramethylene ethylenediamine (TEMED) and 0.000567% by volume of ammonium persulfate (APS) (store TEMED and APS at 4 °C). The stacking (6% polyacrylamide), top part of the gel is

prepared from a stock of 30% by weight of acrylamide to 0.8% by weight of N,N′-bis-methylene acrylamide, in 0.125 M Tris–HCl (pH 6.8) and 0.1% SDS and polymerized with 0.001% by volume of TEMED and 0.01% by volume of APS.

17. SDS sample buffer (2×): 125 mM Tris–HCl pH 6.8, 2% SDS, 20% glycerol, 1.43 M β-mercaptoethanol, 0.2% bromophenol blue.
18. Sample buffer to solubilize bacterial pellet (optional): 100 mM Sodium phosphate pH 6.8, 20 mM Tris–HCl pH 8.0, 8 M urea.
19. Coomassie stain: 10% Acetic acid, 0.2% Sigma brilliant blue, 25% isopropanol, bring to 1 L with ddH_2O, and stir for 30 min. Filter using paper towels and let it sit overnight.
20. Destain: 10% Acetic acid, 10% isopropanol and bring to 1 L with ddH_2O.
21. Beckman Ultracentrifuge with JA 20 and JA 14 rotor (Beckman Coulter), or equivalent.
22. Lysis buffer: 6 M Guanidinium hydrochloride, 1 mM EDTA, 1 mM DTT, 50 mM Tris–HCl pH 7.5.
23. Dialysis buffer 1: 0.1 M NaCl, 50 mM Tris–HCl pH 7.5, 1 mM EDTA, 1mM PMSF (phenylmethylsulfonyl fluoride).
24. Dialysis tubing: 3.5 kDa Molecular weight cutoff (MWCO).
25. Dounce tissue homogenizer 40 ml for cell lysis.
26. Speed vacuum for drying the protein.

2.2 Cation-Exchange Chromatography

1. Cation-exchange column resin: Macro-Prep CM Cation Exchange support (Bio-Rad) and BioRad economo-column of 15 × 1.5 cm.
2. AG 501-X8 (D) (Biorad) resin to deionize urea.
3. 10 M Deionized urea:
   Preparation of 10 M deionized urea stock
   (a) 10 M Deionized urea stock is used to make Buffer B and Buffer C.
   (b) Weigh urea (MW 60.06) for 1 L, add 500 ml of water, and stir until it dissolves completely. It takes approximately 4–5 h to dissolve urea completely (*see* **Note 3**).
   (c) Wrap the volumetric flask with tin foil, or use a brown glass container.
   (d) Once urea starts going into solution, bring it to the final volume by adding ddH_2O.
   (e) Deionize urea (removes all ions from a solution, as they can be detrimental to proteins if present) with resin 5.56 g/100 ml of 10 M urea.

(f) Swirl for at least 2 h with a magnetic stirrer in an Erlenmeyer flask.

(g) Filter 10 M urea using 0.45 μm filter, and store it in a brown bottle at room temperature.

4. Buffer A: 50 mM NaOAC pH 5.2.
5. Buffer B: 7 M Urea, 20 mM NaOAc (pH 5.2), 5 mM β-mercaptoethanol, 1 mM EDTA (*see* **Note 2**).
6. Buffer C: 7 M Urea, 20 mM NaOAc (pH 5.2), 5 mM β-mercaptoethanol, 1 mM EDTA, 1 M NaCl (*see* **Note 2**).
7. Buffer D: 50 mM NaOAc (pH 5.2), 2 M NaCl.
8. Gradient maker.

2.3 High-Performance Liquid Chromatography

1. HPLC system with fraction collector.
2. Nanosep MF 0.45 μm filter (PALL).
3. Solution E: HPLC-grade 100% acetonitrile (ACN). Filter 1 L of acetonitrile using 0.22 μm aqueous filter.
4. Solution F: 0.1% Trifluoroacetic acid (TFA). Add 1 ml of TFA (99%) into 1000 ml distilled water, and filter using 0.45 μm filter.
5. Reversed-phase C_{18} column: C_{18} silica-based column (300 Å pore size, 5 μm particle size).
6. 15% SDS-PAGE.
7. Lyophilizer to freeze-dry the histone fractions.

2.4 Fast Protein Liquid Chromatography

1. Fast protein liquid chromatography (FPLC) AKTA UPC-900 system with fraction collector (GE Healthcare).
2. Unfolding buffer: 7 M Guanidinium hydrochloride, 20 mM Tris–HCl pH 7.5, 10 mM DTT (*see* **Note 4**).
3. Refolding buffer: 2 M NaCl, 10 mM Tris–HCl pH 7.5, 1 mM EDTA, 1 mM DTT.
4. Nanosep MF 0.45 μm filter (PALL).
5. Aqueous filters 0.22 μm.
6. 20% Ethanol, filtered using 0.22 μm filter.
7. Superdex 200 10/300 GL gel filtration column (GE Healthcare).
8. 15% SDS-PAGE.
9. Dialysis tubing 3.5 kDa MWCO.
10. 1 mg/ml BSA in refolding buffer: Filter using 0.45 μm filter.
11. Spectrophotometer, NanoDrop.
12. Syringes: 3 ml and 5 ml.

2.5 Reconstitutions of Nucleosome Core Particles

1. Purified DNA template (147–200 bp, 1–2 mg/ml) and purified histone octamers 1–2 mg/ml (*see* Subheading 3.5).
2. DNA buffer 2×: 4 M NaCl, 20 mM Tris–HCl pH 7.5.
3. Dialysis buffer 2: 2 M NaCl, 10 mM Tris–HCl pH 7.5, 0.5 mM EDTA, 1 mM DTT.
4. Tris–EDTA (TE) buffer: 10 mM Tris–HCl pH 7.5, 0.1 mM EDTA.
5. Dialysis tubing 3.5 kDa MWCO.
6. 15 % SDS-PAGE.

2.6 4 % Native Polyacrylamide Gel Electrophoresis

1. Acrylamide (4 %), with a bis-acrylamide-to-acrylamide ratio of 1:20, in 40 mM Tris–HCl pH 7.5, 1 mM EDTA, 20 mM NaOAc.3H_2O (*see* **Note 5**).
2. 1× E buffer: 40 mM Tris–HCl pH 7.5, 1 mM EDTA, 20 mM NaOAc.3H_2O.
3. Ethidium bromide (EtBr) stain: 0.0001 % Ethidium bromide solution (*see* **Note 6**).
4. Destain: ddH_2O (*see* **Note 6**).
5. Imager: Ultraviolet transilluminator for imaging.

2.7 Analytical Ultracentrifuge

1. AUC XLI/XLA (Beckman Coulter).
2. AUC cells with double-sector aluminum-Epon centerpieces and An 60-Ti and An 55 rotors.
3. UltraScan II version 9.9 (or higher) software available at http://www.ultrascan2.uthscsa.edu/index.php.

3 Methods

3.1 Histone Expression and Isolation

The cDNAs of the human recombinant histones H2A, H2B, H3.1, H3.3, and H4 were cloned into pET 22b expression vector [21] and human histone H2A variants, H2AZ1 and H2AZ2, into pTYB1 and pET 11a expression vector.

3.1.1 Histone Expression

1. Transform *E. coli* BL 21 (DE3) pLyS competent cells (50 μl) with 0.1–1 μg of the pET 22b, pET 11A, and pTYB1 human histone expression plasmids and aseptically streak/spread cultures on agar plate. Incubate overnight at 37 °C.
2. Next day, pick individual colonies and inoculate four 5 ml cultures containing liquid broth (LB). Incubate overnight (~16 h) in a shaking incubator at 280 rpm, at 37 °C.
3. Inoculate 4 × 500 ml cultures with overnight 5 ml aliquot each and keep incubating in a shaking incubator at 37 °C for each

histone H3.1, H3.3, H2A, H2AZ.1, H2AZ.2, and H2B (H4 at 30 °C) at 280 rpm until an $OD_{600} = \sim 0.6$ is reached (takes approximately 2.5–3 h) (*see* **Note 7**).

4. Once the optimal density of cells has been reached, remove 1 ml aliquot and centrifuge at 16,000 × *g* in an Eppendorf tube. Discard the supernatant and keep the uninduced pellet at −80 °C, which will serve as a negative control for expression. Induce the rest of the cultures to a final concentration of 1 mM IPTG. Incubate for another 3 h at 37 °C for human recombinant histones H2A, H2AZ.1, H2AZ.2, H2B, H3.1, and H3.3, and 4 h for human recombinant histone H4 at 30 °C.
5. Harvest cells by centrifugation in 250 ml polycarbonate tubes at 4 °C at 3000 × *g* for 15 min using the JA 20 rotor.
6. At this point you can flash freeze the pellet and keep at −80 °C until use (*see* **Notes 8** and **9**).

3.1.2 HCL Extraction

Histones are basic proteins with Arg- and Lys-rich composition, and can be extracted with dilute acids. The cells expressing specific histones (*see* Subheading 3.1) are first lysed, and then the histones are extracted with 0.5 N HCl and acetone precipitated. The acid extraction method is based on the protocol described by Levenstein and Kadonaga [22].

7. Resuspend each 500 ml bacterial pellet in 10 ml lysis buffer. For a total of 2 L bacterial culture, 40 ml of lysis buffer is required.
8. Homogenize with a 40 ml Dounce (~40 strokes) on ice and dialyze homogenate against 2 L of dialysis buffer 1 using 3.5 kDa MWCO dialysis tubing, for 2 h at 4 °C.
9. Transfer dialyzed solution into a 125 ml flask, add 12 N HCl drop by drop (add very steadily) while stirring, and bring the final HCl concentration to 0.5 N.
10. Incubate at −20 °C for 30 min (this will help solubilize the histone proteins).
11. Centrifuge at 26,000 × *g* at 4 °C for 20 min using JA 20 rotor (use 30 ml polycarbonate bottles with lids).
12. Histone precipitation is achieved by adding 6 volumes of acetone to the collected supernatant. To this end, the supernatant can be split into 4 ml aliquots using 30 ml Corex tubes. Subsequently, add 24 ml of acetone to each tube (*see* **Note 10**), invert a couple of times (*see* **Note 11**), and keep at −20 °C or 4 °C overnight.
13. Next day, spin tubes at 7000 × *g* for 10 min at 4 °C (JA-20 Rotor) and discard acetone supernatant.

14. With the help of a spatula, resuspend the pellet in fresh room-temperature acetone (*see* **Note 12**) and centrifuge again at 7000 × *g* for 10 min 4 °C (JA-20 Rotor). Repeat this step twice.
15. While spinning, get the vacuum pump ready.
16. Discard supernatant (*see* **Note 13**) and start drying the protein (using the device shown in Fig. 1) for 5–10 min while warming the Corex tube with the palm of your hand (*see* **Note 14**) until the tube does not feel cold anymore (*see* **Note 15**).
17. After that, close the three-way valve to break the vacuum, and gently release the Corex tube from the valve. Store the powder thus obtained at −80 °C until use. 2 L Culture yields approximately 250 mg of histone powder.
18. Analyze purified recombinant histone by 15 % SDS-PAGE; the histone bands in the HCl extract should be very prominent (Fig. 2, AS).

3.2 Cationic Exchange Chromatography

3.2.1 Packing the Column

1. Prepare all the buffers and pack the column at least 1 day before you plan to run it.
2. Pour column using 25 ml of the Macro-Prep CM cation-exchange resin stored in 20% ethanol in an Erlenmeyer flask and let it settle for approximately 20 min.
3. Decant the excess 20% ethanol solution, wash resin with approximately 100–150 ml Buffer A, then gently stir the suspension with a glass rod for 1 min, and let it settle for approximately 20 min.
4. Repeat the previous step until the pH of the buffer in the resin is ~5.2 and stable.
5. Pour the resin into a column following the manufacturer's instructions. The resin will span about 7 cm.

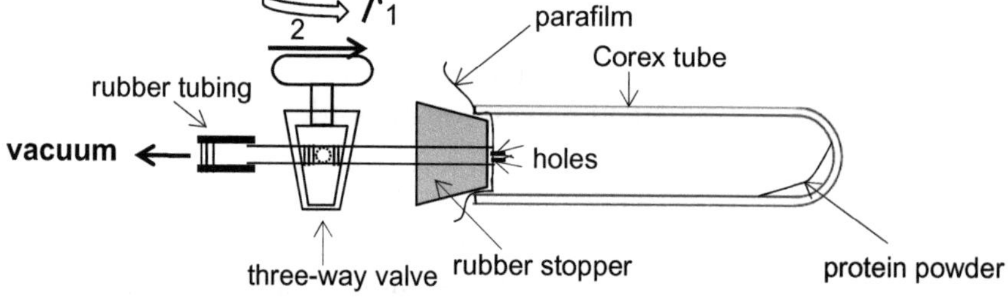

Fig. 1 To dry the protein pellets after precipitation of the acid-extracted proteins with acetone, a very simple device can be improvised from a glass three-way valve and a perforated rubber stopper connected to a vacuum source (vacuum pump or aspirator). The valve should be on its "closed" position (1) when the Corex centrifuge tube—with the precipitated protein pellet (*see* Subheading 3.1, **step 16**)—is attached to the rubber stopper, and in the "open" position (2) for drying. After the pellet is completely dry and powdery, the valve should be returned to position (1) before the centrifuge tube is gently detached

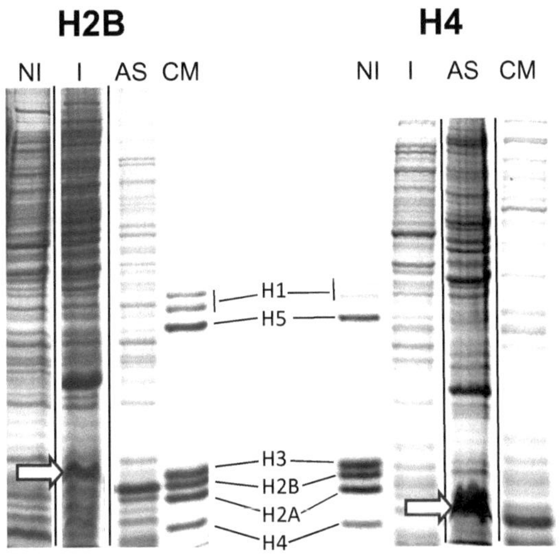

Fig. 2 SDS-PAGE analysis of the recombinant protein from bacteria expressing histone H2B and H4, before IPTG induction (NI), after IPTG induction (I), and after 0.5 N HCl solubilization (*see* Subheading 3.1) (AS). The *arrow* indicates the histone expressed. The slight decrease in its electrophoretic mobility compared to the marker is due to the composition of sample buffer used to solubilize the bacterial pellet. *CM* chicken erythrocyte histone marker. The *vertical lines* are there to indicate that the images were taken from different noncontiguous regions of the same gel

6. Connect the column to peristaltic pump and equilibrate the column with Buffer A for at least 5 h at a flow rate of 20 ml/h.
7. Change to Buffer B, and let the column equilibrate for another 2 h at the same flow rate.

3.2.2 Loading the Sample and Elution

8. Weigh dried recombinant histone powder ~30 mg and dissolve in a 1 ml of Buffer B.
9. Stop the peristaltic pump, remove the top plug of the column, and then carefully overlay the sample on top of resin—drop by drop, very slowly—without disturbing the resin bed. Start peristaltic pump at 20 ml/h flow rate and let the sample descend into the column until a meniscus is formed just above the resin.
10. Rinse the empty sample tube with 1 ml of Buffer B, and load onto the column as in the previous step. Repeat twice with 1 ml of Buffer B.
11. Add Buffer B to the top of the resin, put back the plug of the column, and connect it to a gradient maker.
12. Elute the sample with a linear gradient of 0 M to 1 M NaCl in 7 M urea, 20 mM NaOAc (pH 5.2), 5 mM β-mercaptoethanol, and 1 mM EDTA consisting of approximately 6–7× column volumes (40 ml of Buffer B + 40 ml of Buffer C).

13. Collect fractions at 3-min intervals (approximately 1 ml).
14. At the end of the gradient, wash the column with Buffer D for 2 h to remove any other proteins still attached to the column.
15. Wash column with Buffer A for 5 h and store in Buffer A until next use. This column can be used at least five times. To reuse it, wash the column with Buffer B for 2 h.
16. Read the OD_{230} using a conventional spectrophotometer (*see* **Note 16**) of the collected fractions, using Buffer B as a blank. Plot a graph and store the fractions at 4 °C (Fig. 3 shows examples from elution profiles obtained with this column).
17. Analyze samples using 15 % SDS-PAGE and combine appropriate fractions (*see* **Note 17**) (Fig. 3).

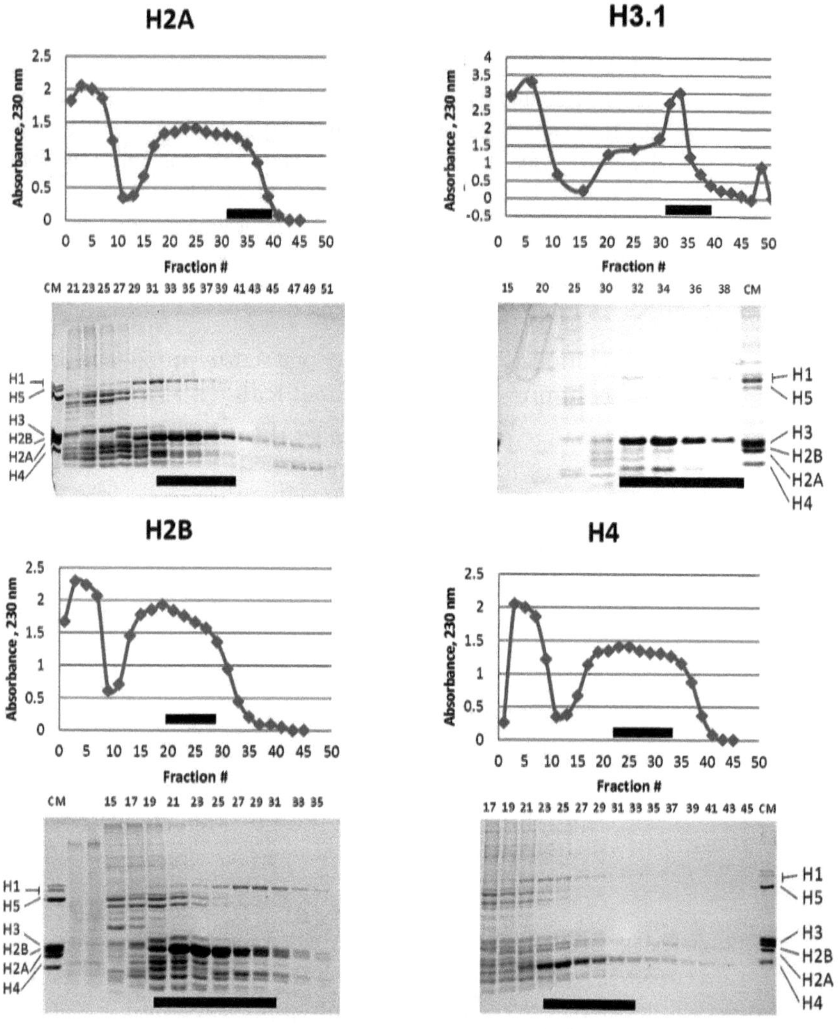

Fig. 3 Ionic exchange fractionation (*see* Subheading 3.2, **step 17**) and SDS-PAGE analysis of the acid-soluble fractions (*see* Subheading 3.1, **step 19**). The fraction numbers are indicated on *top* of the gel. The *bars* indicate the fractions pooled. Elution conditions are as described in the text. *CM* chicken erythrocyte histone marker

18. Dialyze fractions of interest against 2 L of distilled water using 3.5 kDa MWCO dialysis tubing while stirring at 4 °C for 2 h (*see* **Note 17**).
19. Repeat **step 18** one more time.
20. Lyophilize samples overnight, resuspend in 2 ml of ddH_2O, and measure absorbance at 230 nm (OD_{230}). The approximate concentration of the protein can be determined using the average extinction coefficient for the core histones (H2A + H2B + H3 + H4) of 4.2 cm^2/mg [14].
21. Store samples at −80 °C until you are ready to do HPLC (*see* Subheading 3.3).

3.3 Reversed-Phase High-Performance Liquid Chromatography

Recombinant histones can be easily purified using a reversed-phase HPLC system using C_{18} column (300 Å pore size, 5 μm particle size). C_4 columns can also be used, but the elution gradient needs to be modified.

1. In a reversed-phase HPLC, 100% acetonitrile (Solution E) is used as a mobile phase and 0.1% TFA (Solution F) is used as ion pairing reagent to maintain the low pH of the mobile phase.
2. Approximately 1 mg of the samples prepared in Subheading 3.2, **step 21**, dissolved in 0.5–0.8 ml of ddH_2O can be used. The solution needs to be filtered using Nanosep MF 0.45 μm at 780 × *g* for 5 min using a microcentrifuge at 4 °C.
3. Elution is carried out at a 1 ml/min flow rate with an acetonitrile gradient in the presence of 0.1% TFA as indicated in Table 1.

Table 1
Acetonitrile gradient for elution of histones from C_{18} column

Time (min)	Solution E (acetonitrile)	Solution F (0.1 %TFA)	Duration (in min)
0	5	95	1
1	25	75	10
11	30	70	15
26	35	65	20
46	40	60	20
66	43	57	10
76	55	45	50
126	90	10	5
131	100	0	20
151	100	0	5

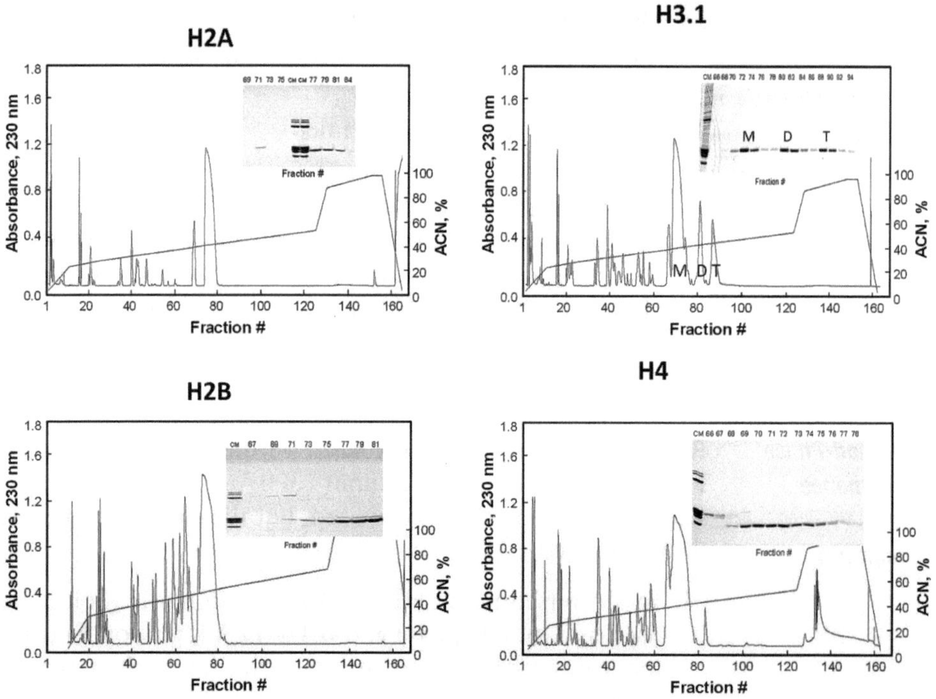

Fig. 4 Reversed-phase HPLC fractionation and SDS-PAGE analysis (*inset*) of the fractions pooled in Fig. 3. The fraction numbers are indicated on *top* of the gel. The elution conditions are indicated in the text (*see* Subheading 3.3, **step 3**). *CM* chicken erythrocyte histone marker

4. Figure 4 shows several examples of the chromatograms and purification achieved at this stage.
5. The fractions of interest (as analyzed with 15 % SDS-PAGE) are pooled, frozen with liquid nitrogen, and lyophilized overnight.
6. Add 1 ml of distilled water to the lyophilized samples, and measure OD_{230} to determine the approximate concentration as in Subheading 3.2, **step 20**.
7. In those instances where the induction of recombinant protein expression provides very clean samples Fig. 5a (this is usually achieved by inducing at OD_{600} 0.5–0.6), it is possible to skip Subheading 3.2 and proceed directly to Subheading 3.3 (Fig. 5b).

3.4 Histone Octamer Preparation

1. Based on OD_{230}, combine appropriate volumes corresponding to stoichiometric amounts of each recombinant histone obtained in Subheading 3.3 and run a 15 % SDS-PAGE, loading an increasing overall amount of combined histones in the range of 0.5–1.5 μg (*see* Fig. 6a). The different loadings will be helpful when visually adjusting the volumes of the different histones to attain an approximate normalization in which all

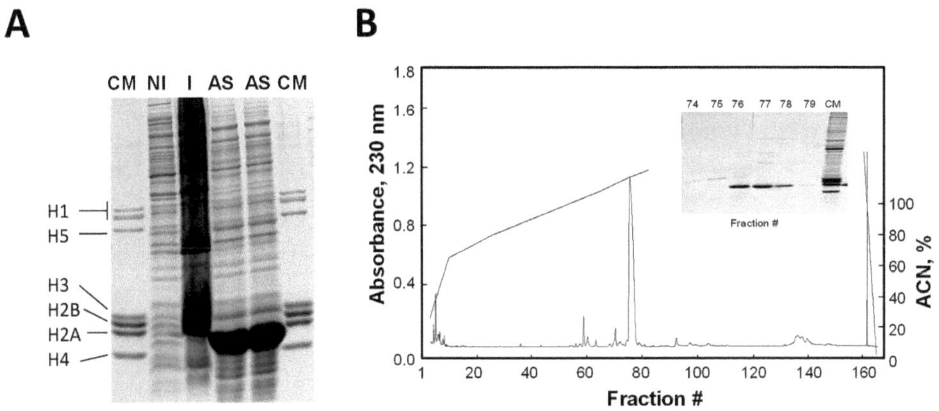

Fig. 5 (**a**) SDS-PAGE analysis of a bacterial culture expressing histone H2A.Z-1, grown for 3 h at 37 °C after induction with IPTG at an $OD_{600} = 0.6$. *NI* protein sample before induction, *I* protein sample after induction, *AS* 0.5 N HCl acid-soluble fraction (*see* Subheading 3.1) (**b**). Reversed-phase HPLC of the sample shown in (**a**) lane AS. *CM* chicken erythrocyte histone marker. The *inset* shows an SDS-PAGE analysis of some of the fractions collected during the elution

the histone bands exhibit a similar staining intensity (*see* **Note 18**). Two or three 15% SDS-PAGE gels may be required to attain this.

2. Once this stoichiometry is achieved, flash freeze the sample with liquid nitrogen and lyophilize overnight.
3. All the processes used to obtain the histones up to this point—bacterial expression, protein solubilization, acid extraction, ionic exchange in the presence of urea, and reversed-phase HPLC—have been of a denaturing nature. Therefore, the histone mixture needs to be renatured as described next. Resuspend the lyophilized sample of Subheading 3.4, **step 2**, in unfolding buffer at an approximate concentration of 2 mg/ml (*see* **Note 19**).
4. Incubate at room temperature for 30 min.
5. Add an equal volume of refolding buffer and immediately dialyze twice at 4 °C against 1 L of refolding buffer for 2 h each time while stirring; the sample is finally dialyzed against 2 L of ice-cold refolding buffer overnight at 4 °C.

3.5 FPLC Fractionation of Histone Octamer

1. After dialysis, remove any precipitated proteins by centrifugation at 16,000 × *g*, 4 °C, in a microcentrifuge, and filter the supernatant using Nanosep MF 0.45 μm filter in the same centrifuge at 700 × *g*, for 5 min at 4 °C.
2. One day ahead of the histone octamer preparation, wash the gel filtration column for 5 min with 20% ethanol at 4 °C at a 0.15 ml/min flow rate, using FPLC AKTA equipment.

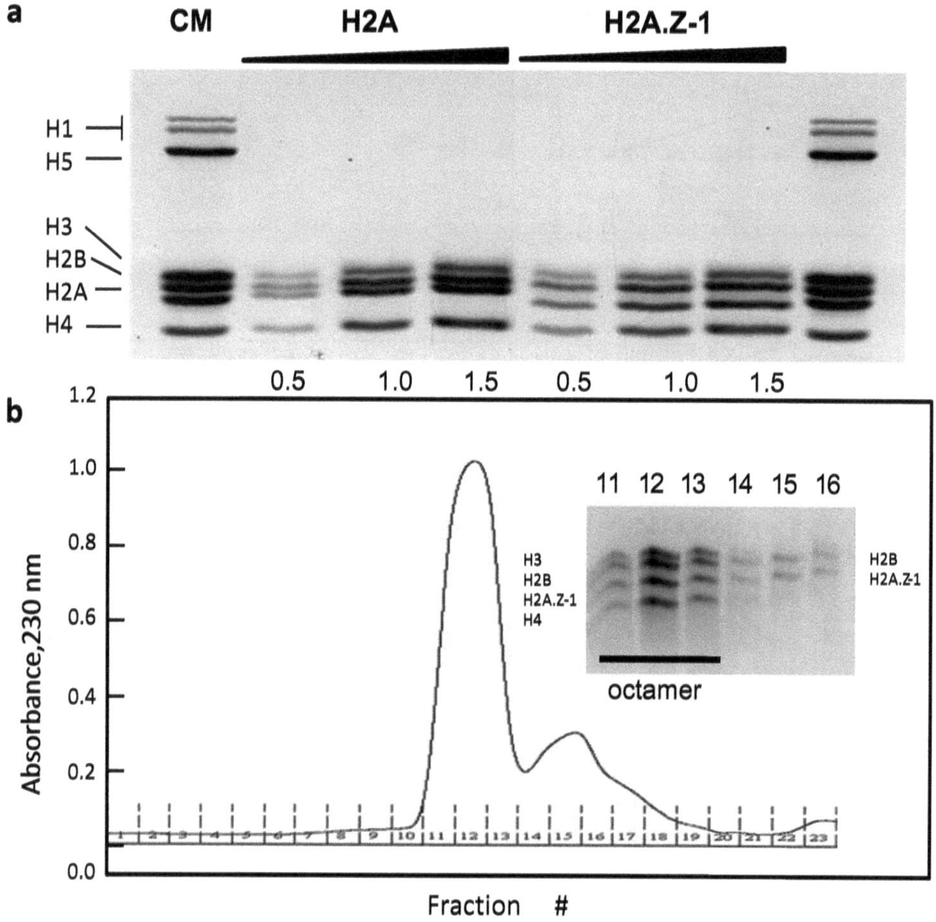

Fig. 6 (**a**) Increasing amount of purified histone mixtures (*see* Subheading 3.4, **step 1**) dissolved in water, consisting of approximately equal amounts of each individual histone as estimated from the concentration determined of their absorbance at 230 nm (*see* Subheading 3.3, **step 6**). (*H3-H4-H2B*)-H2A and (*H3-H4-H2B*)-H2A.Z-1 mixtures are shown. The *numbers below* indicate the approximate number of micrograms loaded in each lane. (**b**) FPLC chromatogram of an (*H3-H4-H2B*)-H2A.Z-1 mixture after octamers (*see* Subheading 3.5, **step 6**). The *inset* shows an SDS-PAGE analysis of some of the fractions collected along the elution profile. The *bar* indicates the fractions corresponding to the histone octamer

3. Wash column with 5 volumes of ddH_2O at a 0.15 ml/min flow rate overnight, followed by a column equilibration run with refolding buffer at a 0.45 ml/min flow rate for about 2 h.
4. After the column has been equilibrated with the running buffer, inject 1 ml of BSA (1 mg/ml in refolding buffer) and elute at a flow rate of 0.45 ml/min using the same buffer.
5. Under these conditions, a distinct peak should appear at around fraction 13 corresponding to the molecular mass of monomeric BSA (66.5 kDa). This may be preceded by a peak at fractions 6–7 which corresponds to protein aggregates.

6. After ensuring that the column is performing as expected (from the previous **step 5**), proceed with the injection of the sample prepared in Subheading 3.4, **step 5** (Fig. 6b provides an example of an elution profile obtained with this column) (*see* **Note 20**).
7. Pool fractions corresponding to the histone octamer (Fig. 6b) and determine the concentration, spectrophotometrically reading the absorbance at 230 nm (*see* Subheading 3.2, **step 20**).

3.6 Preparation of Histone-DNA Complexes (Reconstituted Nucleosomes)

1. A 208 bp sequence-positioning DNA template is generally used in our lab for the reconstitution of chromatin complexes (*see* **Note 21**). It corresponds to a fragment of the 5S rRNA gene from the sea urchin *Lytechinus variegatus* [23].
2. To reconstitute 100 μg of DNA we use 90 μg of histones (*see* **Note 22**).
3. Bring the 208 bp DNA in TE buffer to 2 M NaCl and 10 mM Tris–HCl (pH 7.5) by addition of an equal volume of DNA buffer 2×. Histone octamers prepared from FPLC are already in 2 M NaCl, 10 mM Tris–HCl pH 7.5, 1 mM EDTA, and 1 mM DTT.
4. Using a conventional spectrophotometer, accurately determine the concentration of the DNA and histone solutions (*see* **Note 23**).
5. Mix the appropriate volumes of histone and DNA, and use a stepwise salt (NaCl) gradient dialysis from 2 M NaCl to 0 M NaCl to reconstitute the histone-DNA mixture to nucleosome [8].
6. Start the reconstitution procedure the day before by dialyzing against 2 L of dialysis buffer 2 overnight while stirring.
7. Next day remove 500 ml of the dialysis buffer and transfer to a 500 ml beaker. Place the dialysis tubing containing the sample in this beaker while you add and mix 500 ml of cold TE to the rest (1.5 L) of the buffer in dialysis container. This will bring the NaCl concentration down to 1.5 M. Return the dialysis tubing to the dialysis flask and continue dialysis for another 2 h.
8. Repeat the procedure described in the previous step (Subheading 3.6, **step** 7) every 2 h by removal and addition of the following TE volumes: 400 ml (1.2 M NaCl), 500 ml (0.9 M NaCl), 666 ml (0.6 M NaCl), and 1000 ml (0.3 M NaCl).
9. At this point, transfer the dialysis tubing to a container with 2 L of cold TE buffer, and continue dialysis overnight.
10. Analyze the efficiency of reconstituted nucleosome by running a native 4% polyacrylamide gel electrophoresis (Fig. 7a).

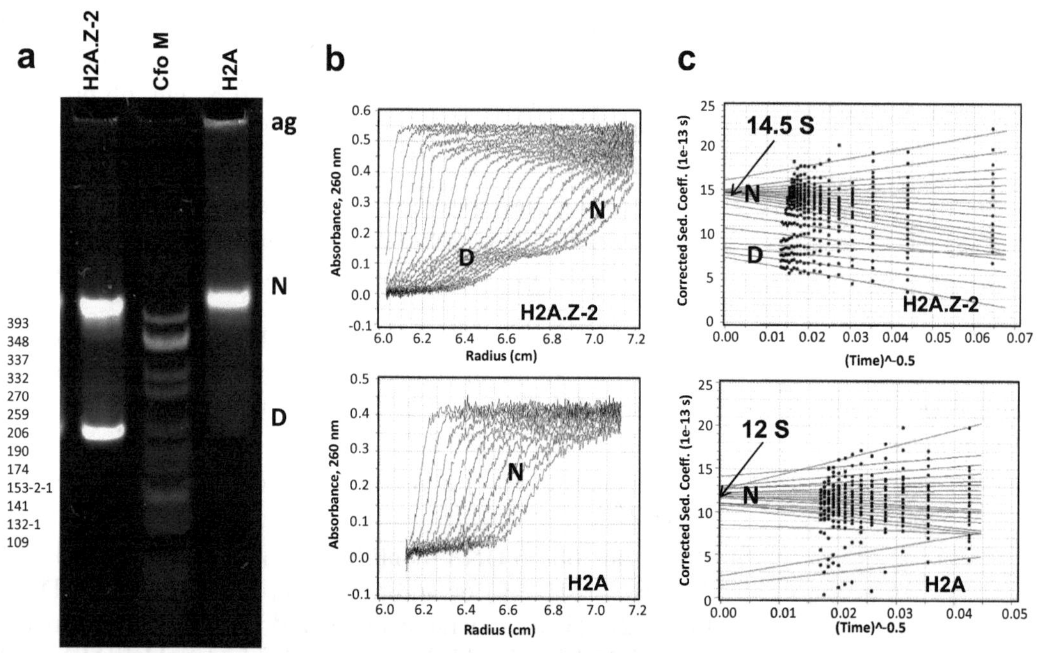

Fig. 7 (**a**) Ethidium bromide-stained native 4 % PAGE of nucleosomes consisting of (*H3-H4-H2B*)-H2A (lane H2A) and (*H3-H4-H2B*)-H2A.Z-2 (lane H2A.Z-2), reconstituted onto a 208 bp DNA template. Notice the downward shift in electrophoretic of the (*H3-H4-H2B*)-H2A.Z-2-containing nucleosome. *ag* aggregates, D = 208 bp DNA; *N* nucleosome, *Cfo M* pBr322 plasmid DNA digested with *CfoI*, used as a marker. The number of bp corresponding to the bands is indicated on the *left*. The amount of free DNA in the H2A.Z-2 nucleosomes (approx. 15 %, *see* Fig. (**b**)—H2A.Z-2) does not reflect any difference in its reconstitution efficiency compared to the H2A nucleosome, but simply indicates a lower amount of histones loaded at the start of the reconstitution. Notice that the brightest appearance of the DNA band (D) is due to the fact that free DNA is able to intercalate three times more ethidium bromide than when reconstituted into nucleosomes (N). (**b**) Sedimentation profiles of the H2A and H2A.Z-2 nucleosome samples in the AUC. (**c**) Analysis of the boundaries shown in (**b**) using the "fan plot" representation of the van Holde and Weischet [25] analysis method. In this representation, the number of lines converging to a point in the "*y*-axis" is proportional to the amount of sample sedimenting, with its corresponding sedimentation coefficient indicated on this axis. Notice that the higher average sedimentation coefficient value of the H2A.Z-2 nucleosomes (14.5 S) indicates a more compact conformation when compared to the H2A nucleosome (12 S), in agreement with the change in electrophoretic mobility observed in (**a**)

3.7 Analytical Ultracentrifuge Analysis of Reconstituted Nucleosome

Analytical ultracentrifuge is the most versatile and accurate means for determining the conformational changes of macromolecules as determined from their sedimentation coefficient. For a more detailed description and theoretical basis, *see* [14, 24]. The nucleosome core particle reconstituted above can then be analyzed using this technique [14]. The data obtained from AUC (Fig. 7b) can then be edited and analyzed using different software such as UltraScan II version 9.9, to determine the sedimentation coefficients of reconstituted nucleosomes.

1. Dialyze the reconstituted samples prepared in Subheading 3.6 overnight against your choice of buffer. Buffers may contain different salt concentrations (monovalent, divalent), different

pHs, or urea—depending on the types of interactions to be analyzed (ionic, hydrophobic ...).

2. Samples should be stored at 4 °C until use.
3. Perform AUC sedimentation velocity analysis as soon as possible after dialysis.
4. Sedimentation velocity analysis is usually carried out in cells consisting of double-sector Epon-filled centerpieces, using either an An 60-Ti Class R or An 55 Class R analytical rotor.
5. Assemble the cells very carefully as per Beckman instructions manual, and load 425 μl of buffer in reference and 395 μl of sample in the sample sector.
6. Load the cells and counterbalance into the rotor (*see* **Note 24**); align them properly and introduce the rotor into the AUC.
7. Assemble the monochromator, close the lid, turn the vacuum on, and start the run when the vacuum is lower than 150 μm. Set up the temperature at which you desire to perform the run (*see* **Note 25**).
8. Perform a pre-run at 3000 rpm. While scanning at 260 nm (or at any other desired wavelength, i.e., 230/280 nm for proteins), check that the vacuum is maintained at 10 microns for indications that cells are not leaking.
9. Increase the speed to the desired speed, normally 40,000 rpm for nucleosomes, and 16,000–20,000 rpm for longer chromatin complexes (oligonucleosomes).
10. Once the rotor has reached the desired temperature for the run, select the number of scans required (usually 20 scans per cell) and start the method scan, following the Beckman manual instructions.

Analysis of reconstituted chromatin complexes

11. There are several methods and programs in use to analyze the AUC-generated data. Our lab carries out this analysis with the UltraScan II software (*see* **Note 26**), which uses the Van Holde and Weischet [25] analysis to analyze the sedimentation velocity data. For more detailed information, *see* [14, 24]. Figure 7b shows some examples of data obtained in this way.

4 Notes

1. Please note that Corning discontinued the production of these tubes, but Pyrex replacements are available from the same company. Otherwise they can be obtained from GENTAUR (Brussels, Belgium).

2. Buffers B and C should be covered with aluminum foil in a brown bottle, because urea is light sensitive.
3. To dissolve urea mixture faster, continue stir on low heat but make sure that it does not get too hot.
4. Make fresh unfolding buffer to get good refolding efficiency.
5. Make fresh 4 % native gel on the day of use (older gels tend to shrink and do not run well). This gel is very thin and fragile and has to be handled very carefully while staining with ethidium bromide and destaining with ddH_2O.
6. Dissolve 10 μl of 10 mg/ml ethidium bromide stock into 100 ml ddH_2O to make 0.0001 % stain, and stain for at least 12 min. Always make fresh ethidium bromide stain to get better resolution of bands. Destain for 12 min using ddH_2O.
7. Induction of each histone expression was carried out at a bacterial culture of OD_{600} ~0.6–0.8 to get optimum results. If the culture is induced at OD_{600} ~0.6, the HCl-purified histone proteins from the bacterial cell extract are usually clean enough for purification by HPLC (*see* Subheading 3.3) without any cationic exchange (*see* Subheading 3.2) re-purification step.
8. Purifying the histone on the same day usually results in higher yields.
9. The protein composition of the sample collected in Subheading 3.1, **step 4**, and the protein obtained after induction in Subheading 3.1, **step 6**, can be analyzed by 15 % SDS-PAGE. For a pellet from 1 ml suspension, resuspend in 50 μl of buffer in Subheading 2.1, **item 18**, and load 10 μl on the SDS-PAGE. Alternatively, straight SDS sample buffer 1× (*see* Subheading 2.1, **item 17**) can also be used.
10. The tube is sealed with a double layer of parafilm.
11. Inverting the tubes after addition of acetone has to be quick, and the top of the tube should be capped with the hand, as acetone can dissolve parafilm upon long exposure.
12. Resuspend the pellet so that it becomes particulate. The particles in the second resuspension should have a very thin powdery appearance, to allow acetone to absorb as much of the protein hydration as possible, and to facilitate the drying in the next step.
13. If the acetone resuspension (Subheading 3.1, **step 14**) has been done properly, this pellet should be very unstable; therefore the acetone supernatant needs to be removed very carefully.
14. Dry the protein using a device shown in Fig. 1. Any vacuum pump can be used for this purpose. During this procedure, it is advisable to break the vacuum once or twice and disperse the pellet into small pieces with a spatula.

15. Evaporation is an endergonic process.
16. NanoDrop will not work properly at these high concentrations of salt and urea.
17. The SDS-PAGE analysis and dialysis of the samples (*see* Subheading 3.2, **steps 17** and **18**) should be performed as soon as possible to avoid the time-dependent deleterious effects of the products of urea decomposition on protein (i.e., formation of isocyanate).
18. The stoichiometry of the bands does not have to be perfect, and usually a visual normalization is good enough. The chromatographic step after refolding will produce an octamer (*see* Subheading 3.4) with stoichiometric amounts of H2A/H2B/H3 and H4. However, the better the accuracy in Subheading 3.4, **step 1**, the higher the octamer yield.
19. Ideally the overall amounts of octamer should be 500 μg or higher.
20. Histone octamers can be perfectly reconstituted from each of their recombinant histones. However, in instances where only the role of the variability of one of the histones is to be analyzed (i.e., H2AZ, H3K27me$_3$), it is possible to use combinations of the recombinant histone with native counterparts. Native H2A/H2B dimers and H3-H4 tetramers can be easily obtained from chicken erythrocytes [26]. Chicken erythrocyte histones, while having an identical sequence to human histones, contain a low level of posttranslational modifications (PTMs) and are very amenable to reconstitution studies [11].
21. DNA fragments of different lengths (146–250 bp) can be used. Fragments of DNA smaller than those of the DNA in the nucleosome core particle (145–147 bp) usually lead to aggregation. The 208 bp fragment of Simpson [23] can be digested to smaller sites with different restriction enzymes. Other DNA sequences of similar sizes can also be frequently used, including the DNA sequence positioning clone 601 of Widom [27].
22. The ratio histone (H)/DNA (D) = 0.9 is calculated from the weight ratio WH:WD for 1 mol of histone octamer (M_r = 108,486):1 mol of 208 bp DNA (M_r: 137,904) × 1.1 (1.1 = empirical factor). Notice that this calculation is for a DNA sequence containing a relatively strong positioning sequence. DNA sequences longer than 200–250 bp may lead to additional complications. For instance for a DNA sequence containing 312 bp, the calculation of the weight H/D ratio for 1 mol histone/1 mol DNA would be (108,486/312 × 663) × 1.1 = 0.58 mg histone/mg DNA. However, sequences of DNA sizes larger than 200 bp may tend to bind histone octamers nonspecifically at sites other than the positioning ones. This may complicate both the analysis of these nucleosomes and the

required weight H/D ratio which would usually be much higher than predicted from these calculations.

23. Use extinction coefficients $A_{260} = 20$ cm^2/mg for DNA and $A_{230} = 4.2$ cm^2/mg for the core histone octamers [28].
24. It is very important that the cells and the counterbalance are properly balanced. The weight of the counterbalance has to be equal to or 0.5 g lighter than the weight of any of the other cells in the rotor.
25. The AUC can be used to perform thermodynamic studies, and hence the samples can be run at different temperatures if desired (within the range of 4–40°C). Samples that can undergo protein degradation are also run at 4 °C. However, the apparent sedimentation coefficient in a given buffer B at a given temperature T ($S_{T,B}^{app}$) should be corrected to standard conditions, in water (w) at 20 °C ($S_{20,w}$). The UltraScan can perform this correction.
26. The latest version (UltraScan III Windows XP/7, version 3.3, release 1884/SOMO 2734) of UltraScan software is available online (http://www.ultrascan.uthscsa.edu/). This website regularly updates the program with more recent versions. For the analysis shown in this protocol chapter we used UltraScan II version 9.9 (Win 32—9.9).

Acknowledgement

This work was supported by a Canadian Institutes of Health Research (CIHR) [MOP-130417] grant.

References

1. van Holde KE (1988) Chromatin. Springer, NY
2. Ausio J (2006) Histone variants—the structure behind the function. Brief Funct Genomic Proteomic 5:228–243
3. Patel DJ, Wang Z (2013) Readout of epigenetic modifications. Annu Rev Biochem 82:81–118
4. Gurard-Levin ZA, Almouzni G (2014) Histone modifications and a choice of variant: a language that helps the genome express itself. F1000Prime Rep 6:76
5. Chen P, Zhao J, Li G (2013) Histone variants in development and diseases. J Genet Genomics 40:355–365
6. Huang H, Sabari BR, Garcia BA, Allis CD, Zhao Y (2014) SnapShot: histone modifications. Cell 159, 458–458.e451
7. Tessarz P, Kouzarides T (2014) Histone core modifications regulating nucleosome structure and dynamics. Nat Rev Mol Cell Biol 15: 703–708
8. Tatchell K, van Holde KE (1977) Reconstitution of chromatin core particles. Biochemistry 16:5295–5303
9. Song F, Chen P, Sun D, Wang M, Dong L et al (2014) Cryo-EM study of the chromatin fiber reveals a double helix twisted by tetranucleosomal units. Science 344:376–380
10. Ausió J, Dong F, van Holde KE (1989) Use of selectively trypsinized nucleosome core particles to analyze the role of the histone "tails" in the stabilization of the nucleosome. J Mol Biol 206:451–463
11. Peterson CL, Hansen JC (2008) Chicken erythrocyte histone octamer preparation. CSH Protoc 2008: pdb.prot5112.
12. Luger K, Mader AW, Richmond RK, Sargent DF, Richmond TJ (1997) Crystal structure of

the nucleosome core particle at 2.8 A resolution. Nature 389:251–260

13. Luger K, Rechsteiner TJ, Richmond TJ (1999) Expression and purification of recombinant histones and nucleosome reconstitution. Methods in Molecular Biology 119:1–16
14. Ausió J, Moore SC (1998) Reconstitution of chromatin complexes from high-performance liquid chromatography-purified histones. Methods 15:333–342
15. Arents G, Moudrianakis EN (1995) The histone fold: a ubiquitous architectural motif utilized in DNA compaction and protein dimerization. Proc Natl Acad Sci U S A 92: 11170–11174
16. Li A, Yu Y, Lee SC, Ishibashi T, Lees-Miller SP et al (2010) Phosphorylation of histone H2A.X by DNA-dependent protein kinase is not affected by core histone acetylation, but it alters nucleosome stability and histone H1 binding. J Biol Chem 285:17778–17788
17. Ishibashi T, Dryhurst D, Rose KL, Shabanowitz J, Hunt DF et al (2009) Acetylation of vertebrate H2A.Z and its effect on the structure of the nucleosome. Biochemistry 48:5007–5017
18. Shogren-Knaak M, Ishii H, Sun JM, Pazin MJ, Davie JR et al (2006) Histone H4-K16 acetylation controls chromatin structure and protein interactions. Science 311:844–847
19. Shogren-Knaak MA, Peterson CL (2004) Creating designer histones by native chemical ligation. Methods Enzymol 375:62–76
20. Simon MD, Chu F, Racki LR, de la Cruz CC, Burlingame AL et al (2007) The site-specific installation of methyl-lysine analogs into recombinant histones. Cell 128:1003–1012
21. Thakar A, Gupta P, Ishibashi T, Finn R, Silva-Moreno B et al (2009) H2A.Z and H3.3 histone variants affect nucleosome structure: biochemical and biophysical studies. Biochemistry 48:10852–10857
22. Levenstein ME, Kadonaga JT (2002) Biochemical analysis of chromatin containing recombinant Drosophila core histones. J Biol Chem 277:8749–8754
23. Simpson RT, Thoma F, Brubaker JM (1985) Chromatin reconstituted from tandemly repeated cloned DNA fragments and core histones: a model system for study of higher order structure. Cell 42:799–808
24. Ausió J (2000) Analytical ultracentrifugation and the characterization of chromatin structure. Biophys Chem 86:141–153
25. van Holde KE, Weischet WO (1978) Boundary analysis of sedimentation velocity experiments with monodisperse and paucidisperse solutes. Biopolymers 17:1387–1403
26. Simon RH, Felsenfeld G (1979) A new procedure for purifying histone pairs H2A+H2B and H3+H4 from chromatin using hydroxyapatite. Nucleic Acids Res 6:689–696
27. Thastrom A, Lowary PT, Widlund HR, Cao H, Kubista M et al (1999) Sequence motifs and free energies of selected natural and non-natural nucleosome positioning DNA sequences. J Mol Biol 288:213–229
28. Ausió J, Seger D, Eisenberg H (1984) Nucleosome core particle stability and conformational change. Effect of temperature, particle and NaCl concentrations, and crosslinking of histone H3 sulfhydryl groups. J Mol Biol 176:77–104

Chapter 7

SILAC-Based Quantitative Strategies for Accurate Histone Posttranslational Modification Profiling Across Multiple Biological Samples

Alessandro Cuomo*, Monica Soldi*, and Tiziana Bonaldi

Abstract

Histone posttranslational modifications (hPTMs) play a key role in regulating chromatin dynamics and fine-tuning DNA-based processes. Mass spectrometry (MS) has emerged as a versatile technology for the analysis of histones, contributing to the dissection of hPTMs, with special strength in the identification of novel marks and in the assessment of modification cross talks. Stable isotope labeling by amino acid in cell culture (SILAC), when adapted to histones, permits the accurate quantification of PTM changes among distinct functional states; however, its application has been mainly confined to actively dividing cell lines. A spike-in strategy based on SILAC can be used to overcome this limitation and profile hPTMs across multiple samples. We describe here the adaptation of SILAC to the analysis of histones, in both standard and spike-in setups. We also illustrate its coupling to an implemented "shotgun" workflow, by which heavy arginine-labeled histone peptides, produced upon Arg-C digestion, are qualitatively and quantitatively analyzed in an LC-MS/MS system that combines ultrahigh-pressure liquid chromatography (UHPLC) with new-generation Orbitrap high-resolution instrument.

Key words Histone posttranslational modifications, SILAC, SILAC spike-in, Ultrahigh-pressure liquid chromatography, High-resolution mass spectrometry, Epigenetics

1 Introduction

Histones are the basic protein constituents of chromatin and prototype of hyper-modified proteins, being decorated by several distinct posttranslational modifications (PTMs) [1, 2]. Different hPTM combinations are hypothesized to generate a code (the "histone code") that orchestrates the functional state of the underlying genome [3]. Lysine acetylation and lysine/arginine methylations are among the best characterized PTMs on histones, although other modification types and modified residues have been described over the last two decades [4].

*These authors contributed equally to this work.

Benoît Guillemette and Luc Gaudreau (eds.), *Histones: Methods and Protocols*, Methods in Molecular Biology, vol. 1528,
DOI 10.1007/978-1-4939-6630-1_7,

Mass spectrometry (MS) is a powerful technology for unbiased analysis of posttranslational modifications and has emerged as the elective strategy for the in-depth characterization of hyper-modified proteins, including histones [5, 6]. Different MS approaches, from peptide-centric/bottom-up to intact-protein/top-down analyses, have been designed to gain a comprehensive view on composite histone modification patterns [7–10]. Standard bottom-up analysis combines the high sensitivity of detection at the MS level with the high efficiency of classical fragmentation techniques, such as collision-induced dissociation (CID) [11]. More recently, higher energy collisional dissociation (HCD) fragmentation has been recognized to be advantageous for hPTM analysis [12, 13].

In the last decade, the standard shotgun proteomics workflow has been ad hoc tailored to the analysis of histone microheterogeneity, with a number of methodological implementations that take into account the peculiar chemo-physical properties of these molecules [14]. A representative scheme of the current workflow for MS-based hPTMs analysis is illustrated in Fig. 1 and comprises distinct steps. The extraction of histones from fractionated nuclei by strong acid is followed by protein digestion using a protease different from trypsin, such as Arg-C, to generate suitable peptides for the subsequent LC-MS/MS detection [15]. Alternatively, longer peptides (>2 kDa) can be analyzed by middle-down MS [16], which provides information about long-distance modification connectivity and thus helps assessing the combinatorial feature of the histone code [17, 18].

In addition to the qualitative annotation of hPTMs, MS can also inform on quantitative aspects, such as their relative abundance and dynamic changes [19]. To this aim, two main approaches are generally employed: label-free and stable isotope-based quantitation. Relative quantification of modified peptide by label-free MS proteomics can be achieved using the chromatographic ion intensity of each modified peptide peak [9]. In this approach, all samples to be compared are prepared separately and analyzed in individual LC-MS/MS runs.

Alternatively, stable isotope labels can be introduced, either by chemical reactions or in vivo metabolic labeling [20]. Stable isotope labeling by amino acids in cell culture (SILAC) is a very accurate and reproducible strategy for comparative analysis of proteins or peptides. In SILAC, up to three individual cell populations are grown in media supplemented with either natural ("light") or isotope-encoded ("heavy" and "medium") amino acids. When the cell populations are mixed, corresponding peptides have the same

Fig. 1 (continued) presence of D_3-alkylation on lysine residues). (**d**) Digested peptides are separated via nano(n)-UHPLC system on a shallow gradient of ACN in 100 min and then analyzed on a benchtop Q Exactive instrument. (**e**) Histone PTM identification and site-specific assignment are performed semi-automatically, with the assistance of a dedicated software; subsequently, quantitation is performed manually

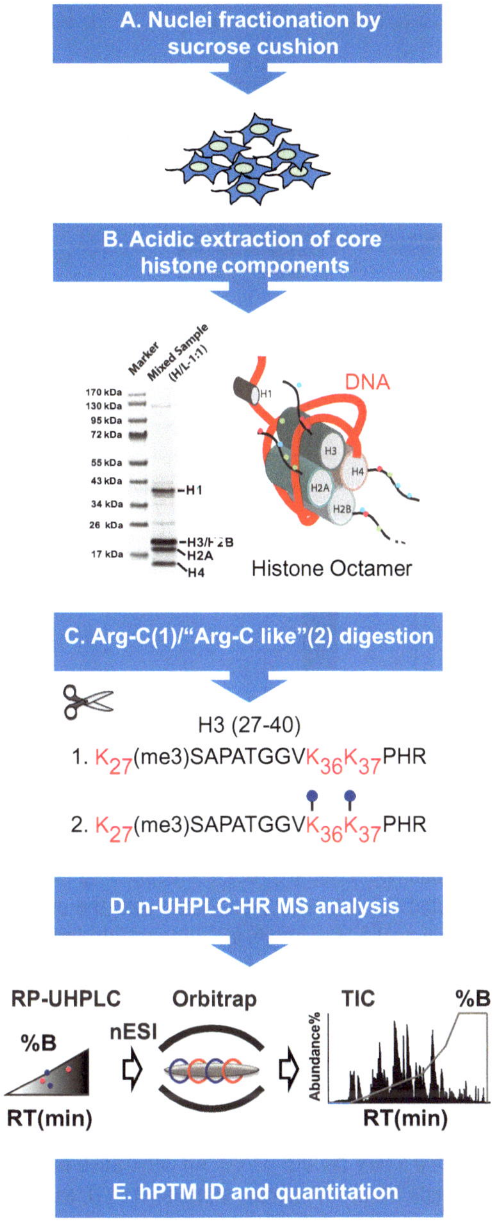

Fig. 1 Schematic representation of the workflow for MS-based analysis of histone posttranslational modifications. (**a**) Nuclei are fractionated using sucrose cushion. (**b**) Core histones are extracted with strong acids from fractionated nuclei. Yield and purity of the core histones are evaluated on a Coomassie Blue-stained 4–12% SDS-PAGE gradient gel. The four core histones (H3, H2A, H2B, and H4) are visible in a mass range around 17 kDa, together with the co-purified linker histones H1. A scheme of the nucleosome, with the modified histone tails protruding from the globular core-octamer, the DNA wrapped around (*red line*) and the linker H1, is displayed. (**c**) Histones are digested following either in-solution Arg-C or in-gel "Arg-C-like" protocols. As an example, the H3 (27–40) peptide bearing K27me3 and generated from the two digestions is reported (*blue circles* indicate the

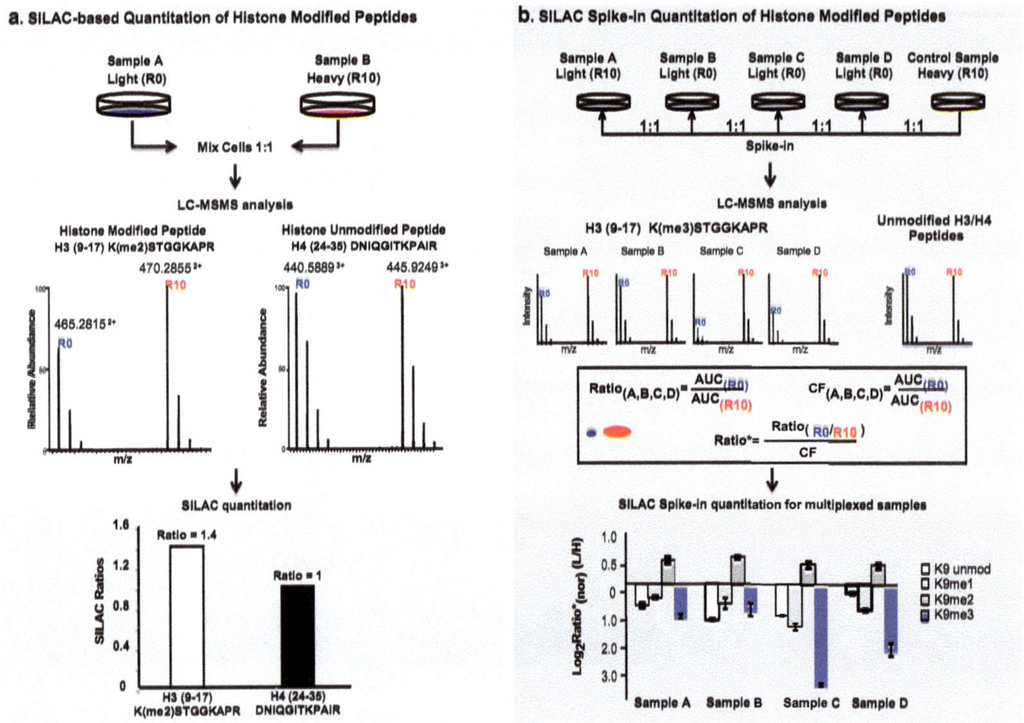

Fig. 2 Quantitative analysis of hPTMs using SILAC and SILAC spike-in strategies. (**a**) In the standard SILAC setup, cells are grown in light (L, Arg0, *blue*) and heavy (H, Arg10, *red*) media, harvested upon complete labeling and mixed in equal amounts (1:1) (*upper panel*). MS-based SILAC quantitation of the peptide pair at *m/z* 465.2815 (L, *blue*) and 470.2855 (H, *red*), corresponding to di-methylated H3 (9–17) peptide (*middle panel*) results in an estimated 1.4 H/L ratio, normalized over the unmodified H4 (24–35) peptide, which has an H/L ratio equal to 1 (linear scale) (*bar graph, lower panel*). (**b**) In the SILAC spike-in approach, the heavy (H, Arg10, *red*) cell line used as reference for quantitation is serially mixed in equal amounts (1:1:1:1 ratio) with other cell lines, cultured in light medium (L, Arg0, *blue*) (*upper panel*). Changes in the abundance of the tri-methylated H3 (9–17) peptide across the A-to-D samples are profiled based on the corresponding SILAC H/L ratio calculated from the relative intensity of each peptide pair (MS spectra in the *upper central panel*). The correction factor (CF) is calculated using the average ratio from a pool of non-modified histone peptides. As such, the adjusted values (*asterisk*) include the correction for possible changes in intact protein level or mixing error (*boxed formulas in the lower central panel*). An illustrative bar chart displays the relative changes for methylated species of peptide H3 (9–17) across samples A to D, as $\log_2$ ratio values (*lower panel*)

chemo-physical properties but present specific molecular mass difference ("delta mass"), which makes them distinguishable in MS [21–23]. The very early mixing step reduces the noise due to experimental variation and, consequently, increases the accuracy in detecting mild protein/PTM fluctuations (Fig. 2a). Standard SILAC protocol is typically based on the employment of both heavy lysine and heavy arginine, whereas the quantitative analysis of hPTMs is typically based on labeling histones with only heavy arginine [15], as a consequence of the use of Arg-C/"Arg-C like" digestion strategies.

SILAC has two main limitations: first, only a maximum of three distinct functional states can be directly compared within a single experiment; second, the strategy cannot be applied to samples not amenable to metabolic labeling, such as the clinical ones. To overcome these restraints, a modified version of SILAC is based on a spike-in strategy, where a SILAC-labeled sample is added to a sample series and used as reference for multiplexed relative quantitation [24] (Fig. 2b). For example, we adapted the spike-in SILAC strategy to histones in order to profile hPTM patterns across different breast cancer cell lines, with the aim of outlining a "breast cancer-specific epigenetic signature" to be used as biomarker for tumor prognosis and patient stratification [15].

In this chapter, we offer a detailed description of the protocol for SILAC labeling of histones from cell lines, to carry out accurate hPTM quantification; furthermore, we illustrate the spike-in SILAC strategy to accurately profile such modifications across multiple samples.

In a bottom-up MS setup, the majority of isobaric modified histone peptides can be generated by two complementary digestion protocols (in-solution Arg-C and in-gel "Arg-C like") and separated using a newly established analytical platform that combines ultra-HPLC with high-resolution MS (Fig. 3). The quantification is carried out at the peptide level, taking advantage of SILAC and normalizing the obtained modified peptide ratios with the SILAC ratio from the unmodified histone peptides, which serve as protein loading control (correction factor). The illustrated workflow enables assessing comprehensively and accurately the changes of complex hPTM patterns across multiple biological conditions, with applications in both basic and clinical research.

2 Materials

2.1 Cell Culture

2.1.1 Standard Culture

1. Dulbecco's modified Eagle medium (DMEM).
2. Fetal bovine serum (FBS), heat inactivated.
3. 50 units/mL Penicillin.
4. 50 mg/mL Streptomycin.
5. Cell culture dishes.
6. Incubator at 37 °C, with 5% CO_2 supply.
7. Phosphate-buffered saline (PBS) pH 7.4: 0.8% NaCl, 0.02% KCl, 0.02% KH_2PO_4, 0.2% Na_2HPO_4 ($7H_2O$).

2.1.2 SILAC Labeling

1. Dulbecco's modified Eagle medium (DMEM) lacking arginine and lysine.
2. L-Lysine hydrochloride (light, K0).

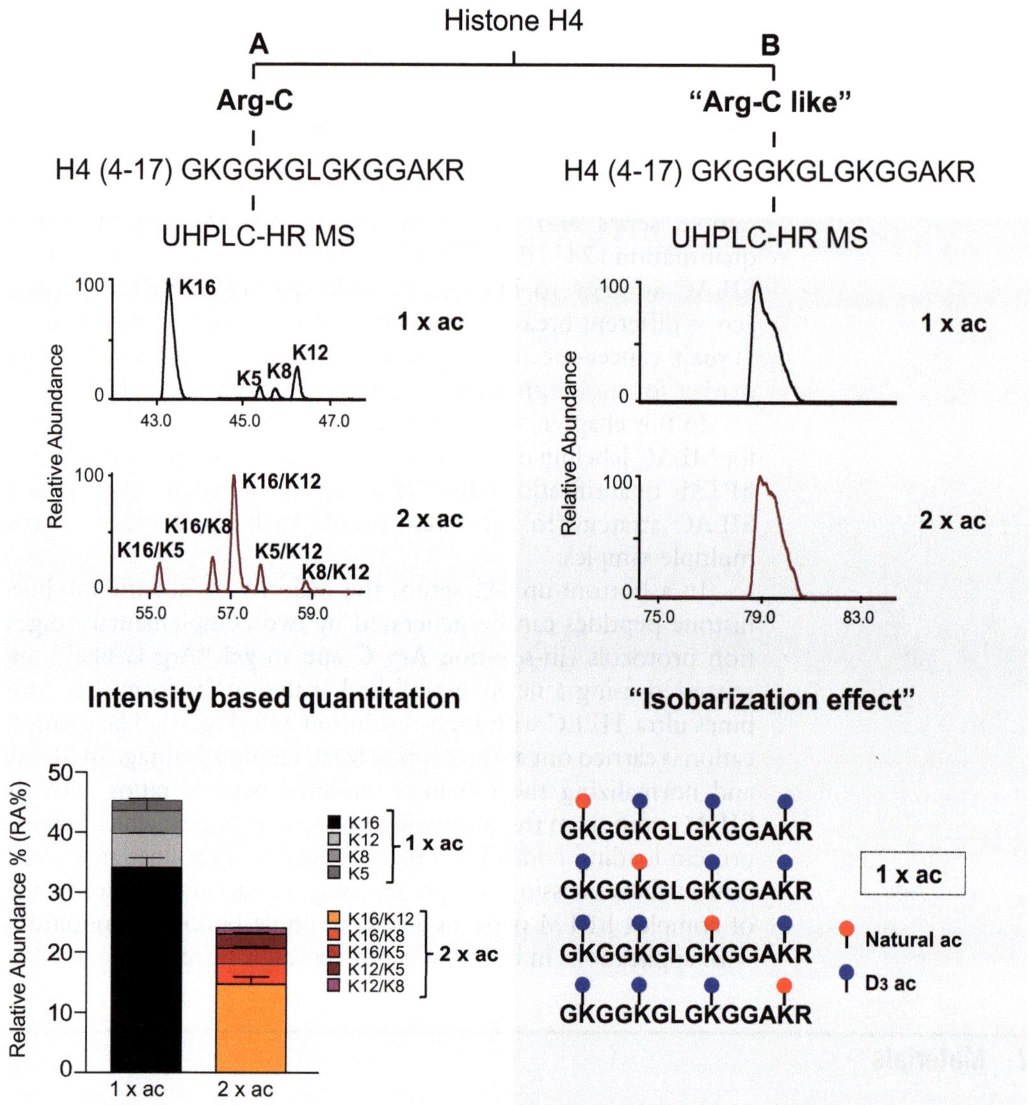

Fig. 3 Arg-C and "Arg-C-like" strategies lead to different resolution of H4 (4–17) peptide isoforms in nano(n)-UHPLC-HR MS. (**a**) Extracted ion chromatograms (XIC) of 2H+ modified H4 (4–17) peptide, corresponding to mono- (1 × ac) and di-acetylated (2 × ac) isoforms generated upon in-solution Arg-C digestion (*upper panel*). The UHPLC configuration permits the separation of all modified isobaric species corresponding to mono- and di-acetylated H4 (4–17) peptide isoforms, thus allowing their quantitation (*lower panel*). (**b**) Extracted ion chromatograms (XIC) of 2H+ H4 (4–17) peptide, corresponding to mono- and di-acetylated forms (*upper panel*). In this case, upon the "Arg-C-like" protocol, the chemical D_3-acetylation of the unmodified lysines is summed to the native acetylations, causing the elution of the different modified isoforms within the same retention time window, the so-called isobarization effect (*upper panel*). Schematic representation of the "isobarization effect," illustrating how the four distinct isoforms of native mono-acetylated peptide at different positions (*red circles*) are combined with the chemical ones (*blue circles*) to generate a hybrid tetra-acetylated species (*lower panel*)

3. L-Arginine hydrochloride (light, R0).
4. L-Arginine $^{13}C_6{}^{15}N_4$ hydrochloride (heavy, R10).
5. Dialyzed fetal bovine serum (D-FBS), heat inactivated.
6. 50 units/mL Penicillin.
7. 50 mg/mL Streptomycin.
8. Cell culture dishes.
9. Incubator at 37 °C, with 5 % CO_2 supply.
10. PBS pH 7.4: 0.8 % NaCl, 0.02 % KCl, 0.02 % KH_2PO_4, 0.2 % Na_2HPO_4 ($7H_2O$).

2.2 Sample Preparation

2.2.1 Histone Purification

1. N-Buffer solution: 10 % Sucrose, 0.5 mM EGTA pH 8.0, 15 mM NaCl, 60 mM KCl, 15 mM HEPES pH 7.5, 0.5 mM PMSF, 1 mM DTT, 5 mM NaF, 5 mM Na_3VO_4, 5 mM Na butyrate, cocktail of protease inhibitors (Protease Inhibitor Cocktail, Roche Applied Science).
2. PBS pH 7.4: 0.8 % NaCl, 0.02 % KCl, 0.02 % KH_2PO_4, 0.2 % Na_2HPO_4 ($7H_2O$).
3. Triton X-100 solution: 8 % Triton X-100 in N-Buffer solution.
4. Sucrose cushions: 10 % Sucrose in 20 mL of N-Buffer solution.
5. 37 % Hydrochloric acid (HCl).
6. 99 % Acetic acid (CH_3COOH).
7. Dialysis tubing with a 6–8 kDa cutoff.
8. SpeedVac vacuum centrifuge.

2.2.2 SDS-PAGE

1. NuPAGE® Novex 4–12 % Bis-Tris gel system with NuPAGE® MES SDS buffer kit (Invitrogen).
2. NuPAGE® LDS Sample Buffer (4×) (Invitrogen) (*see* **Note 1**).
3. Colloidal Blue Staining Kit (Invitrogen).
4. Lysis buffer: 8 M Urea, 20 mM HEPES pH 7.6, cocktail of protease inhibitors (Protease Inhibitor Cocktail, Roche Applied Science) and 1000 U/μL of Benzonase.

2.2.3 In-Gel Digestion with Trypsin

1. Ammonium hydrogen carbonate (AmBic).
2. Acetonitrile (ACN) (LC-MS grade).
3. Ethanol.
4. Destaining buffer: 50 % Ethanol in 25 mM AmBic.
5. Reduction buffer: 10 mM Dithiothreitol (DTT) in 50 mM AmBic.
6. Alkylation buffer: 55 mM Iodoacetamide in 50 mM AmBic.
7. Extraction buffer: 30 % ACN, 3 % TFA in ddH_2O.
8. Digestion buffer: 50 mM AmBic in ddH_2O (pH 8.0).
9. SpeedVac vacuum centrifuge.

2.2.4 "Arg-C-Like" In-Gel Digestion

1. Acetonitrile (ACN) (LC-MS grade).
2. Ammonium hydrogen carbonate (AmBic).
3. Destaining buffer: 50% ACN/50% ddH_2O.
4. Acetic anhydride-d_6.
5. Saturated sodium acetate (CH_3COONa).
6. Digestion buffer: 50 mM AmBic in ddH_2O (pH 8.0).
7. Trypsin solution: 100 ng/μL Sequencing-grade trypsin (Promega Corporation, Madison, WI) in 50 mM AmBic.
8. Extraction buffer: 5% Formic acid (Sigma-Aldrich)/95% ddH_2O.
9. SpeedVac vacuum centrifuge.
10. StageTips [25]: Empore SPE C18 disks (3 M); Empore SPE Carbon disks (3 M); Empore SPE strong cation-exchange (SCX) disks (3 M).
11. 100% Ammonium hydroxide (NH_4OH).
12. 100% Methanol.

2.2.5 Arg-C in Solution Digestion

1. Digestion buffer: 50 mM AmBic in water (pH 8.0).
2. Arginine-C solution: 100 ng/μL Sequencing-grade Arginine-C in 50 mM AmBic.
3. 50% Trifluoroacetic acid (TFA) in ddH_2O.

2.3 Ultrahigh-Performance Liquid Chromatography–Tandem Mass Spectrometry

1. Trifluoroacetic acid (TFA).
2. Formic acid (FA) 100%.
3. Acetic acid (AA) 100%.
4. Acetonitrile (ACN) (LC-MS grade).
5. UHPLC solvent "A": 0.1% FA in ddH_2O.
6. UHPLC solvent "B": 0.1% FA/80% ACN in ddH_2O.
7. Reversed-phase material for nano-ultra HPLC (n-UHPLC) column: Reprosil-Pur C18-AQ, 1.9 μm (Dr. Maisch).
8. EASY-nLC 1000 system (Thermo Fisher Scientific).
9. 25 cm Analytical capillary column (75 μm inner diameter, 350 μm outer diameter) (Microcolumn).
10. Q Exactive mass spectrometer (Thermo Fisher Scientific).
11. Nanospray Flex Ion Source (Thermo Fisher Scientific).

2.4 Data Acquisition and Analysis

1. Thermo Fisher Scientific Xcalibur software 2.2 SP1.48.
2. MaxQuant (MQ) software 1.5.1.2.
3. PC with 2 GB RAM and a 64-bit Windows operating system is required.

4. MQ-supported versions: Windows Vista SP2, Windows 7, Windows 8, Windows Server 2008, and Windows Server 2012.
5. QualBrowser (Thermo Fisher Scientific).

3 Methods

3.1 Standard Culture

1. Prepare the standard medium according to the specific cell type. For the breast cancer cell lines used in the illustrative study of a SILAC spike-in approach for multiplexed hPTM profiling [15], DMEM medium containing 10% fetal bovine serum (FBS), 1% glutamine, and 1% Pen/Strep was employed.
2. Thaw the cells in standard medium and keep them in culture for at least 2 days for adaptation, before moving to the SILAC medium. This time may change, depending on the cell type.

3.2 SILAC Labeling

1. Prepare the amino acid stock solution (*see* **Note 2**) and the SILAC medium (*see* **Note 3**).
2. Aliquot equal amounts of cells previously thawed in standard medium into two tubes and wash cell pellets with ice-cold PBS. After centrifugation, discard PBS and resuspend cells from each tube in the media containing heavy and light amino acids, respectively (*see* **Note 4**).
3. Pass the cells according to the cell type-specific procedure.
4. Grow cells in SILAC media for up to eight to ten doublings, to ensure the complete incorporation of isotope-encoded amino acids (*see* **Note 5**).
5. Freeze a small aliquot of heavy-labeled cells to assess the labeling efficiency (*see* **Note 6**) (*see* Subheading 3.3).
6. In parallel, compare the labeled cells with those growing in standard medium (*see* **Note 7**), to detect any alteration from physiology caused by the poorer growth conditions; for instance:
   (a) Inspect cells at the microscope to uncover any morphological alterations.
   (b) Count cells in SILAC versus standard medium, and plot and compare the respective growth curves.
   (c) Assess cell-specific phenotypes that may be diagnostic of possible functional alterations (i.e., activation of specific genes, or expression of specific marker proteins).

3.3 Evaluation of Labeling Efficiency

1. After growing cells in heavy medium for eight to ten replication cycles, wash the cell pellet (5×10^6 cells) in ice-cold PBS.
2. Add to the cellular pellet 100 μL of lysis buffer, supplemented with 1 μL of Benzonase (100,000 U), for 30 min on a rotating

wheel at room temperature (RT). Centrifuge at 16,000 × *g* for 30 min, at 4 °C, and recover the supernatant.

3. Resolve proteins from heavy-labeled whole-cell extracts by SDS-PAGE.
4. Perform the in-gel digestion of proteins with trypsin [26], as follows:
   (a) Excise two to three gel slices (5 mm high) from each lane with a scalpel, cut them into 1 mm^3 cubes, and put the cubes into an Eppendorf tube.
   (b) Add an excess (200 μL) of destaining buffer to the gel pieces; incubate for 20 min at RT with strong shaking. Repeat twice and at each wash discard the supernatant carefully, avoiding picking up the gel pieces.
   (c) Add an excess (200 μL) of ethanol and incubate for 10 min at RT, until gel pieces shrink.
   (d) Repeat **steps** (**b**) and (**c**) until the gel pieces are completely destained (they become opaque white and sticky).
   (e) Add an excess of the reduction buffer, to fully cover the gel pieces. Incubate for 1 h at 56 °C.
   (f) Remove the reduction buffer and immediately add the alkylation buffer to cover the gel pieces; incubate for 45 min at RT, in the dark.
   (g) Wash twice the gel pieces with the digestion buffer for 10 min at RT, with strong shaking; at each wash discard the supernatant carefully, avoiding picking up the gel pieces.
   (h) Add ACN for 10 min at RT to shrink the gel pieces. Discard the supernatant and dry the gel cubes in a vacuum centrifuge.
   (i) Incubate the gel pieces with 40 μL of ice-cold 12.5 ng/μL trypsin solution on ice, until they get fully rehydrated.
   (j) Remove the trypsin solution in excess and add 100 μL of digestion buffer to thoroughly cover the gel pieces. Incubate overnight at 37 °C.
   (k) Collect the supernatants and add the extraction buffer to the gel pieces. Incubate with strong agitation for 10 min at RT. Repeat twice and pool supernatants.
   (l) Incubate the gel pieces in ACN for 10 min with strong agitation. Repeat twice and pool all supernatants.
   (m) Lyophilize the peptides in a vacuum centrifuge. Resuspend the dried peptides in 100 μL of 0.5% acetic acid/0.1% TFA.

(n) Desalt and concentrate peptides on reversed-phase C_{18} StageTips (*see* Subheading 3.7).

(o) Elute peptides from the C_{18} StageTip using 20 μL of 80% ACN/0.5% TFA. Repeat twice (final elution volume 40 μL). Remove the organic solvent by evaporation in a vacuum centrifuge, resuspend the peptides in 0.1% FA (typically 5–10 μL), and proceed with RP-LC/MS analysis.

Ideally, peptides identified from this pool should contain only heavy amino acids, with no detectable signal at *m/z* values corresponding to light peptides; however, a small percentage of light peptides may remain and non-normalized H/L ratios can be measured (*see* **Notes 8** and **9**).

3.4 Differential Sample Mixing in SILAC and Spike-in SILAC Strategies

1. Harvest all differentially labeled cells, according to the cell type-specific procedure (*see* **Note 10**).
2. Count carefully all cell populations.
3. For standard SILAC experiments, mix light- and heavy-labeled cells in equal amount (one-to-one). Typically, the ideal number of starting cells for a robust histone preparation following the described protocol is 2×10^7 cells *per* sucrose cushion (*see* **Note 11**).
4. In a spike-in SILAC experiment using the heavy-labeled sample as internal standard, mix each of the multiple light-labeled cellular populations in one-to-one ratio with the reference cell line. Different spike-in SILAC experimental setups can be designed, depending on the isotopic state of the sample used as standard (*see* **Notes 12** and **13**) and the type and stage at which the mixing is performed (*see* **Note 14**).
5. From this step on, the two cellular populations—differentially labelled and mixed in equal amount—will be treated as one sample.

3.5 Histone Extraction

Purification of core histones from the total cellular lysate reduces sample complexity, improving the identification and quantification of PTMs by LC-MS/MS (*see* **Note 15**).

1. Wash cells (*see* Subheading 3.4) twice with ice-cold PBS.
2. Resuspend the cell pellet in 7.5 mL of N-Buffer solution.
3. Add 500 μL of Triton X-100 solution (to reach a 0.5% final Triton X-100 concentration); incubate at 4 °C for 10 min on a rotating wheel.
4. Pour carefully each lysate on a 20 mL of sucrose cushion (*see* **Note 16**).
5. Centrifuge in a swing-out rotor at 3695 × *g* for 20 min at 4 °C, to separate the nuclei (pellet) from the cytoplasm (*see* **Note 17**). Remove supernatant.

6. Wash twice the pellet containing the nuclei in ice-cold PBS (*see* **Note 18**). Centrifuge at 3695 ×*g* for 15 min at 4 °C (*see* **Note 17**).
7. Resuspend the nuclear pellet in 2 mL of ice-cold PBS; add an equal volume of HCl 0.8 N (final concentration of HCl 0.4 N) and roll for 5–6 h at 4 °C, on a rotating wheel (*see* **Note 19**).
8. Centrifuge at 16,000 ×*g* for 10 min at 4 °C. Collect the supernatant (*see* **Note 20**).
9. Repeat the extraction using 1 mL of 0.4 N HCl for 1–2 h, at 4 °C, on a rotating wheel (*see* **Note 19**).
10. Spin at 16,000 ×*g* for 10 min at 4 °C and pool the second supernatant with the first one.
11. Dialyze overnight at 4 °C the pooled supernatants in precooled 100 mM acetic acid (at least 100–200 volumes), using 6–8 kDa cutoff dialysis tubing (*see* **Note 21**).
12. Collect the dialyzed histones and make 1 mL aliquots in Eppendorf tubes.
13. Lyophilize histones in vacuum centrifuge and store them at -80 °C.

3.6 SDS-PAGE Separation of Histones

1. Measure the concentration of the extracted core histones using the Bradford protein assay.
2. Add NuPAGE® LDS buffer (*see* **Note 1**) to 5 μg of extracted core histones and boil the sample for 10 min at 70 °C.
3. Load the protein sample onto 4–12 % gradient NuPAGE® gel. Perform the electrophoresis separation using MES SDS buffer and applying 90 V constant for 10 min (stacking gel) followed by 150 V constant for about 1 h (running gel) (*see* **Note 22**).
4. Stain gel with Coomassie Blue enabling to assess the purity of the preparation.

3.7 "Arg-C-Like" Digestion

1. Lay the gel on a clean glass surface and cut the gel bands corresponding to individual core histones with a scalpel.
2. Cut each gel slice in small cubes (around 1 mm^3 in size) and transfer to an Eppendorf (*see* **Note 23**).
3. Wash twice the gel pieces with water, for 10 min at RT. Discard the supernatant carefully at each wash using a pipette, avoiding picking up the gel pieces that remain at the bottom of the Eppendorf tube.
4. Destain the gel pieces alternating washes with the following solutions: 100 % ddH_2O, 50 % ACN in ddH_2O, and, finally, 100 % ACN to shrink the gel pieces. At each wash, discard the supernatant carefully using a pipette without picking up the gel pieces at the bottom of the tube. Repeat the washes with strong shaking at RT in a Thermomixer until the gel pieces are completely destained.

5. Dry the gel pieces in a vacuum centrifuge.
6. Add acetic anhydride-d_6 1:9 v/v in 1 M AmBic (typically in a final volume of 80 μL) and 3 μL of saturated sodium acetate, as catalyzer. Incubate for 3–4 h at 37 °C with strong shaking (*see* **Note 24**).
7. Rinse the gel pieces several times with 50 mM AmBic, alternated with ACN at increasing percentages (from 50 to 100), in order to completely eliminate acetic anhydride-d_6. At each wash, discard carefully the supernatant without picking up the gel pieces.
8. Shrink the gel pieces in 150 μL of 100% ACN for 15 min.
9. Dry them in a vacuum centrifuge, to ensure complete dehydration.
10. Rehydrate the gel pieces with 2 μL of ice-cold 100 ng/μL trypsin solution in 50 mM AmBic. Add 10 μL of 50 mM AmBic and incubate in ice until the gel pieces are fully swollen (*see* **Notes 25** and **26**).
11. Add 30 μl of 50 mM AmBic to fully cover the gel pieces. Incubate overnight at 37 °C.
12. Add one volume of ACN 100% and shake for 20 min.
13. Collect the supernatant in a new Eppendorf.
14. Add one volume of ACN 100%, shake for 20 min at RT, collect the supernatant, and pool it with the previous one.
15. Add one volume of extraction buffer to the gel pieces; incubate in a Thermomixer, with strong agitation, for 45 min at RT.
16. Add one volume of ACN 100%, shake for 20 min, collect, and pool the supernatant.
17. Add one volume of ACN 100% to shrink the gel pieces.
18. Lyophilize the peptide mixture from the pooled supernatant in a vacuum centrifuge.
19. Reconstitute the dried samples in 0.1% TFA/0.5% AA in ddH_2O.
20. Prepare the StageTip microcolumns by putting one plug of Teflon meshwork disks containing immobilized reversed-phase C_{18}, carbon, and strong cation-exchange chromatography (SCX) beads in distinct 200 μL pipette tips [25].
21. Assemble the C_{18}/carbon "sandwich StageTip" by loading the C_{18} microcolumn on top of a second tip, loaded with the carbon filter plug.
22. Desalt and concentrate 50% of the peptides on the C_{18}/carbon "sandwich StageTip" and the remaining 50% on the SCX microcolumn (*see* **Note 27**).

23. Load 50% peptides onto the C_{18}/carbon "sandwich StageTip" and 50% onto the SCX StageTip.
24. Elute the peptides from the C_{18}/carbon Tip using 80% ACN/0.5% TFA in ddH_2O and from the SCX Tip with 5% NH_4OH/30% methanol in ddH_2O.
25. Remove the organic solvent by evaporating in a vacuum centrifuge and resuspend the peptides in a suitable injection volume (typically 5–10 μL) of 1% TFA, pooling the two eluates.
26. Inject approximately 1 μg of peptides into the RP nano(n) column for nLC–MS analysis.

3.8 Arg-C In-Solution Digestion

1. Add the endoproteinase Arg-C to 5–10 μg of core histones, in a 1:100 enzyme/substrate ratio (*see* **Note 28**).
2. Reach a final reaction volume of 40 μL using 50 mM AmBic in ddH_2O.
3. Incubate overnight at 37 °C.
4. The day after stop the reaction adding 2 μL of 50% TFA in ddH_2O.
5. Desalt and concentrate the digested peptides on a sandwich/SCX StageTip system, as described in Subheading 3.7, **steps 19–23**.
6. Elute the peptides prior to LC-MSMS, as described in Subheading 3.7, **steps 24** and **25**.

3.9 Liquid Chromatography Analysis

This step allows separating the peptides from in-gel or in-solution digestion by ultrahigh-pressure liquid chromatography (UHPLC) and ionizing them via an online nano-electrospray (nESI) system.

1. Pack an analytical column (25 cm fused silica emitter) using reverse-phase (RP) C_{18} 1.9 μm resin in methanol, at a constant helium pressure (50/80 bar) using a bomb-loader device, as previously described [27] (*see* **Note 29**).
2. Load the digested peptide mixture onto C_{18} column at a constant flow of 500 nL/min.
3. After sample loading, apply a gradient of 0–40% solvent B over 100 min followed by a gradient of 40–60% solvent B in 5 min and 60–95% solvent B over 5 min at a flow rate of 250 nL/min, for peptide elution (*see* **Note 30**).

3.10 Mass Spectrometry Analysis

Mass spectrometry analysis is performed on a Q Exactive mass spectrometer.

1. Operate the mass spectrometer in the data-dependent mode (DDA) (*see* **Note 31**), to automatically switch between full scan MS and MS/MS acquisition using the Xcalibur 2.2 software package (*see* **Note 32**).

2. Acquire survey full-scan MS spectra from *m/z* 300 to 1150 in the Orbitrap with a resolution of $R = 70{,}000$ at *m/z* 200, after accumulation of ions at 3×10^6 target value (maximum filling time of 20 ms).
3. Isolate and fragment the top five most intense multiple-charged ions ($z \geq 2$) by HCD fragmentation (maximum filling time of 50 ms), with a resolution of $R = 17{,}500$ (*see* **Note 33**).
4. Standard acquisition method settings:
   (a) Electrospray voltage, 2.4 kV.
   (b) No sheath and auxiliary gas flow.
   (c) Ion transfer (heated) capillary temperature: 250 °C.
   (d) Dynamic exclusion of 20 s upon MS/MS.
   (e) MS/MS ion selection threshold: 2×10^4 counts (underfill ratio 0.1 %) and 2.0 Da isolation width.
   (f) Normalized collision energy: 25 % (*see* **Note 34**).

3.11 Data Analysis

Acquired raw data are analyzed by the integrated MaxQuant software [28] v.1.5.1.2, using the Andromeda search engine. The data are processed according to the following parameters.

1. Group-specific parameters:
   (a) Variable modifications: N-acetyl (protein), oxidation (M), acetyl (K), methyl (K, R), dimethyl (K, R), trimethyl (K) (*see* **Notes 35** and **36**).
   (b) Missed cleavages: up to 2.
   (c) Enzyme: Arg-C.
   (d) Multiplicity: 2, set labeled amino acid (Arg10) (*see* **Note 37**).
2. Parameters for identification and quantification:
   (a) Peptide false discovery rate (FDR): 0.01.
   (b) Protein FDR (*see* **Note 38**): 0.01.
   (c) Minimum peptide length: 6.
3. At the end of the analysis, open the evidence.txt of specific modification "site.txt". Remove peptides with score less than 60, with more than five putative PTMs and/or localization probability score lower than 0.75.
4. Validate each identified peptide containing modifications by visual inspection of the MS/MS spectra using the QualBrowser and/or Viewer.exe in the MQ environment (*see* **Note 39**).
5. Construct the extracted ion chromatograms (XIC) for each precursor ion, based on the *m/z* value, specifying the mass tolerance of 10 ppm with mass precision up to four decimals, using the QualBrowser.

6. Calculate SILAC ratios using the peak areas of the heavy and light version of each histone peptide, measured within the same retention time window (*see* **Note 40**).
7. Divide the SILAC ratio of each modified peptide pair by the estimated correction factor (CF), calculated using the unmodified peptides, to correct either for potential inaccuracy in light and heavy cell mixing or for protein expression changes between different functional states, which may cause systematic errors in the SILAC-based quantification of modified peptides (*see* **Note 41**).

4 Notes

1. Add DTT to a final concentration of 100 mM just before boiling.
2. To prepare amino acid stock solutions, dissolve arginine and lysine in PBS to a concentration of 84 g/L and 146 g/L, respectively.
3. Add to the SILAC medium dialyzed serum, antibiotics, L-glutamine, and any other supplements that particular cell lines may need, at the same concentration used for growing cells in standard conditions. Add the light/heavy arginine and lysine at a final concentration of 84 mg/L and 146 mg/L, respectively, which is ideal for most cell lines. For SILAC-based histone quantification, add only the heavy arginine, since lysine is not used for hPTM profiling.
4. To increase the confidence in the detection of peptide/protein changes, it is advisable to perform biological replicates adopting the "labeling swap" strategy, which consists in running one experiment in the "forward" setup, followed by at least one replicate in the "reverse" setup, where the heavy and light media are "swapped" between the two biological conditions to compare. As such, the SILAC ratio readout corresponding to a defined change in PTM/protein abundance is inverted between the two experiments, facilitating the discrimination of genuine variations.
5. The number of cell doublings required to achieve full incorporation is cell type specific and must be assessed case by case.
6. An incorporation test on the heavy-labeled sample is advisable before starting any functional experiment and particularly critical when working with a new cell line, or medium formulation.
7. Building a growth curve is a useful way to evaluate the growth of a cell line. For some cell types a slightly reduced growth in SILAC medium—relative to standard conditions—may be observed, even when significant reduction in viability or other morphological alterations are not evident.

8. The minimal recommended incorporation of heavy amino acids that ensure accurate SILAC-based quantification is 95%. At lower incorporation efficiency, the maximum measurable ratio decreases accordingly (i.e., 80% incorporation results in a maximum SILAC ratio equal to 4 upon H/L mixing).
9. In some cell types the metabolic interconversion between arginine and proline can occur when arginine is in excess. Similarly, the metabolic proline-to-arginine conversion can occur when cells are provided with insufficient arginine [29]. Errors in quantification, especially for peptides containing more than one proline, are a consequence of this reaction. Hence, when assembling a SILAC medium for a new cell type, the optimal arginine concentration should be determined experimentally, assessing the frequency of heavy proline in the heavy SILAC cells. This is crucial in the case of histones, since arginine is the sole amino acid employed for the labeling.
10. Trypsin or PBS-EDTA can be used to detach adherent cells.
11. Increasing cell number *per* sucrose cushion results in a less pure sample after nuclei fractionation and acidic extraction.
12. Typically, heavy-labeled cells are used as spike-in and mixed in equal amounts with all other light-labeled samples, thus enabling to perform the multiplexed comparison also for samples not compatible with metabolic labeling, for example clinical samples. However, in a previous study conducted in our group, we adopted light-labeled normal breast cell line (MCF10A) as spike-in reference for a set of breast cancer cell lines to be profiled [15]. This peculiar experimental design was imposed by the fact that MCF10A cells exhibited a lower SILAC labeling efficiency than the tumor counterparts.
13. A pool of different cell lines, cultured in heavy-isotope media, can be used as spike-in reference instead of a single cell line. This strategy is named "super-SILAC" and is an extension of the SILAC spike-in concept, aimed to improve the accuracy in quantification of heterogeneous samples, such as clinical ones; in fact the super-SILAC mixture is intended to better recapitulate the heterogeneity in cell types and functional states occurring in a tissue biopsy [30, 31].
14. Although mixing the SILAC samples at the level of the differentially labeled cellular populations is most advantageous to minimize the experimental variability arising from the subsequent sample preparation, it is also possible to mix the samples at the protein level, upon an accurate quantification of the purified histones by means of a spectrophotometric assay (i.e., Bradford), in combination with SDS-PAGE inspection. This is the only choice when analyzing clinical samples.

15. Further modification-specific biochemical enrichment of distinct protein/peptide isoforms prior to MS (for example achieved by immunoprecipitation using antibodies against a specific histone modification) is a useful option to improve the detection of specific marks [32].
16. When pouring the cellular lysate onto the sucrose cushion, make sure to avoid mixing the two phases, which would reduce nuclei purity.
17. If your swing-rotor centrifuge cannot reach the indicated speed, the centrifugation time can be increased (i.e., 30 min at $3273 \times g$).
18. Nuclear pellets can be more or less visible as brilliant white, according to their purity.
19. The final volume of the strong acid buffer solution depends on the amount of nuclear pellet.
20. Upon incubation with a strong acid, core histones remain in solution together with linker histones and high-mobility-group (HMG) proteins, whereas most of the remaining proteins precipitate.
21. Replacing the buffer several times can reduce the dialysis time.
22. Handmade gels, at a concentration of 17.5 % PAA, can also be used to separate individual histone molecules (H3, H2A, H2B, and H4). Alternatively, reversed-phase high-performance liquid chromatography (RP-HPLC) and hydrophilic interaction liquid chromatography (HILIC) can be adopted to obtain purified histone forms [33]. Recently a combination of weak-cation exchange (WCX) and HILIC has been shown to improve the dissection of isobaric modified histone species prior to MS [18].
23. Take care to minimize keratin contaminations that interfere with LC-MS/MS during the steps of protein digestion and peptide extraction [34].
24. The density of acetic anhydride-d_6 is 1.143 g/mL. The reaction may generate bubbles in the very first minutes after the assembly; therefore handle the tubes with caution and open them from time to time during the incubation, to release the gas generated.
25. Use 200 ng of trypsin for each core histone band (typically corresponding to 1 μg of histone molecule).
26. Trypsin is not the ideal protease for MS analysis of histones, because it produces several short hydrophilic peptides that are poorly retained on reverse-phase (RP) columns and less efficiently detected with conventional LC-MS setups. The endoproteinase Arg-C generates instead peptides of suitable length for MS and is the best choice for bottom-up hPTM analysis.

However, Arg-C is poorly efficient in gel. A chemical derivatization procedure, followed by in-gel trypsin digestion, has been designed to overcome this issue, generating "Arg-C-like" histone peptides for subsequent MS analysis. This strategy has multiple advantages, extensively discussed elsewhere [35], for instance, the increase in separation of some hyper-modified histone peptides upon UHPLC-HR MS detection [36].

27. Both hydrophilic and very short peptides that are not retained by the C_{18} filter remain in the flow-through from the upper column and are loaded directly on the carbon tip. In parallel, the SCX StageTips can enrich other peptide subsets, not efficiently retained by RP chromatography.

28. Arg-C in-solution digestion results as the most efficient strategy to discriminate and quantify the majority of H4 (4–17) acetylated isoforms by UHPLC-MSMS, outperforming the "Arg-C-like" protocol. The chemical derivatization using deuterated anhydride, in fact, leads to the "isobarization effect," consisting in the co-elution of differently acetylated peptide species, as recently described [36] (*see* also **Note 40** and Fig. 3).

29. Small-particle-sized C_{18} columns (<2 μm internal diameter) in the ultrahigh-pressure regime lead to very high peak capacities, increasing peptide separation [37].

30. Further adjustments of the gradient might be tuned to maximize LC performance. Recently, our group proved that optimal steepness (slope) of the gradient allows the efficient separation of histone peptides even in case of isobaric hyper-modified species. For instance, with a 100-min gradient (0.4% (B)/min) in UHPLC, 13 distinct methylated isoforms of the H3 (27–40) peptide were separated. Instead, an extended gradient of 200 min using 0.2% (B)/min led to only a marginally better separation [36].

31. The Q Exactive instrument, working in DDA mode, typically completes a cycle of ten MS/MS events in about 1 s (10 Hz). However, in LC-MS/MS many nearly isobaric peptides with several magnitudes of intensity (abundance) difference elute in a very narrow time window, even with optimal chromatographic setups. In this condition the less abundant peptides, which would require longer injection times (>250 ms), generate poor-quality MS/MS spectra, irrespective of the speed achieved by the instrument (*see* also **Note 32**).

32. The sensitivity issue mentioned above (*see* **Note 31**) is an intrinsic limitation of the current analytical strategy to identify and quantify hPTMs that relies on a "shotgun" approach with data-dependent acquisition (DDA) method. To overcome the limit of the "intensity-dependent" selection by DDA, it would be interesting to investigate whether data-independent acqui-

sition (DIA) methods [38], recently designed to maximize the MS acquisition time, can increase the sensitivity in MS characterization of low-abundant histone modifications.

33. With longer histone peptides non-ergodic fragmentations, such as ETD and ECD, should be used instead of CID because of the high charge state of histones in acidic conditions [39, 40]. Although the analysis of longer peptides increases the global view of hPTMs, the analysis of raw data is more challenging, due to the higher complexity of both the MS and the MS/MS spectra.
34. HCD increases the frequency of modification-specific fragmentation; this raises the confidence in the identification of isobaric modifications in high-resolution MS/MS spectra [41].
35. Other modification types, such as arginine-methylation, lysine ubiquitylation, crotonylation, and succinylation [4, 42], could be added as variable modifications in the search. It is however worth noting that increasing the number of variable modifications causes the expansion of the search space, which reduces the global accuracy and the statistical confidence in the identification process. As such, dividing the search into multiple parallel jobs, in which a reduced set of variable PTMs is included, is desirable to limit the number of false-positive modification IDs.
36. In the case of "Arg-C-like" in-gel digestion, the composition and mass of unmodified lysine to be added in the Andromeda configuration are H(-1) C(2) O D(3) and 45.0294 Da, respectively. In the case of mono-methylated lysine, the corresponding composition and mass are H C(3) O D(3) and 59.0450 Da, respectively.
37. The modification corresponding to arginine labeled with heavy stable isotopes is treated as fixed in the Andromeda search.
38. False-positive rates for peptides are calculated as described [43]. In general, we estimate 1% FDR, both at the level of peptides and proteins.
39. Typically, an MS-based proteomics workflow relies on database search engines to obtain peptide sequence identification. These scoring-based identification methods may generate wrong assignment of modified sites, especially for more complex cases, such as the MS/MS spectra of hyper-modified peptides. In these circumstances, the in-depth manual inspection of raw data assisted by dedicated software is crucial to eliminate incorrect assignments and reduce false positives. Notably, a lack of sequence annotation for one histone peptide can be complemented by its retention time (RT) information. In fact, robust LC offers high RT reproducibility, even when comparing

different runs across distinct acquisitions [44, 45]. As such, RT values can be definitely used as "diagnostic markers" for specific modified peptide, thus strengthening the ID information and permitting comparison among multiple samples. Indeed, new generation software for quantitative proteomics allow the automatic RT alignment and the ID transfer across multiple runs in order to minimize missing values in DDA.

40. Notably, the use of "Arg-C like" for SILAC-labeled samples is unsuitable for acetylated H4 (4–17) quantitation. In fact, for this peptide the interference due to the overlap between the isotopes of the light and the heavy peptide (i.e., m/z 2+ light-mono-acetylated peptide = 724.4381, m/z 2+ heavy-fully natural acetylated peptide = 724.9129) impairs the quantification of these forms ("isobarization effect"). In-solution Arg-C digestion, eventually coupled with pre-fractionation of intact histone molecules by semi-preparative RP-HPLC, is the solution to thoroughly separate and quantify all differently acetylated isoforms of this peptide (*see* also **Note 28** and Fig. 3).
41. The normalization of hPTM variations over total protein abundance is achieved by using a "correction factor" (CF), calculated as the average of the light and heavy peak areas (SILAC ratio) for at least three distinct unmodified peptides of H4 (24–35), H3 (41–49), and H2A (82–88), which are robustly detected in LC-MS/MS. Although some modifications have been reported for these peptides (acetylation and methylation of H4K31 and H3R42 methylation), to our experience they occur with an extremely low frequency and their abundance is typically below 1 %. As such, the effect of such modifications can be considered negligible for the protein normalization procedure. A similar step of SILAC ratio normalization is carried out by default in the MQ data processing, when the quantitation feature is enabled. This MQ feature uses all unmodified peptides from all histone molecules identified in the sample to normalize actual ratios. To our experience, MQ normalization is comparable with the one obtained by manual CF calculation, as described above.

Acknowledgments

Research in TB group is supported by grants from the Giovanni Armenise-Harvard Foundation Career Development Program, the Italian Association for Cancer Research (AIRC), the Italian Ministry of Health and CNR-EPIGEN flagship project. We would like to thank R. Noberini and A. Silvola for critical reading of the manuscript and fruitful discussion.

References

1. Bannister AJ, Kouzarides T (2011) Regulation of chromatin by histone modifications. Cell Res 21:381–395
2. Kouzarides T (2007) Chromatin modifications and their function. Cell 128:693–705
3. Jenuwein T, Allis CD (2001) Translating the histone code. Science 293:1074–1080
4. Tan M, Luo H, Lee S et al (2011) Identification of 67 histone marks and histone lysine crotonylation as a new type of histone modification. Cell 146:1016–1028
5. Garcia BA, Shabanowitz J, Hunt DF (2007) Characterization of histones and their post-translational modifications by mass spectrometry. Curr Opin Chem Biol 11:66–73
6. Sidoli S, Cheng L, Jensen ON (2012) Proteomics in chromatin biology and epigenetics: elucidation of post-translational modifications of histone proteins by mass spectrometry. J Proteomics 75:3419–3433
7. Britton LM, Gonzales-Cope M, Zee BM, Garcia BA (2011) Breaking the histone code with quantitative mass spectrometry. Expert Rev Proteomics 8:631–643
8. Pesavento JJ, Bullock CR, LeDuc RD, Mizzen CA, Kelleher NL (2008) Combinatorial modification of human histone H4 quantitated by two-dimensional liquid chromatography coupled with top down mass spectrometry. J Biol Chem 283:14927–14937
9. Pesavento JJ, Mizzen CA, Kelleher NL (2006) Quantitative analysis of modified proteins and their positional isomers by tandem mass spectrometry: human histone H4. Anal Chem 78:4271–4280
10. Plazas-Mayorca MD, Zee BM, Young NL et al (2009) One-pot shotgun quantitative mass spectrometry characterization of histones. J Proteome Res 8:5367–5374
11. Shukla AK, Futrell JH (2000) Tandem mass spectrometry: dissociation of ions by collisional activation. J Mass Spectrom 35:1069–1090
12. Jung HR, Pasini D, Helin K, Jensen ON (2010) Quantitative mass spectrometry of histones H3.2 and H3.3 in Suz12-deficient mouse embryonic stem cells reveals distinct, dynamic post-translational modifications at Lys-27 and Lys-36. Mol Cell Proteomics 9:838–850
13. Olsen JV, Macek B, Lange O et al (2007) Higher-energy C-trap dissociation for peptide modification analysis. Nat Methods 4:709–712
14. Soldi M, Cuomo A, Bremang M, Bonaldi T (2013) Mass spectrometry-based proteomics for the analysis of chromatin structure and dynamics. Int J Mol Sci 14:5402–5431
15. Cuomo A, Moretti S, Minucci S, Bonaldi T (2011) SILAC-based proteomic analysis to dissect the "histone modification signature" of human breast cancer cells. Amino Acids 41:387–399
16. Sidoli S, Schwammle V, Ruminowicz C et al (2014) Middle-down hybrid chromatography/tandem mass spectrometry workflow for characterization of combinatorial post-translational modifications in histones. Proteomics 14:2200–2211
17. Jung HR, Sidoli S, Haldbo S et al (2013) Precision mapping of coexisting modifications in histone H3 tails from embryonic stem cells by ETD-MS/MS. Anal Chem 85:8232–8239
18. Young NL, DiMaggio PA, Plazas-Mayorca MD et al (2009) High throughput characterization of combinatorial histone codes. Mol Cell Proteomics 8:2266–2284
19. Zee BM, Young NL, Garcia BA (2011) Quantitative proteomic approaches to studying histone modifications. Curr Chem Genomics 5:106–114
20. Eberl HC, Mann M, Vermeulen M (2011) Quantitative proteomics for epigenetics. Chembiochem 12:224–234
21. Ong SE, Blagoev B, Kratchmarova I et al (2002) Stable isotope labeling by amino acids in cell culture, SILAC, as a simple and accurate approach to expression proteomics. Mol Cell Proteomics 1:376–386
22. Ong SE, Mann M (2006) A practical recipe for stable isotope labeling by amino acids in cell culture (SILAC). Nat Protoc 1:2650–2660
23. Ong SE, Mann M (2007) Stable isotope labeling by amino acids in cell culture for quantitative proteomics. Methods Mol Biol 359:37–52
24. Geiger T, Wisniewski JR, Cox J et al (2011) Use of stable isotope labeling by amino acids in cell culture as a spike-in standard in quantitative proteomics. Nat Protoc 6:147–157
25. Rappsilber J, Mann M, Ishihama Y (2007) Protocol for micro-purification, enrichment, pre-fractionation and storage of peptides for proteomics using StageTips. Nat Protoc 2:1896–1906
26. Olsen JV, Ong SE, Mann M (2004) Trypsin cleaves exclusively C-terminal to arginine and lysine residues. Mol Cell Proteomics 3:608–614
27. Thakur SS, Geiger T, Chatterjee B et al (2011) Deep and highly sensitive proteome coverage by LC-MS/MS without prefractionation. Mol Cell Proteomics 10:M110.003699
28. Cox J, Mann M (2008) MaxQuant enables high peptide identification rates, individualized p.p.b.-range mass accuracies and proteome-wide

protein quantification. Nat Biotechnol 26: 1367–1372
29. Kirchner M, Selbach M (2012) In vivo quantitative proteome profiling: planning and evaluation of SILAC experiments. Methods Mol Biol 893:175–199
30. Pozniak Y, Geiger T (2014) Design and application of super-SILAC for proteome quantification. Methods Mol Biol 1188:281–291
31. Shenoy A, Geiger T (2015) Super-SILAC: current trends and future perspectives. Expert Rev Proteomics 12:1–7
32. Bremang M, Cuomo A, Agresta AM et al (2013) Mass spectrometry-based identification and characterisation of lysine and arginine methylation in the human proteome. Mol Biosyst 9:2231–2247
33. Boersema PJ, Mohammed S, Heck AJ (2008) Hydrophilic interaction liquid chromatography (HILIC) in proteomics. Anal Bioanal Chem 391:151–159
34. Shevchenko A, Tomas H, Havlis J, Olsen JV, Mann M (2006) In-gel digestion for mass spectrometric characterization of proteins and proteomes. Nat Protoc 1:2856–2860
35. Soldi M, Bonaldi T (2014) The ChroP approach combines ChIP and mass spectrometry to dissect locus-specific proteomic landscapes of chromatin. J Vis Exp. doi:10.3791/51220
36. Soldi M, Cuomo A, Bonaldi T (2014) Improved bottom-up strategy to efficiently separate hypermodified histone peptides through ultra-HPLC separation on a bench top Orbitrap instrument. Proteomics 14:2212–2225
37. Nagaraj N, Kulak NA, Cox J et al (2012) System-wide perturbation analysis with nearly complete coverage of the yeast proteome by single-shot ultra HPLC runs on a bench top Orbitrap. Mol Cell Proteomics 11:M111.013722
38. Egertson JD, Kuehn A, Merrihew GE et al (2013) Multiplexed MS/MS for improved data-independent acquisition. Nat Methods 10:744–746
39. Zubarev RA, Horn DM, Fridriksson EK et al (2000) Electron capture dissociation for structural characterization of multiply charged protein cations. Anal Chem 72:563–573
40. Syka JE, Coon JJ, Schroeder MJ, Shabanowitz J, Hunt DF (2004) Peptide and protein sequence analysis by electron transfer dissociation mass spectrometry. Proc Natl Acad Sci U S A 101:9528–9533
41. Michalski A, Neuhauser N, Cox J, Mann M (2012) A systematic investigation into the nature of tryptic HCD spectra. J Proteome Res 11:5479–5491
42. Xie Z, Dai J, Dai L et al (2012) Lysine succinylation and lysine malonylation in histones. Mol Cell Proteomics 11:100–107
43. Cox J, Neuhauser N, Michalski A et al (2011) Andromeda: a peptide search engine integrated into the MaxQuant environment. J Proteome Res 10:1794–1805
44. Tyanova S, Mann M, Cox J (2014) MaxQuant for in-depth analysis of large SILAC datasets. Methods Mol Biol 1188:351–364
45. Schilling B, Rardin MJ, MacLean BX et al (2012) Platform-independent and label-free quantitation of proteomic data using MS1 extracted ion chromatograms in skyline: application to protein acetylation and phosphorylation. Mol Cell Proteomics 11:202–214

Chapter 8

Characterization of Individual Histone Posttranslational Modifications and Their Combinatorial Patterns by Mass Spectrometry-Based Proteomics Strategies

Simone Sidoli and Benjamin A. Garcia

Abstract

Histone posttranslational modifications (PTMs) play an essential role in chromatin biology, as they model chromatin structure and recruit enzymes involved in gene regulation, DNA repair, and chromosome condensation. Such PTMs are mostly localized on histone N-terminal tails where, as single units or in a combinatorial manner, they influence chromatin reader protein binding and fine-tune the abovementioned activities. Mass spectrometry (MS) is currently the most adopted strategy to characterize proteins and protein PTMs. We hereby describe the protocols to identify and quantify histone PTMs and their patterns using either bottom-up or middle-down proteomics. In the bottom-up strategy we obtain 5–20 aa peptides by derivatization with propionylation followed by trypsin digestion. The newly generated N-termini of histone peptides can be further derivatized with light or isotopically heavy propionyl groups to increase chromatographic retention and allow multiplexed analyses. Moreover, we describe how to perform derivatization and trypsin digestion of histones loaded into a gel, which is usually the final step of immunoprecipitation experiments. In the middle-down strategy we obtain intact histone tails of 50–60 aa by digestion with the enzyme GluC. This allows characterization of combinatorial histone PTMs on N-terminal tails.

Key words Histones, Mass spectrometry, Proteomics, Bottom-up, Middle-down

1 Introduction

Epigenetics is defined as the study of inheritable changes in the phenotype of an organism caused by mechanisms other than changes in the underlying DNA sequence [1]. The phenotype of a complex organism changes dramatically during development, from the embryo to the adult form, even though its DNA remains mostly unaltered. The epigenetic machinery involves different cellular biomolecules, including histone posttranslational modifications (PTMs), histone variants, noncoding RNAs, DNA methylation, and DNA-binding factors [2]. While DNA methylation is known as irreversible modification that inactivates chromatin regions from

Benoit Guillemette and Luc Gaudreau (eds.), *Histones: Methods and Protocols*, Methods in Molecular Biology, vol. 1528, DOI 10.1007/978-1-4939-6630-1_8, © Springer Science+Business Media New York 2017

being translated [3], histone variants and histone PTMs are more dynamic units that influence chromatin-related functions. PTMs are mostly localized in the N-terminal tails, as it is the region of the histones most exposed and flexible. Even though histone marks have been extensively characterized in the last decade, many links between known histone marks and their function are still missing. This is mostly due to the peculiarity of histone PTMs to shuffle in a large variety of combinations, modifying dramatically the affinity with histone-interacting proteins and thus their role in the chromatin. The presence of sequence variants also contributes to increase the complexity of histone analysis, as histone isotypes are generally highly similar in sequence, but they might have different roles in the chromatin; for example H2A.x has a C-terminal sequence which is more easily phosphorylated in case of DNA damage compared to canonical H2A [4] and it is required for inactivation of sex chromosomes in male mouse meiosis [5], while CENP-A substitutes canonical histone H3 in centromeres [6].

Antibody-based techniques such as western blotting have been extensively adopted to characterize histones. However, this approach is limited for the following reasons: (1) antibodies only work as confirmation and they cannot identify unknown PTMs; (2) they are biased through presence of coexisting marks, which might influence binding affinity; (3) they cannot identify combinatorial marks, as only very few antibodies are available for such purpose; and (4) they happen to cross-react between highly similar histone variants or multiple PTMs (e.g., di- and trimethylation of lysine residues). Egelhofer et al. described that more than 25% of commercial antibodies fail specificity tests by dot blot or western blot, and among specific antibodies more than 20% fail in chromatin immunoprecipitation experiments [7]. Mass spectrometry (MS) is currently the most suitable analytical tool to study novel and/or combinatorial PTMs, and it has been extensively implemented for histone proteins (reviewed in [8]). This is mostly due to MS high sensitivity, high mass accuracy, and the possibility to perform large-scale analyses. In this chapter, we describe the workflow to purify histones and prepare them for PTM analysis via bottom-up or middle-down proteomics (Fig. 1). Both strategies achieve quantification of single-histone marks. While bottom-up is more sensitive and requires less advanced instrumentation, middle-down is more suitable to characterize distant coexisting marks and their respective histone variants. An overview of the major differences between the bottom-up and the middle-down proteomics strategies is illustrated in Table 1. We recommend consulting the table prior to deciding which strategy to follow for histone PTM analysis, as it specifies both the requirements and the different types of results that can be achieved.

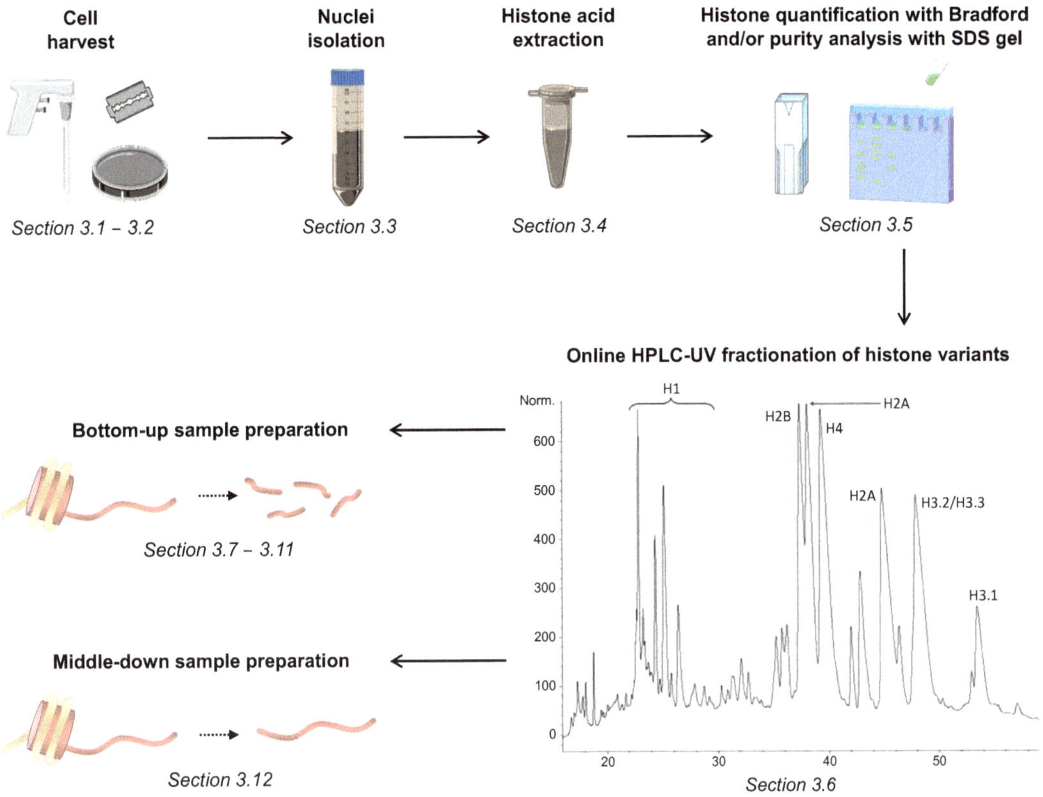

Fig. 1 Workflow for histone sample preparation. After cell harvesting, nuclei are extracted with acid precipitation (TCA). The yield of histone extraction can be verified by using protein quantification methods such as Bradford, and the purity of the sample by SDS gel. When histone amount is sufficient it is possible to fractionate the different histone variants by reversed-phase HPLC. The different fractions (or the crude histone extract in case of low sample amounts) can then be digested for bottom-up or middle-down analysis

2 Materials

2.1 Reagents and Abbreviations

1. Bradford protein assay reagent.
2. Ammonium hydroxide (NH_4OH), 28% NH_3 in water.
3. Trichloroacetic acid (TCA).
4. Trifluoroacetic acid (TFA).
5. Propionic anhydride (D0 and D10) and 2-propanol for propionylation mixture.
6. Kasil®#1 (PQ corporation, Valley Forge, PA, USA) and formamide to prepare frits for nano-columns.

2.2 Buffers

1. Phosphate-buffered saline (PBS): 137 mM NaCl, 2.7 mM KCl, 10 mM Na_2HPO_4, 2 mM KH_2PO_4.
2. Nuclei isolation buffer (NIB-250): 15 mM Tris–HCl (pH 7.5), 15 mM NaCl, 60 mM KCl, 5 mM $MgCl_2$, 1 mM $CaCl_2$, 250 mM sucrose.

Table 1
Overview of bottom-up and middle-down strategies

Step of the workflow	Bottom-up	Middle-down
Sample preparation	Medium-easy: Derivatization with propionic anhydride is required if trypsin is used	Easy
Fractionation of histone isotypes and separate analysis	Does not provide significant gain in sensitivity as compared to the LC-MS/MS analysis of the crude histone mixture	Provides significant gain in sensitivity as compared to the LC-MS/MS analysis of the crude histone mixture (ref. [11])
Enzyme for digestion	Trypsin (recommended), ArgC	GluC (recommended), AspN
Possibility of multiplexing	Yes	Not at the moment
Chromatography	C18 reversed phase. C18-AQ material is recommended, as it can tolerate 100 % H_2O	Weak cation exchange—hydrophilic interaction chromatography (WCX-HILIC). Polycat A (PolyLC, USA) is recommended
HPLC	Two channels required	Three channels required
	• Buffer A/loading buffer: 0.1 % formic acid	• Loading buffer: 0.1 % formic acid
	• Buffer B: 95 % acetonitrile, 0.1 % formic acid	• Buffer A: 75 % acetonitrile, 20 mM propionic acid, pH 6
		• Buffer B: 25 % acetonitrile, 0.1 % formic acid, pH 2.5
Mass spectrometer	Must provide at least high-resolution full MS and collision-induced dissociation (CID) as fragmentation method for MS/MS	Must provide high-resolution full MS and MS/MS, with electron transfer dissociation (ETD) as fragmentation method
MS acquisition method	Data-dependent acquisition. Targeted MS/MS for isobaric peptides (identical precursor mass) required	Data-dependent acquisition. No targeted MS/MS required. Dynamic exclusion disabled to allow selection of isobaric peptides
Database searching	Mascot (Matrix Science, UK) is recommended. Other tools can be used. However, some database searching engines do not properly deal with many dynamic modifications	Mascot is required, due to the optimization of the following bioinformatics steps
Quantification	Extracted ion chromatogram, which can be performed manually or with software	Total MS/MS ion intensity and fragment ion relative ratio calculated by isoScale (ref. [10]). Too demanding to be done manually
Results		
Quantification of single PTMs	Yes, but not on arginines (target of digestion enzyme)	Yes (only on histone tails)
Quantification of combinatorial PTMs	No, unless both sites of interested are on the same peptide	Yes (only on histone tails)
Interplay evaluation between coexisting PTMs	No, unless both sites of interested are on the same peptide	Yes (only on histone tails)
Discrimination of the histone isotype where the PTM resides	Very limited for histones with highly homolog sequence	Possible for several histone isotypes

3. Ammonium bicarbonate (NH_4HCO_3): 50 mM NH_4HCO_3, pH 8.0.
4. In-gel digestion buffer: 50 mM NH_4HCO_3, 12.5 ng/μL trypsin (sequencing grade).
5. HPLC-UV buffer A: 5% Acetonitrile, 0.1% TFA in HPLC-grade water.
6. HPLC-UV HPLC buffer B: 95% Acetonitrile, 0.1% TFA in HPLC-grade water.
7. Stage-tip loading and wash buffer: 0.1% TFA.
8. Stage-tip elution buffer: 75% Acetonitrile, 0.025% TFA.
9. Bottom-up online HPLC loading buffer and buffer A: 0.1% Formic acid in HPLC-grade water.
10. Bottom-up online HPLC buffer B: 0.1% Formic acid, 95% HPLC-grade acetonitrile, in HPLC-grade water.
11. Middle-down online HPLC loading buffer: 0.1% Formic acid in HPLC-grade water.
12. Middle-down online HPLC buffer A: 75% HPLC-grade acetonitrile, 20 mm propionic acid, adjusted to pH 6.0 with NH_4OH in HPLC-grade water.
13. Middle-down online HPLC buffer B: 25% HPLC-grade acetonitrile adjusted to pH 2.5 with formic acid in HPLC-grade water.

2.3 Solutions

1. Protease inhibitors (add fresh to buffers prior to use): 1 M Dithiothreitol (DTT) in ddH_2O (1000×); 200 mM AEBSF in ddH_2O (400×).
2. Phosphatase inhibitor (add fresh to buffers prior to use): 2.5 μM Microcystin in 100% ethanol (500×).
3. HDAC inhibitor (add fresh to buffers prior to use): 5 M Sodium butyrate, made by titration of 5 M butyric acid using NaOH to pH 7.0 (500×).
4. 10% (v/v) NP-40 Alternative in ddH_2O.
5. 0.2 M H_2SO_4 in ddH_2O.
6. 100% TCA (w/v) in ddH_2O.

2.4 Equipment

1. Tissue and cell homogenizers (optional).
2. Glass Pasteur pipettes.
3. pH indicator strips (pH 0–14).
4. Liquid nitrogen.
5. Razor blades.
6. 0.5 and 1.5 mL microcentrifuge tubes.
7. 15 and 50 mL conical tubes.
8. Pipettes from P10 to P1000 range with respective tips.

9. –80 °C Refrigerator.
10. Heat blocks or water baths.
11. HPLC-UV (~0.1–1 mL/min flow range), equipped with C_{18} 5 μm particle commercial column (size 4.6×250 mm or 2.1×250 mm) (optional).
12. 3 M Empore™ Solid Phase Extraction Disks C_{18}.
13. 75 and 100 μm internal diameter fused silica tubings.
14. Micro-stir magnets.
15. C_{18}-AQ 3 μm bulk resin with 200–300 Å pore size for trap column and analytical column for nanoLC (bottom-up strategy).
16. Polycat A bulk resin with 1500 Å pore size (PolyLC, Columbia, MD, USA) for analytical column for nanoLC (middle-down strategy).
17. Pressure cell for capillary column packing with respective compressed gas bomb (either helium, nitrogen, or air).
18. Appropriate nanoLC-MS setup: Bottom-up analysis requires HPLC with at least two channels (one for buffer A/loading buffer and one for buffer B) and high-resolution MS. High-resolution MS/MS is optional. Middle-down analysis requires HPLC with at least three channels (one for loading buffer, one for buffer A, and one for buffer B) and high-resolution MS and MS/MS with electron transfer dissociation (ETD) as fragmentation technique.

3 Methods

Carry out all procedures at room temperature, unless otherwise specified.

3.1 Cell Harvest from Tissue Culture

1. In case cells in suspension are grown, centrifuge cells at 300 rcf for 5–10 min. In case attached cells are grown, trypsinize cells, stop the trypsinization, and centrifuge at 300 rcf for 5–10 min.
2. Remove supernatant.
3. Resuspend cells in PBS and transfer them in a 15 or 50 mL conical tube, depending on the volume of the suspension.
4. Centrifuge cells at 300 rcf for 5–10 min and remove supernatant.
5. Add PBS for a second wash and repeat **step 4.**
6. Estimate the volume of cell pellets
   Pause—Sample can be frozen into liquid nitrogen and stored at –80 °C.
7. Continue with cell nuclei isolation (Subheading 3.3).

3.2 Cell Harvest from Tissue (Alternative to Subheading 3.1).

1. Dissect out desired tissue and rinse with ice-cold PBS.
2. Mince fresh or frozen tissue with a razor blade into small pieces to increase surface contact for nuclei isolation.
3. Collect minced tissue in microcentrifuge tubes or 15 mL conical tubes and estimate the volume of tissue.
   Pause—Sample can be frozen into liquid nitrogen and stored at −80 °C.
4. Continue with cell nuclei isolation (Subheading 3.3).

3.3 Cell Nuclei Isolation

1. Add protease inhibitors and other inhibitors to NIB-250 buffer. For 1 mL of cell pellet, approximately 50 mL of NIB-250 buffer is prepared. Add to 50 mL NIB-250 buffer 50 μL of 1 M DTT, 125 μL of 200 mM AEBSF, 100 μL of 2.5 μM microcystin, and 100 μL of 5 M sodium butyrate.
2. Lyse the cell pellet with 10:1 (v/v) ratio of NIB-250+ inhibitors + 0.2 % NP-40 Alternative.
3. Homogenize with the appropriate instrument. For instance, liver samples can be homogenized using pestles or Dounce homogenizers. Tissue culture cells can be homogenized by gentle pipetting.
4. Incubate homogenized cells on ice for 5–10 min; the cells will lyse and release nuclei.
5. Centrifuge at 1000 rcf for 5–10 min at 4 °C. The pellet contains mostly cell nuclei, while the supernatant contains mostly cytoplasmic components.
6. Wash the nuclei pellet by gently resuspending with 10:1 (v/v) NIB-250 + inhibitors. Do not add NP-40 Alternative anymore, as detergents should be removed prior to histone extraction.
7. Centrifuge at 1000 rcf for 5 min at 4 °C and remove supernatant.
8. Repeat **steps 6** and 7 from two to four times to completely remove NP-40 Alternative. Removal of NP-40 Alternative is evident as gentle pipetting during the washing step does not form bubbles anymore.
   Pause—Optionally, sample can be resuspended in the minimum volume possible of NIB-250 + inhibitors + 5 % glycerol, and stored at −80 °C.
9. Continue to purification of histone proteins (Subheading 3.4). Alternatively, nuclei can be used to perform immunoprecipitation of nucleosomes or other chromatin-binding proteins. For analysis of histone PTMs enriched from immunoprecipitation experiments skip directly to derivatization and proteolytic digestion of histones (Subheading 3.9).

3.4 Purification of Histone Proteins

Histones are highly enriched in basic amino acid residues. This property highly facilitates their interaction with DNA, which has a backbone containing phosphoric acid. The described histone extraction protocol is based on their acid solubility (with 0.2 M H_2SO_4) followed by precipitation with highly concentrated TCA (33%) (*see* **Note 1** for alternative protocol).

1. Resuspend cell nuclei with 0.2 M H_2SO_4 with about five times the volume of the nuclei pellet by gentle pipetting.
2. Incubate the sample with constant rotation or gentle shaking for 2–4 h at 4 °C. For samples with more than 500 μL cell pellet, a 2-h extraction is enough incubation time. Longer incubation is not recommended, as other basic proteins will also be extracted. For small sample size (<200 μL cell pellet), 4-h extraction gives a better yield.
3. Centrifuge at 3400 rcf for 5 min.
4. Transfer the supernatant to a new 1.5 or 15 mL tube, depending on the sample volume.
5. Repeat **steps 3** and **4**.
6. Add 100% TCA to the sample solution with a ratio of 1:3 (v/v), in order to obtain a final TCA concentration of 33%. This step will precipitate histones.
7. Let the mixture precipitate on ice for at least 1 h. Do not disturb the precipitation. For samples that start with small cell numbers, overnight precipitation is recommended.
8. Centrifuge at 3400 rcf for 5 min. Remove the supernatant by aspiration without touching the precipitated proteins. In particular, do not touch the white layer condensed around the bottom of the tube, as these are the histones. The pellet in the very bottom of the tube contains mostly other proteins or other biomolecules.
9. By using a glass Pasteur pipette rinse the tube with acetone + 0.1% HCl to cover the precipitated proteins.
10. Centrifuge at 3400 rcf for 2 min and discard supernatant.
11. Repeat **steps 9** and **10** using acetone without 0.1% HCl.
12. Dry pellet with airflow or with a SpeedVac centrifuge, or just by leaving the tube open. Acetone evaporates quickly.
13. Dissolve the histones with ddH_2O in the minimum volume possible to dissolve completely the white layer. For pellets in a 1.5 mL microcentrifuge tube, 100 μL ddH_2O is usually enough.
14. Centrifuge at 3400 rcf for 2 min and transfer the supernatant to a new tube. Discard the pellet at the very bottom of the tube, as it contains mostly nonhistone proteins and other biomolecules. *Pause*—Sample can be stored at −80 °C. Before freezing collect few μL for quality control ofhistone extraction (Subheading 3.5).

3.5 Quality Control of Histone Extraction

1. Measure protein concentration. BCA, Bradford protein assay, or amino acid analysis (AAA) are recommended. Do not use techniques that adopt absorbance at 280 nm, as histones are poor in aromatic amino acid residues.
2. Verify the purity of extracted histones with SDS gel and Coomassie staining (optional).
3. If high purity of single-histone variants is desired continue to HPLC-UV fractionation of histone variants (Subheading 3.6). Alternatively, skip directly to sample preparation for either bottom-up or middle-down histone PTM analysis (Subheadings 3.7 and 3.12, respectively).

3.6 Online HPLC-UV Fractionation of Histone Variants (Optional)

The crude histone mixture can be fractionated with reversed-phase HPLC coupled to a UV detector. This step allows for purification of histone variants and thus leads to more sensitive analyses for single histones as compared to the analysis of the crude histone mixture (*see* **Note 2** for recommendations on when to perform this step).

1. Connect a C_{18} 5 μm column to an HPLC. We recommend either a 4.6×250 mm or a 2.1 mm×250 mm column. For the first, use a flow rate of ~0.8 mL/min, and for the second a flow rate of ~0.2 mL/min. Use buffer A and buffer B as described in Buffers (Subheading 2.2, **items 5** and **6**)
2. Connect the other extreme of the column to a UV detector, and set the absorbance to 210–220 nm.
3. Acidify the histone sample dissolved in water with 100% TFA to achieve a final concentration of 0.1–1 % TFA.
4. Equilibrate the column with 100% buffer A for at least 15 min at the recommended flow rate, which corresponds approximately to three column volumes. Use this signal to set the zero-absorbance level of the UV detector.
5. Prepare 1.5 or 15 mL tubes to collect fractions or, if available, an automatic sample collector.
6. Inject sample at the concentration of ~1 μg/μL or higher, if using 2.1 mm ID column, or at the concentration of ~0.75 μg/μL or higher, if using 4.6 mm ID column. Samples dissolved in larger volumes might alter the equilibration of the column during loading and lead to lower retention.
7. Run the gradient, programmed as follows: from 0 to 30% B in 1 min, 30 to 60% B in 50 min, and 60 to 90% B in 1 min.
8. Collect 1–2-min fractions between 10 and 50 min (*see* **Note 3** for further instructions). Elution of histone variants is displayed in Fig. 1.
9. Merge fractions containing the same chromatographic peak and dry down in a SpeedVac centrifuge

Pause—Sample can be stored at −80 °C as dry or reconstituted in ddH_2O.

10. Continue to sample preparation for bottom-up or middle-down histone PTM analysis (Subheadings 3.7 and 3.12, respectively).

3.7 Propionic Anhydride Derivatization Prior to Histone Digestion forc Bottom-Up Analysis

The bottom-up strategy is the most commonly used MS-based proteomics strategy for histone characterization, as it is based on histone digestion into short peptides (5–20 aa), which facilitates both HPLC separation and MS detection (Fig. 2). Masses in the range of 600–2000 Da are commonly more easily ionized, and identified with higher mass accuracy and resolution than larger masses. Smaller masses are instead hard to retain by chromatography. MS/MS fragmentation is also facilitated, as short peptides are generally well suited for collision-induced dissociation (CID). However, histones are highly enriched in basic amino acid residues such as lysine and arginines. Therefore, trypsin digestion leads to the generation of too short peptides for HPLC retention and unambiguous localization of the PTMs. Our protocol includes a step of lysine and peptide N-terminal chemical derivatization [9]. We use propionic anhydride for this purpose, as we recently proved that it is the most suitable anhydride for the purpose [10]. Such derivatization blocks the ε-amino groups of unmodified and monomethyl lysine residues, allowing trypsin to perform proteolysis only at the C-terminal of arginine residues. Moreover, derivatized lysine residues, contrary to unmodified ones, cannot exchange protons with the solution and thus the peptides are generally only doubly or triply charged, facilitating MS and MS/MS detection. N-terminal derivatization increases peptide hydrophobicity and thus reversed-phase chromatographic retention (*see* **Note 4** for alternative protocol).

1. Dissolve histone samples in 30 μL of 50 mM NH_4HCO_3, pH 8.0 (recommended amount: 50–100 μg). If samples were in pure ddH_2O, add concentrated NH_4HCO_3 until reaching 50 mM.
2. Dip a P10 pipette tip into the sample and touch with this a pH indicator strip to monitor the pH. It should be enough to have an idea of the current pH without having sample losses. NH_4OH and glacial acetic acid can be used to adjust the pH (*see* **Note 5** for safety instructions).
3. Prepare propionylation reagent by mixing propionic anhydride with 2-propanol in the ratio 1:3 (v/v); for example for three samples that have the volume of 20–30 μL, mix 15 μL of propionic anhydride and 45 μL of 2-propanol. This reagent must be made fresh every three to four samples (*see* **Note 6** for details regarding reagents' reactivity).
4. Add rapidly the propionylation reaction to the histone sample with a ratio of 1:2 (v/v), e.g., 15 μL propionylation reaction for 30 μL sample.

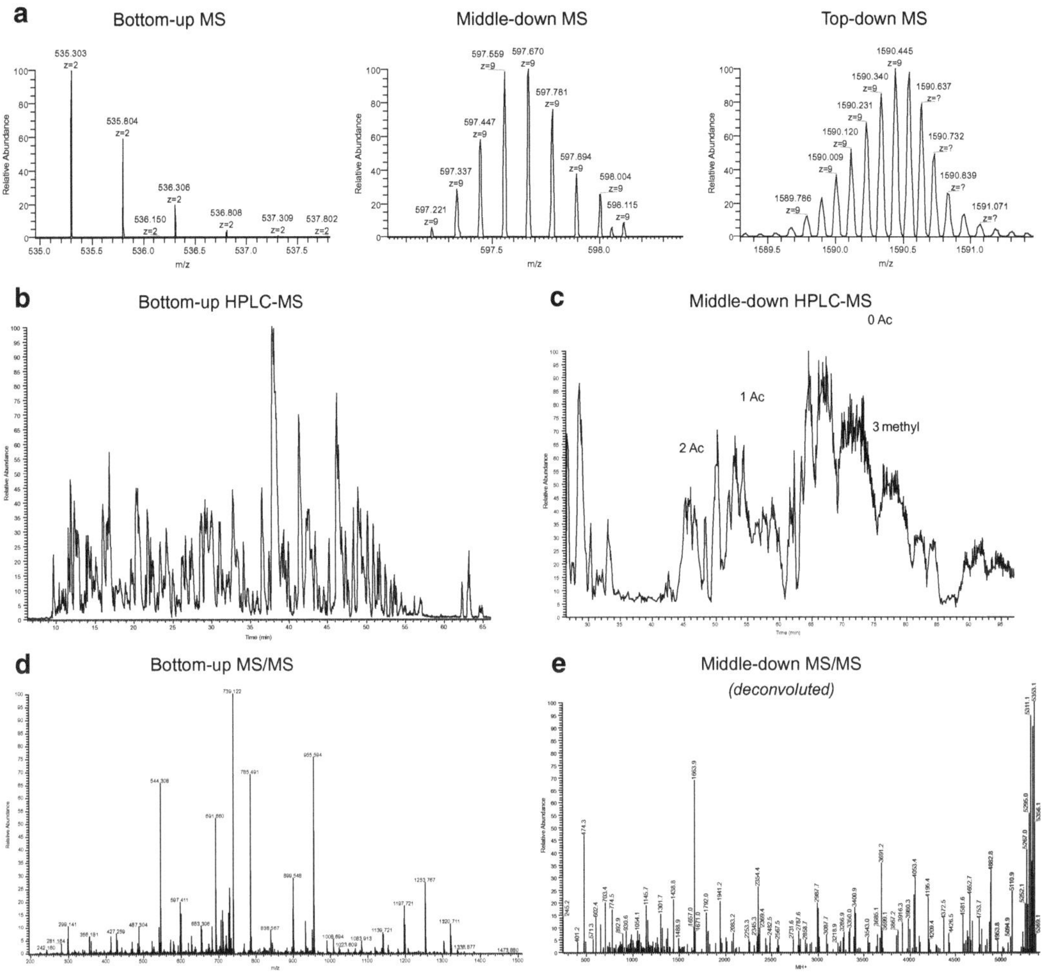

Fig. 2 Comparison of HPLC-MS performance for bottom-up and middle-down analysis. (**a**) Full MS scan performed at 60,000 resolution with an Orbitrap Fusion (Thermo Fisher Scientific) of a bottom-up-like peptide (1 kDa), a middle-down-like peptide (5 kDa), and a mass comparable to an intact histone (14–15 kDa). Higher masses lead to lower efficiency in resolving the isotopic distribution of the analyzed molecule. (**b**) Reversed-phase HPLC separation of a histone mixture digested with trypsin after derivatization with propionic anhydride. Peptides are eluted in sharp, mostly baseline separated, peaks. (**c**) WCX/HILIC separation of histone H3 N-terminal tails. The different peaks correspond to the same peptide sequence with different number of PTM equivalents. Heavily methylated and acetylated peptides elute first, while poorly modified tails elute in the final part of the chromatography. (**d**) MS/MS spectrum of a bottom-up-like peptide (1 kDa). (**e**) Deconvoluted MS/MS spectrum of a middle-down-like peptide (5 kDa), where ETD fragmentation was used. On the right side of the spectrum the precursor mass and its respective neutral losses are the most abundant species in the spectrum, indicating that the fragmentation is not complete

5. Add rapidly NH_4OH to reestablish pH 8.0 to the solution. Propionic anhydride reacting with the free amines of the peptides produces propionic acid that decreases pH. Usually, adding NH_4OH to the sample with a ratio of 1:5 (v/v) is appropriate to reestablish pH 8.0, e.g., 6 μL of NH_4OH to 30 μL of sample.

Warning—When pH is larger than 10.0, labeling of other amino acid residues with higher pKa is possible.

6. Mix immediately by vortex.
7. Check pH with the same procedure as **step 2**.
8. Briefly centrifuge and incubate samples at 37 °C on a heat block or in a water bath for 15 min.
9. Repeat **steps 3–8**, always taking care not to perform the reaction for more than three to four samples per batch of propionylation reagent.
10. Dry samples down to 5–10 μL in a SpeedVac centrifuge. This evaporates unreacted propionic anhydride, 2-propanol, acetic acid, and ammonia gas released from NH_4OH. If samples dry out completely, no significant sample losses occur.
11. Resuspend or dilute samples with ddH_2O until achieving 30 μL of final volume.
12. Repeat **steps 2–10**. A double round of histone propionylation ensures >95 % of reaction completion.
    Pause—Sample can be stored at −80 °C as dry or reconstituted in ddH_2O.
13. Continue with proteolytic digestion with trypsin (Subheading 3.8).

3.8 Proteolytic Digestion with Trypsin (In Solution)

1. Resuspend histones in 50 mM NH_4HCO_3 to achieve a concentration of 1 μg/μL or higher. More diluted samples lead to lower trypsin efficiency.
2. Verify that pH is about 8.0.
3. Add trypsin to histone samples at a 1:20 ratio, e.g., 5 μg of trypsin for 100 μg of histones.
4. Incubate at 37 °C for 6 h.
5. Stop the digestion by adding 2–5 μL (or more) of glacial acetic acid to reach pH 3.0, or 1–2 μL of TFA.
6. Dry down the sample to 5–10 μL in a SpeedVac centrifuge.
    Pause—Sample can be stored at −80 °C.
7. Continue with propionylation of histone peptides at N-termini after trypsin digestion (Subheading 3.10).

3.9 Derivatization and Proteolytic Digestion of Histones (In Gel—Alternative to Subheadings 3.7 and 3.8)

This part of the protocol should be used for histones loaded in gel. Protein separation using SDS-PAGE is an efficient technique to achieve both sample fractionation and removal of detergents (e.g., if sample is an elution from immunoprecipitation). This part is alternative to Subheadings 3.7 and 3.8. Subheading 3.10 is common for both in-solution and in-gel histone derivatization and digestion.

1. Excise the histone fraction from the polyacrylamide gel. Cut as close to the protein band as possible to reduce the amount of background.
2. Cut the excised piece into ~1 mm^3 cubes and transfer them to a clean 1.5 mL or 0.5 mL microcentrifuge tube.
3. Wash the gel pieces with ddH_2O corresponding to five times gel volume. For washing the tubes can be left for 15 min on a shaker or vortex.
4. Remove water, and replace it with the same volume of 50% acetonitrile. Repeat the wash.
5. Remove the solution. If gel bands are still heavily stained repeat **steps 3** and **4.**
6. Remove the solution and replace it with the same volume of 100% acetonitrile. Repeat the wash.
7. Remove acetonitrile, taking care not to aspirate shrunk gel pieces.
8. Add 50 μL (or sufficient volume to cover gel pieces) of 100 mM NH_4HCO_3, immediately followed by 100 μL of propionic anhydride. Quickly vortex and incubate for 20 min at room temperature.
9. Spin down and remove supernatant.
10. Wash with 500 μL of 100 mM NH_4HCO_3.
11. Repeat **step 10** once or twice. Check pH of the second wash; if it is not ~8.0 repeat the wash.
12. Remove supernatant and add five times gel volume of acetonitrile to shrink gel pieces.
13. Remove acetonitrile.
14. Repeat **steps 8–13** to assure completion of derivatization.
15. On ice (4 °C) rehydrate gel particles with digestion buffer (50 mM NH_4HCO_3, and 12.5 ng/μL trypsin). Add enough digestion buffer to cover the gel pieces. If after 2 min all the initially added volume has been absorbed by the gel pieces add 20 μL more digestion buffer.
16. Incubate at room temperature overnight.
17. The next day transfer supernatant with peptides to a clean 1.5 mL tube. The supernatant contains the digested peptides eluted from gel bands. To increase peptide recovery (e.g., low sample amounts), follow **steps 18–21**.
18. Add 20 μL of ddH_2O and wash gel pieces for 15 min.
19. Add the same volume of acetonitrile and wash gel pieces for 15 min.
20. Transfer the supernatant to the same 1.5 mL tube of **step 17.**

21. Repeat **steps 18–20**.
22. Dry the sample in a SpeedVac centrifuge.
    Pause—Sample can be stored dry at −80 °C.
23. Continue with propionylation of histone peptides at N-termini after trypsin digestion (Subheading 3.10).

3.10 Propionylation of Histone Peptides at N-termini After Trypsin Digestion

This section describes the derivatization of peptide N-termini. Such procedure is not essential for most of the histone peptides, but it facilitates the HPLC retention of the shortest ones (e.g., aa 3–8 histone H3), as the propionyl group increases peptide hydrophobicity (*see* **Note 7** for modifications of the protocol that includes multiplexing).

1. Resuspend samples in 30 μL of 100 mM NH_4HCO_3.
2. Repeat **steps 2–12** of Subheading 3.7, also if in-gel digestion was performed. In case light and heavy anhydride is used, perform propionylation with the light form in one sample, and with the heavy form in the other sample.
3. Resuspend or dilute samples with 50–100 μL ddH_2O + 0.1 % TFA. If propionylation with light and heavy anhydride was performed, samples can now be mixed together.
   Pause—Sample can be stored at −80 °C.
4. Continue to sample desalting with StageTips (Subheading 3.11). If your HPLC-MS setup is equipped with a trap column skip directly to preparation of the nano-HPLC setup for online HPLC-MS analysis (Subheading 3.13).

3.11 Sample Desalting with StageTips (This Step Can be Omitted if Using Trap Column in HPLC-MS)

The protocol we describe leads to the presence of salts in the sample at this stage of the preparation. Salts are detrimental for HPLC-MS analysis. First, ionized salts are also injected into the mass spectrometer, contributing in suppressing the signal of the peptides and contaminating the instrument. Moreover, salts might form ionic adducts with peptides, reducing the signal intensity of the "clean" peptide, as a percentage of such peptide would be detected with a different molecular weight. This prevents efficient identification and quantification of the given peptide. Desalting can be performed off-line with StageTips or online when the HPLC-MS setup consists of a two-column system (Fig. 3). In this section we describe the off-line protocol.

1. By using a P1000 pipette tip cut a disk of C_{18} material from a 3 M Empore™ Solid Phase Extraction Disk C_{18}, and deposit this minidisk to the bottom of a P100/200 pipette tip. You can push the minidisk out of the P1000 tip by using a fused silica capillary. Ensure that the disk is securely wedged in the bottom of the tip.

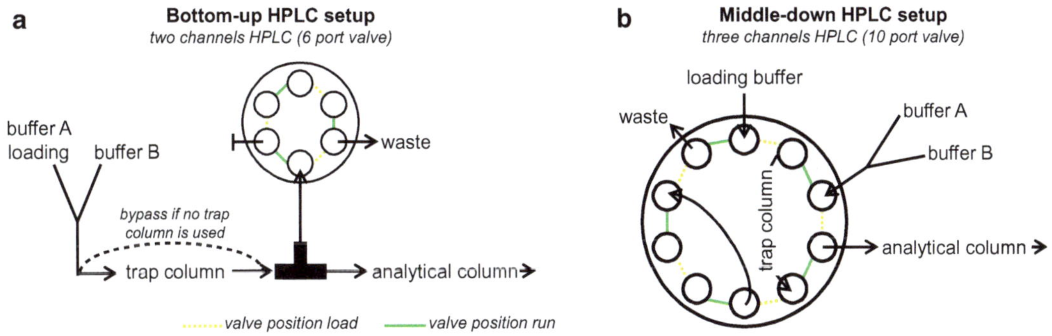

Fig. 3 Column configuration for online HPLC-MS. (**a**) Connection of trap and analytical column for bottom-up analysis. The *arrows* represent the direction of the flow. During loading the loading buffer, which is also the buffer A, is pumped through the trap column and to the waste, as the valve leaves open the waste line. During the run the valve blocks the tee and the gradient flows through trap and analytical column. If trap column is omitted both load and run are performed with the valve in the position run, while the valve in position load is used only to rapidly change buffer composition without system backpressure. (**b**) Connections for middle-down analysis. During loading the loading buffer is pumped through the trap column, connected to generate a loop within the valve, while the analytical column can be equilibrated with buffer A

2. Repeat **step 1** in the same P100/200 tip if you are desalting more than 25 μg of sample.
3. Wash disk by flushing 100 μL of 75 % acetonitrile and 0.025 % TFA with air pressure, e.g., using a syringe (*see* **Note 8** for alternative procedure).
4. Equilibrate disk by flushing 50 μL of 0.1 % TFA by air pressure.
5. Load sample onto the disk by applying air pressure.
6. Wash sample by flushing 50 μL of 0.1 % TFA by applying air pressure.
7. Elute sample by flushing 50 μL 75 % acetonitrile and 0.025 % TFA by air pressure. Collect the sample in a 0.5 or 1.5 mL tube.
8. Dry sample in a SpeedVac centrifuge to ~5 μL.
   Pause—Sample can be stored at −80 °C.
9. Continue to preparation of the nano-HPLC setup for online HPLC-MS analysis (Subheading 3.13).

3.12 Sample Preparation for Middle-Down Histone PTM Analysis (Alternative to Subheadings 3.7–3.11)

The middle-down strategy takes advantage of the fact that the N-terminal tail of the histones can be proteolytically digested by GluC, an enzyme that cleaves at the C-terminal of the glutamic acid residue. This generates a polypeptide of 40–50 aa residues (5–6 kDa) that contain the majority of histone PTMs. For instance, histone H3 isotypes in mammals and many model organisms contain the first glutamic acid in position 50. This strategy is an effective compromise between bottom-up and top-down (intact protein analysis), as it allows precise mapping and quantification of single PTMs, still technically challenging with top-down and combinatorial PTMs, not possible with bottom-up.

1. Resuspend histones in 50 mM NH_4HCO_3 (pH 8.0) to achieve a concentration of 1 µg/µL or higher. More diluted samples lead to lower GluC efficiency. Alternatively, histones can be resuspended in 50 mM ammonium acetate ($NH_4C_2H_3O_2$, pH 4.0). At pH >10 deamidation of glutamine has high kinetics; this is an issue as glutamine is present on all histone tails and, if deamidated, it is converted into glutamic acid.
2. Add GluC to the sample at a 1:20 enzyme:sample ratio (w/w), e.g., 5 µg of GluC for 100 µg of histones.
3. Incubate at room temperature for 6 h. Higher temperatures increase deamidation kinetics.
4. Stop the digestion by adding 2–5 µL (or more) of glacial acetic acid to reach pH 3.0, or 1–2 µL of TFA.

   Pause—Freeze samples at −80 °C until analysis.
5. Continue with preparation of the nano-HPLC setup for online HPLC-MS analysis (Subheading 3.13).

3.13 Preparation of the Nano-HPLC Setup for Online HPLC-MS Analysis

Online HPLC-MS in proteomics is commonly performed using nano-liquid chromatography. This is because nano-HPLC allows for loading of low amounts of material, and guarantees high-sensitivity analyses. However, particular attention must be used when preparing the HPLC setup, as small errors in column cuts or connections highly affect chromatographic performance. Here we describe how to prepare nano-HPLC columns (**steps 1–10** can be omitted if using commercial columns) and how to configure the HPLC setup for bottom-up and middle-down analysis.

1. Cut ~30 cm of fused silica capillary in which you wish to make a frit. We recommend the use of 75 µm internal diameter (ID) capillaries for analytical columns and 100 µm ID for trap columns
2. Transfer 88 µL Kasil® to a 0.5 mL tube
3. Add 16 µL formamide to the tube and vortex quickly for 10–15 s. Formamide is toxic, and all the necessary safety precautions should be taken.
4. Dip 1 cm of fused silica into the mixture and remove quickly. The mixture will enter into the capillary for about 1–2 cm by capillarity.
5. Leave the fused silica overnight for polymerization. Alternatively, polymerization can be catalyzed by placing the capillary in a heater at ~110 °C for 3–4 h.
6. Cut the frit to leave no more than 3–4 mm at the top of the column. The frit should appear bright below illumination. The fused silica is now ready for packing (*see* **Note 9** for alternative procedure).
7. Prepare in a clean HPLC glass vial the resin slurry for column packing in 100% methanol or any other organic solvent and

add a micro-stir magnet. Use C_{18}-AQ reversed-phase 3 μm particles for trap columns and bottom-up analytical column, and Polycat A resin 3 μm 1500 Å pore size for middle-down analytical column.

8. Place the resin slurry in a pressure bomb and turn on magnetic stirring.
9. Place the Kasil-fritted fused silica in the pressure bomb. Pressure is delivered by a gas bomb, containing helium, nitrogen, or air. Traditional pressure bombs cannot stand pressures above 100–150 bars. Verify that value on the pressure-limiting valve placed on the gas bomb.
10. Turn on pressure and leave the column packing. The recommended lengths for columns are as follows:
    (a) 15–18 cm for C_{18} analytical column for bottom-up analysis: The column can be packed indefinitely, and then cut in the desired length.
    (b) 10–12 cm for Polycat A analytical column for middle-down analysis: The column can be packed indefinitely, and then cut in the desired length.
    (c) 1.5 cm for trap columns: Make sure that you turn off the pressure at the desired length, as it is not possible to cut a capillary so short and then make the HPLC connections. Leave at least 5 cm of empty capillary for bottom-up analysis, and 8 cm for middle-down.
11. Connect the trap and the analytical column as indicated in Fig. 3. For the bottom-up analysis the trap column can be omitted if StageTips desalting was performed (Subheading 3.11).
12. Continue to bottom-up or middle-down analysis of histone peptides (Subheadings 3.14 and 3.15, respectively).

3.14 Bottom-Up Analysis of Histone Peptides

At this stage the histone sample and the HPLC setup are ready. It is possible now to proceed to the HPLC-MS/MS analysis. The method described is meant to be used for the columns we previously recommended (Subheading 3.13).

1. Program the HPLC method as follows: from 0 to 30% buffer B in 30 min, from 30 to 100% B in 5 min, and 8 min at isocratic 100% B. If the HPLC is not programmed for automated column equilibration before sample loading then include this part in the method: from 100 to 0% B in 1 min and isocratic flow at 0% B for 10 min. The flow rate of the analysis should be 250–300 nL/min.
2. Program the MS acquisition method to perform MS/MS data-dependent acquisition. With C_{18} chromatography the average

baseline peak width is about 30 s for the gradient we described. Make sure that the MS duty cycle allows one full MS scan every ~2 s, in order to have enough data points to draw accurately the peak shape of the eluted peptides.

3. Include in the MS acquisition method targeted MS/MS for peptides that have isobaric species (displayed with an asterisk in Table 2). These peptides need to be selected for fragmentation through their entire elution, as the discrimination of the relative abundance of the isobaric species is performed by monitoring the elution profile of the fragment ions. All the other settings are common to other standard proteomics experiments.
4. Load ~1 μg of sample onto the HPLC column.
5. Run the HPLC-MS/MS method as programmed.
6. Perform label-free quantification by extracting the area below the curve of the chromatographic peak for each peptide. This step can be performed manually or with dedicated software. The area of the chromatographic peak should be calculated for the $[M+H]^{+}$, $[M+2H]^{2+}$, and $[M+3H]^{3+}$ ions of the same peptide, even though in most cases the $[M+2H]^{2+}$ is the prevalent form (*see* **Note 10** for further instructions on how to discriminate the differently modified peptides).
7. Calculate the relative abundance of each PTM by calculating the sum of all different modified forms of a histone peptide (100%), and divide the area of the particular peptide by the total histone peptide. When isobaric species are present, e.g., K18ac and K23ac, MS/MS information is used to find the ratio between the two species (Fig. 4). This ratio is used to divide the area of the chromatographic peak between the two species.

3.15 Middle-Down Analysis of Intact Histone Tails (Alternative to Subheading 3.14)

WCX/HILIC is currently the best suited column material to online separate histone tails. Large basic and hydrophilic polypeptides bind efficiently in highly organic solvent (75%) and near-neutral pH (6.00), since the hydrophilic stationary phase contains glutamic acid that deprotonates and generates ionic bonds with positively charged polypeptides. Elution is performed with a gradient of water and decreasing pH, avoiding the use of salts that are potentially detrimental for MS. Detection is performed with high-resolution MS/MS and ETD fragmentation. Afterwards, database searching is mandatory to follow our workflow, which is currently the only one publicly available to map and quantify precisely single and combinatorial histone PTMs with middle-down proteomics. The method described is meant to be used with the column type and configuration we described previously (Subheading 3.13).

1. Program the HPLC method as follows: from 0 to 55% buffer B in 1 min, from 55 to 85% B in 160 min, and from 85 to 100% in 5 min. If the HPLC is not programmed for

Table 2

Peptides of most common interest in bottom-up histone analysis. The table displays the histone variant and the peptide position in the protein sequence. Each peptide is then present in all most common modified states, and we calculated their respective *m*/*z* signal for singly, doubly, triply, and, where possible, quadruply charged forms

Histone	Peptide position	Modified peptide	(+1)	(+2)	(+3)	(+4)	Histone	Peptide position	Modified peptide	(+1)	(+2)	(+3)	(+4)
Histone H3	1 8	ARTKQTAR	1043.596	522.302	348.537		Histone H3	41 49	YRPGTVALR	1088.622	544.815	363.546	
		A**Rme1**TKQTAR	1057.612	529.310	353.209								
								54 63	YQKSTELLIR	1362.763	681.886	454.926	
	3 8	TKQTAR	816.458	408.733					YQ**Kac**STELLIR	1348.747	674.877	450.254	
		T**Kme1**QTAR	830.474	415.741									
		T**Kme2**QTAR	788.463	394.735				73 83	EIAQDFKTDLR	1447.743	724.376	483.253	
		T**Kme3**QTAR	802.478	401.743					EIAQDF**Kme1**TDLR	1461.759	731.383	487.925	
		TphosKQTAR	896.424	448.716					EIAQDF**Kme2**TDLR	1419.748	710.378	473.921	
									EIAQDF**Kme3**TDLR	1433.763	717.386	478.593	
	9 17	KSTGGKAPR	1069.601	535.304	357.205								
		Kme1STGGKAPR	1083.616	542.312	361.877		Histone H4	1 17	**ac**SGRGKGGKGL GKGGAKR	1837.040	919.024	613.019	460.016
		Kme2STGGKAPR	1041.606	521.307	347.874				**ac**SG**Rme1**GKGGK GLGKGGAKR	1851.056	926.032	617.690	463.520
		Kme3STGGKAPR	1055.621	528.314	352.545								
		[*]**Kac**STGGKAPR	1055.584	528.296	352.533			4 17	GKGGKGLGKGGAKR	1550.902	775.955	517.639	388.481
		[*]KSTGG**Kac**APR	1055.584	528.296	352.533				[*]GKGGKGLGKGG A**Kac**R	1536.886	768.947	512.967	384.977
		Kme1STGG**Kac**APR	1069.600	535.304	357.205				[*]GKGGKGLG**Kac**GGAKR	1536.886	768.947	512.967	384.977
		Kme2STGG**Kac**APR	1027.589	514.299	343.202				[*]GKGG**Kac**GLGKGGAKR	1536.886	768.947	512.967	384.977
		Kme3STGG**Kac**APR	1041.605	521.306	347.873				[*]G**Kac**GGKGLGKGGAKR	1536.886	768.947	512.967	384.977
		KacSTGG**Kac**APR	1041.568	521.288	347.861				[*]G**Kac**GGKGLG KGGA**Kac**R	1522.869	761.939	508.295	381.473
		K**Sphos**TGGK APR	1149.566	575.287	383.861				[*]GKGG**Kac**GLGKGGA**Kac**R	1522.869	761.939	508.295	381.473

(continued)

Table 2
(continued)

Histone	Peptide position	Modified peptide	(+1)	(+2)	(+3)	(+4)	Histone	Peptide position	Modified peptide	(+1)	(+2)	(+3)	(+4)
		Kme1SphosTGGKAPR	1163.582	582.295	388.533				*GKGGKGLGKacGGAKacR	1522.869	761.939	508.295	381.473
		Kme2SphosTGGKAPR	1121.571	561.290	374.529				*GKacGGKGLGKacGGAKR	1522.869	761.939	508.295	381.473
		Kme3SphosTGGKAPR	1135.587	568.297	379.201				*GKGGKacGLGKacGGAKR	1522.869	761.939	508.295	381.473
									*GKacGGKacGLGKGGAKR	1522.869	761.939	508.295	381.473
	18 26	KQLATKAAR	1154.690	577.849	385.568				*GKGGKacGLGKacGGAKacR	1508.853	754.930	503.623	377.969
		*Kme1QLATKAAR	1168.705	584.857	390.240				*GKacGGKGLGKacGGAKacR	1508.853	754.930	503.623	377.969
		*KQLATKme1AAR	1168.705	584.857	390.240				*GKacGGKacGLGKGGAKacR	1508.853	754.930	503.623	377.969
		*KacQLATKAAR	1140.674	570.841	380.896				*GKacGGKacGLGKacGGAKR	1508.853	754.930	503.623	377.969
		*KQLATKacAAR	1140.674	570.841	380.896				GKacGGKacGLGKacGGAKacR	1494.837	747.922	498.951	374.465
		KacQLATKacAAR	1126.657	563.833	376.224								
								20 23	KVLR	627.419	314.214		
	27 36	KSAPATGGVKKPHR	1657.939	829.473	553.318	415.241			Kme1VLR	641.435	321.221		
		*Kme1SAPATGGVKKPHR	1671.955	836.481	557.990	418.745			Kme2VLR	599.424	300.216		
		*Kme2SAPATGGVKKPHR	1629.944	815.476	543.987	408.242			Kme3VLR	613.440	307.224		
		*Kme3SAPATGGVKKPHR	1643.959	822.483	548.658	411.746							
		KacSAPATGGVKKPHR	1643.923	822.465	548.646	411.737		68 78	DAVTYTEHAKR	1402.697	701.852	468.237	

		*KSAPATGGVKme1KPHR	1671.955	836.481	557.990	418.745							
		*KSAPATGGVKme2KPHR	1629.944	815.476	543.987	408.242	Histone H2A	82 88	HLQLAIR	906.552	453.780		
		*KSAPATGGVKme3KPHR	1643.959	822.483	548.658	411.746							
		Kme1SAPATGGVKme1KPHR	1685.970	843.489	562.662	422.248		4 11	GKQGGKAR	969.548	485.278		
		*Kme1SAPATGGVKme2KPHR	1643.959	822.483	548.658	411.746			*GKQGGKacAR	955.532	478.270		
		*Kme1SAPATGGVKme3KPHR	1657.975	829.491	553.330	415.250			*GKacQGGKAR	955.532	478.270		
		*Kme2SAPATGGVKme1KPHR	1643.959	822.483	548.658	411.746			GKacQGGKacAR	941.516	471.262		
		Kme2SAPATGGVKme2KPHR	1601.949	801.478	534.655	401.243							
		*Kme2SAPATGGVKme3KPHR	1615.964	808.486	539.327	404.747	H2A.Z	1 19	AGGKAGKDSGKAKTKAVSR	2153.193	1077.100	718.403	539.054
		*Kme3SAPATGGVKme1KPHR	1657.975	829.491	553.330	415.250			*AGGKacAGKDSGKAKTKAVSR	2139.177	1070.092	713.731	535.550
		*Kme3SAPATGGVKme2KPHR	1615.964	808.486	539.327	404.747			*AGGKAGKacDSGKAKTKAVSR	2139.177	1070.092	713.731	535.550
		Kme3SAPATGGVKme3KPHR	1629.979	815.494	543.998	408.251			*AGGKAGKDSGKacAKTKAVSR	2139.177	1070.092	713.731	535.550
		KSphosAPATGGVKKPHR	1737.905	869.456	579.973	435.232			*AGGKAGKacDSGKacAKTKAVSR	2125.160	1063.084	709.059	532.046
		Kme1SphosAPATGGVKKPHR	1751.920	876.464	584.645	438.736			*AGGKacAGKDSGKacAKTKAVSR	2125.160	1063.084	709.059	532.046
		Kme2SphosAPATGGVKKPHR	1709.910	855.459	570.642	428.233			*AGGKacAGKacDSGKAKTKAVSR	2125.160	1063.084	709.059	532.046
		Kme3SphosAPATGGVKKPHR	1723.925	862.466	575.314	431.737			AGGKacAGKacDSGKacAKTKAVSR	2111.144	1056.076	704.387	528.542
H2A.X	4 11	GKTGGKAR	942.537	471.772			macro H2A	15 26	SAKAGVIFPVGR	1313.758	657.383	438.591	

a

Sequence	H3K18ac *b*	H3K18ac *y*	H3K23ac *b*	H3K23ac *y*
K	227.15	1140.71	241.19	1140.71
Q	355.21	914.57	369.25	900.53
L	468.30	786.51	482.34	772.47
A	539.33	673.42	553.37	659.38
T	640.38	602.39	654.42	588.35
K	824.53	501.34	824.53	487.30
A	895.56	317.19	895.56	317.19
A	966.60	246.16	966.60	246.16
R	1122.70	175.12	1122.70	175.12

b

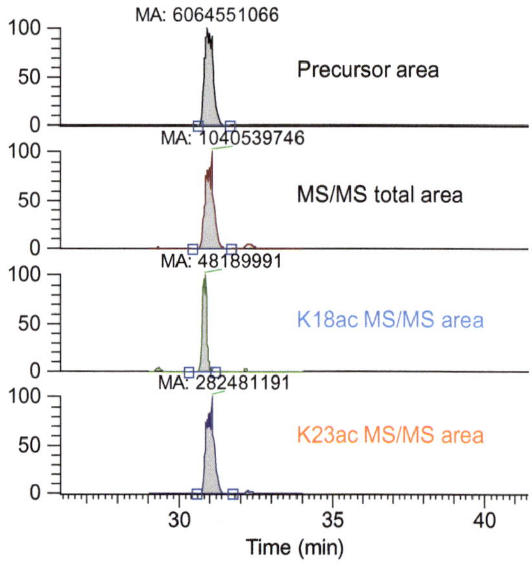

c

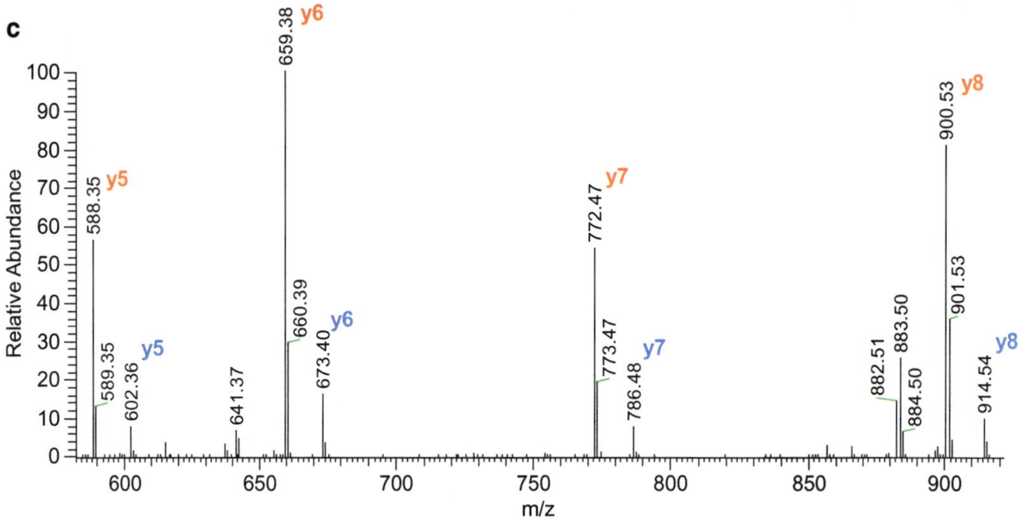

Fig. 4 Quantification of two co-eluted isobaric peptides. (**a**) The histone H3 peptide KQLATKAAR (aa 18–26) was found acetylated in both K18 and K23 residues. The two peptides generate different MS/MS fragments. We used the fragments y5-8 to calculate their relative abundance, as they were the most intense ones (*highlighted*). (**b**) Extracted ion chromatogram of the precursor mass corresponding to the peptide sequence + one acetyl group (*top*) and the ion chromatography of the targeted MS/MS scans (*top-middle*). The extracted MS/MS ion chromatography generates a smaller area than the extracted precursor mass, as the fragmented peptide has a lower signal than the precursor mass. Below, extracted MS/MS ion chromatography of the fragments y5-8 of the K18ac (*bottom-middle*) and the K23ac peptide (*bottom*). The area of the K23ac chromatogram is about six times higher than K18ac. The total precursor area should then be divided by the two species according to their calculated ratio. (**c**) MS/MS spectrum of co-fragmented K18ac and K23ac peptides. Also from the single MS/MS spectrum it is possible to calculate the ratio between K23ac and K18ac, which is about sixfold

automated column equilibration before sample loading then include this part in the method: switch the valve in position load (Fig. 3), from 100 to 0% B in 1 min and isocratic flow at 0% B for 10 min. The flow rate of the analysis should be 250–300 nL/min.

2. Program the MS acquisition method to perform MS/MS data-dependent acquisition of the six to eight most abundant precursor masses without dynamic exclusion. The full MS scan range should be 450–750 *m/z* to avoid repetitive selection of the same peptides in multiple charge states. If only histone H3 is analyzed, the window can be narrowed to 660–720 *m/z*.
3. Program the MS/MS acquisition to be performed with ETD at a resolution of ~30,000. The reaction time should be around 20 ms for polypeptides with eight to ten charges. Include three microscans to improve the quality of the MS/MS spectra acquired, as ETD spectra are overall less reproducible than CID. If using a trapping mass analyzer, i.e., Orbitrap, please note that the automatic gain control should be increased of about one order of magnitude as compared to traditional peptide fragmentation. This is because the number of histone tails accumulated is underestimated by the ion counter, as each of them is heavily charged. All the other settings are common to other standard proteomics experiments.
4. Load ~2 μg of sample onto the HPLC trap column.
5. Run the HPLC-MS/MS method as programmed. The HPLC elution profile should look like in Fig. 2c if only histone H3 is analyzed.
6. Continue to middle-down data processing (Subheading 3.16).

3.16 Middle-Down Data Processing

While bottom-up LC-MS runs do not necessarily need a proper database searching, in the case of middle-down it is mandatory with our developed workflow. In middle-down each precursor mass might easily correspond to more than 30 isobaric peptides (value estimated in [11]), which should be discriminated at the MS/MS level. Peptide-spectrum match validation and peptide quantification are performed with our in-house-developed bioinformatics tools, freely available at http://middle-down.github.io/Software. However, they require the result file of Mascot (Matrix Science, UK) as input.

1. Collect all raw files and submit them to a deconvolution tool. MS/MS spectra ions should be all singly charged previous Mascot database searching. We recommend Xtract as deconvolution algorithm if Thermo Fisher Scientific instrument is used (e.g., LTQ-Orbitrap). Xtract can be directly used in Proteome Discoverer (Thermo Fisher Scientific,

Bremen, Germany) as workflow node. Alternatively, any other deconvolution algorithm that generates Mascot Generic Format (.mgf) files can be used.

2. Perform database searching using the following parameters: MS mass tolerance: 2.2 Da, to include possible errors of the deconvolution algorithm in isotopic recognition. MS/MS mass tolerance: 0.01 Da. Enzyme: GluC with no missed cleavages allowed. No static modifications. Variable modifications: mono- and dimethylation (KR), trimethylation (K), acetylation (K), and, optionally, phosphorylation (ST). The sequence database should contain only histones; large databases increase dramatically searching time.
3. Export Mascot results in .csv file extension. In the export include the following information to the file: all query-level information, and all the default information (already ticked when export page is opened).
4. Import the .csv file in isoScale slim, which you can find at http://middle-down.github.io/Software. Select the tolerance for the search (recommended: 30 ppm) and the type of fragmentation adopted. The result table is in the same folder where the software is located. This table contains the list of peptides that passed the site determining ion validation and their absolute and relative intensity. isoScale outputs the calculated relative abundance for each combinatorial PTM identified and validated (software principle described in [11]).
5. The output table contains duplicates (peptides with the same sequence and PTM combination). Remove them by using the "Remove duplicates" option in Excel.

Middle-down is used to study PTM coexistence, but also to compare multiple conditions. However, all samples need to be run separately (multiplexing is still not possible). From the relative abundance of the combinatorial marks it is possible to extract the relative abundance of single marks simply by summing all relative abundances of peptides that contain the given PTM. In Fig. 5 we display some examples of how to represent middle-down data.

To estimate which histone marks tend to coexist with each other with high or low frequency it is possible to calculate the interplay score [12, 13]. This score is calculated as

$$I_{\mathrm{xy}} = \log 2\left(F_{\mathrm{xy}} / \left(F_{\mathrm{x}} \times F_{\mathrm{y}}\right)\right)$$

where I_{xy} is the interplay score between the marks X and Y, F_{xy} is the coexistence frequency (or relative abundance) of the two marks, and F_x and F_y are the frequencies of the single marks in the dataset. In other words, F_{xy} is the observed coexistence frequency, while $F_x \times F_y$

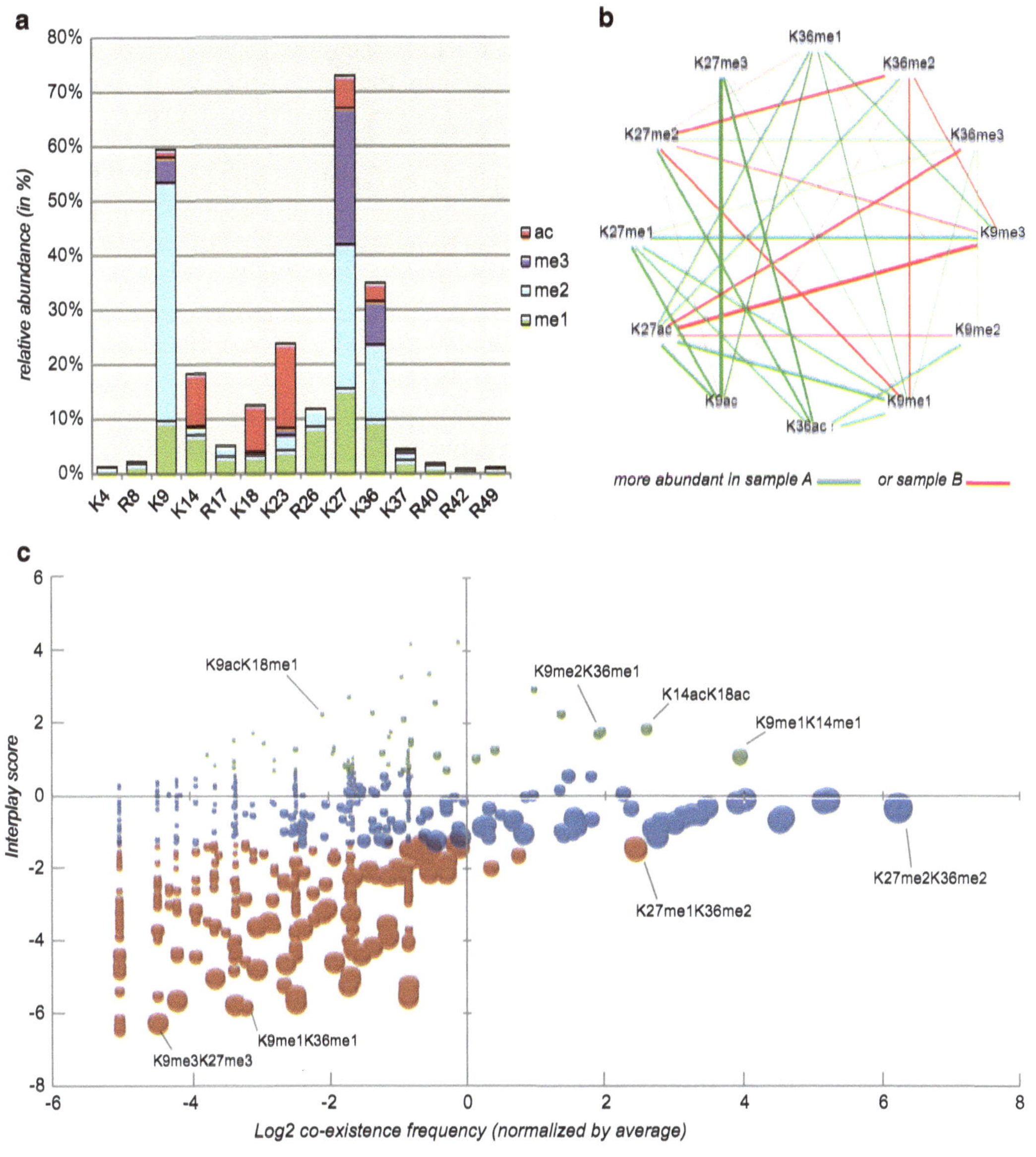

Fig. 5 Examples of middle-down data representation. (**a**) Bar plot of the relative abundance of single PTMs. Middle-down analysis allows for quantification of arginine methylations, while with bottom-up arginine is the cleavage site of the proteolytic enzyme. (**b**) Comparison of coexistence frequency of binary marks in two conditions. The graph displays the relative abundance of coexisting marks in sample A (*green*) and sample B (*red*); for example the combination K9acK27me3 is the binary PTM with the highest A/B ratio. (**c**) Bubble plot of binary PTMs. The graph displays three levels of information: the observed coexistence frequency of the binary marks (*X*-axis), the interplay score of the two marks (*Y*-axis) and the relative abundance of the single marks summed together (bubble size). The colors *green*, *blue*, and *red* represent binary marks with interplay score higher than 1, between 1 and −1, and below −1, respectively. Single marks with high relative abundance (large bubble size) and with low coexistence frequency have intuitively low interplay score, as they are abundant marks that rarely share the same histone tail

is the theoretical coexistence frequency, calculated based on the relative abundance of single PTMs. The interplay score provides how much two marks "like" to share the same histone tail. Positive values indicate tendency to coexist higher than if the two marks were completely independent from each other, while negative values indicate the opposite. The interplay score calculated for binary PTMs quantified in middle-down experiments could be used as an indicator to predict cross talk between histone marks.

4 Notes

1. The high-salt extraction protocol can be used to purify histones [14] in alternative to TCA precipitation. Such protocol is intrinsically milder, as it does not use strong acid. This preserves acid-labile PTMs and increases the yield of extracted histones, as TCA precipitation co-precipitates many other chromatin-binding proteins. However, high-salt extraction leads to samples containing too concentrated salt for HPLC-MS/MS. The StageTip step (Subheading 3.11) is often not sufficient in this case. Salt removal might generate consistent sample losses, making high-salt extraction procedure not favorable for our workflow.
2. Fractionation of intact histone variants ideally requires at least 100 μg of starting material (if 2.1 mm ID column is used), or 300 μg (if 4.6 mm ID column is used). In case the available sample is less than 25% of these references, we recommend avoiding HPLC-UV fractionation.
3. The time and time windows of fraction collection might vary depending on your HPLC system and the collecting tubes in use. Perform a test run at first.
4. Histone digestion for bottom-up proteomics analysis can be performed without propionylation, for instance by reducing trypsin incubation time and the enzyme/substrate ratio [15] or using ArgC as digestion enzyme [16–18]. However, we recommend our described protocol, as it leads to the generation of more hydrophobic peptides which are better retained during liquid chromatography.
5. Propionic anhydride, NH_4OH, and acetic acid should be handled in the fume hood. After its use, the bottle of propionic anhydride should be filled with argon to prevent slow conversion to propionic acid due to water vapor from air.
6. The propionylation mixture rapidly becomes inefficient, due to the conversion of propionic anhydride to propionic acid. For this reason, it is highly recommended to perform the reaction rapidly and for a limited number of samples for each batch.

7. In bottom-up sample preparation it is possible to differentially derivatize peptide N-termini of two samples with light and heavy propionic anhydride. One sample can be modified with a D0 propionic anhydride $(CH_3CH_2CO)_2O$, while the other with a D10 propionic anhydride $(CD_3CD_2CO)_2O$ at this step. This leads to a delta mass between the light- and heavy-labeled peptides of +5 Da and multiplexing analysis can be performed. The two samples should be mixed in equal amounts to obtain the least variation in ionization efficiency. The procedure to extract the area of heavy-labeled peptides is equal to the light version.
8. To desalt a large number of samples, centrifugation can be used instead of air pressure. Do not simply place the tips into 1.5 mL microcentrifuge tubes, as they might break. Use appropriate holders, or drill a hole on the top of the tube using a suitable size screw driver or a mini drill.
9. Alternatively to preparing frit for in-house-packed column, it is possible to pull a tip from one extreme of the capillary, if the lab is equipped with a laser tip puller. Such procedure can be performed only for the analytical column, as it can be directly placed at the front of the mass spectrometer.
10. For the bottom-up analysis lysine acetylation (+42.011 Da) can be discriminated from the nearly isobaric trimethylation (+42.047 Da) by using high-resolution MS acquisition (>30,000). Moreover, acetylation is more hydrophobic than trimethylation, leading to later elution of acetylated peptides as compared to the respective trimethylated ones. The unmodified form of the same peptide elutes even later, due to the fact that the lysine is propionylated. In summary, the order of hydrophobicity for a peptide with one modifiable site is di- and trimethylated < acetylated < unmodified (propionylated) < monomethylated (propionylated).

References

1. Waddington CH (1942) Canalization of development and the inheritance of acquired characters. Nature 150:563–565. doi:10.1038/150563a0
2. Sharma S, Kelly TK, Jones PA (2010) Epigenetics in cancer. Carcinogenesis 31:27–36. doi:10.1093/carcin/bgp220
3. Reik W, Dean W, Walter J (2001) Epigenetic reprogramming in mammalian development. Science 293:1089–1093. doi:10.1126/science.1063443
4. van Attikum H, Gasser SM (2009) Crosstalk between histone modifications during the DNA damage response. Trends Cell Biol 19:207–217. doi:10.1016/j.tcb.2009.03.001
5. Fernandez-Capetillo O, Mahadevaiah SK, Celeste A, Romanienko PJ, Camerini-Otero RD, Bonner WM, Manova K, Burgoyne P, Nussenzweig A (2003) H2AX is required for chromatin remodeling and inactivation of sex chromosomes in male mouse meiosis. Dev Cell 4:497–508. doi:10.1016/S1534-5807(03)00093-5
6. Santaguida S, Musacchio A (2009) The life and miracles of kinetochores. EMBO J 28:2511–2531. doi:10.1038/emboj.2009.173
7. Egelhofer TA, Minoda A, Klugman S, Lee K, Kolasinska-Zwierz P, Alekseyenko AA, Cheung MS, Day DS, Gadel S, Gorchakov AA, Gu TT, Kharchenko PV, Kuan S, Latorre I, Linder-Basso D, Luu Y, Ngo Q, Perry M, Rechtsteiner

A, Riddle NC, Schwartz YB, Shanower GA, Vielle A, Ahringer J, Elgin SCR, Kuroda MI, Pirrotta V, Ren B, Strome S, Park PJ, Karpen GH, Hawkins RD, Lieb JD (2011) An assessment of histone-modification antibody quality. Nat Struct Mol Biol 18:91–93. doi:10.1038/Nsmb.1972

8. Sidoli S, Cheng L, Jensen ON (2012) Proteomics in chromatin biology and epigenetics: elucidation of post-translational modifications of histone proteins by mass spectrometry. J Proteomics 75:3419–3433. doi:10.1016/j.jprot.2011.12.029

9. Plazas-Mayorca MD, Zee BM, Young NL, Fingerman IM, LeRoy G, Briggs SD, Garcia BA (2009) One-pot shotgun quantitative mass spectrometry characterization of histones. J Proteome Res 8:5367–5374. doi:10.1021/Pr900777e

10. Sidoli S, Yuan ZF, Lin S, Karch K, Wang X, Bhanu N, Arnaudo AM, Britton LM, Cao XJ, Gonzales-Cope M, Han Y, Liu S, Molden RC, Wein S, Afjehi-Sadat L, Garcia BA (2015) Drawbacks in the use of unconventional hydrophobic anhydrides for histone derivatization in bottom-up proteomics PTM analysis. Proteomics 15:1459–1469. doi:10.1002/pmic.201400483

11. Sidoli S, Schwammle V, Ruminowicz C, Hansen TA, Wu X, Helin K, Jensen ON (2014) Middle-down hybrid chromatography/tandem mass spectrometry workflow for characterization of combinatorial post-translational modifications in histones. Proteomics 14:2200–2211. doi:10.1002/pmic.201400084

12. Schwammle V, Aspalter CM, Sidoli S, Jensen ON (2014) Large-scale analysis of co-existing post-translational modifications on histone tails reveals global fine-structure of crosstalk. Mol Cell Proteomics 13(7):1855–1865. doi:10.1074/mcp.O113.036335

13. Jung HR, Sidoli S, Haldbo S, Sprenger RR, Schwammle V, Pasini D, Helin K, Jensen ON (2013) Precision mapping of coexisting modifications in histone H3 tails from embryonic stem cells by ETD-MS/MS. Anal Chem 85:8232–8239. doi:10.1021/ac401299w

14. Vonholt C, Brandt WF, Greyling HJ, Lindsey GG, Retief JD, Rodrigues JD, Schwager S, Sewell BT (1989) Isolation and characterization of histones. Methods Enzymol 170:431–523

15. Zhang KL, Tang H, Huang L, Blankenship JW, Jones PR, Xiang F, Yau PM, Burlingame AL (2002) Identification of acetylation and methylation sites of histone H3 from chicken erythrocytes by high-accuracy matrix-assisted laser desorption ionization-time-of-flight, matrix-assisted laser desorption ionization-postsource decay, and nanoelectrospray ionization tandem mass spectrometry. Anal Biochem 306:259–269. doi:10.1006/abio.2002.5719

16. Jufvas A, Stralfors P, Vener AV (2011) Histone variants and their post-translational modifications in primary human fat cells. PLoS One 6, ARTN e15960. doi:10.1371/journal.pone.0015960

17. Bonaldi T, Imhof A, Regula JT (2004) A combination of different mass spectroscopic techniques for the analysis of dynamic changes of histone modifications. Proteomics 4:1382–1396. doi:10.1002/pmic.200300743

18. Zhao XL, Sidoli S, Wang LL, Wang WJ, Guo L, Jensen ON, Zheng L (2014) Comparative proteomic analysis of histone post-translational modifications upon ischemia/reperfusion-induced retinal injury. J Proteome Res 13:2175–2186. doi:10.1021/Pr500040a

Chapter 9

Production and Purification of Antibodies Against Histone Modifications

Benoit Guillemette, Ian Hammond-Martel, Hugo Wurtele, and Alain Verreault

Abstract

Antibodies that recognize specific histone modifications are invaluable tools to study chromatin structure and function. There are numerous commercially available antibodies that recognize a remarkable diversity of histone modifications. Unfortunately, many of them fail to work in certain applications or lack the high degree of specificity required of these reagents. The production of affinity-purified polyclonal antibodies against histone modifications demands a little effort but, in return, provides extremely valuable tools that overcome many of the concerns and limitations of commercial antibodies. We present a series of protocols and guidelines for the production and use of large amounts of polyclonal antibodies that recognize modifications of canonical histones. Our protocols can be applied to obtain antibodies that occur in histone variants and proteins other than histones. In addition, some of our protocols are compatible with the production of monoclonal or recombinant antibodies.

Key words Chromatin, Nucleosome, Histone modification, Antibodies, Polyclonal, Acetylation, Methylation

1 Introduction

Antibodies raised against specific histone posttranslational modifications (PTMs) or combinatorial patterns or PTMs are extremely valuable tools for studies of chromatin and epigenetics. Carefully validated PTM-specific antibodies are key for immunoblotting, immunohistochemistry, immunofluorescence, immunoprecipitation, and chromatin immunoprecipitation (ChIP). Despite remarkable advances in studies of histone PTMs by mass spectrometry [1], some of the aforementioned techniques cannot be performed without PTM-specific antibodies.

Antibodies raised against a wide range of histone PTMs are available from numerous commercial suppliers at a fairly expensive price for small amounts of antibody. Regrettably, a major proportion of commercial antibodies lack the high degree of specificity

Benoit Guillemette and Luc Gaudreau (eds.), *Histones: Methods and Protocols*, Methods in Molecular Biology, vol. 1528, DOI 10.1007/978-1-4939-6630-1_9, © Springer Science+Business Media New York 2017

required of these reagents [2–5]. A large number of these antibodies completely lack the desired PTM specificity or target residue specificity. For instance, several commercial antibodies that were claimed to bind specifically to histone PTMs turned out to recognize the wrong core histone or core histones expressed in *E. coli* [6], which are devoid of PTMs. In many cases, adequate specificity is observed in one lot of a given antibody, but not in others. This lot-to-lot variation in antibody quality curtails the ability of researchers to obtain reproducible results and the ability of different research teams to compare their results. Moreover, although commercially available antibodies are often claimed to work in several species and applications (e.g., immunoblotting, chromatin immunoprecipitation, and immunohistochemistry), their "degree of specificity" in each application is rarely tested by commercial suppliers. In some systems, such as mouse and human cells, the ideal control would be a cell line where the only source of a given histone carries a mutation of the residue that bears the PTM for which the antibody is supposedly specific. Unfortunately, because there are multiple copies of the genes encoding each of the mouse and human canonical histones, there is no such cell line currently available and, even if it were, some histone residues are likely essential for cell viability. Communal resources for systematic commercial histone antibody validation are available online (www.histoneantibodies.com/ [7]) but, because of the lack of human and mouse cell lines where histone residues are mutated, the information provided in these databases should be taken with extreme caution.

In our experience, monoclonal and polyclonal antibodies are never entirely specific. They only exhibit specificity within a range of experimental conditions that are often very narrow. Importantly, at least in our hands, this is the case for ChIP assays [8]. In addition, antibodies that are specific in immunoblots of whole-cell lysates often lose their specificity when used to probe immunoblots prepared from acid-extracted histones, where the histones are detectable by Ponceau S staining of the nitrocellulose or PVDF membrane. Commercial suppliers claim that many of their antibodies react against specific histone PTMs from several species despite the fact that appropriate specificity controls could not be performed. Because large differences in the abundance of specific PTMs can occur between species [9] or even between different cell types from a single species, specificity controls must be performed in the species or cell type in which a given antibody is intended to be used. Despite increasing general awareness about these issues, most laboratories are reluctant in investing time and money to perform specificity controls on commercial antibodies [10].

Another reason why antibodies present off-target binding is that canonical histones carry several residues that bear the same PTMs within similar sequence contexts. Some examples are shown in Fig. 1. There can be large differences in the abundance of PTM

A

	me1/me2/me3
H3 K9me	5-QT***ARKS***TG-12
H3 K27me	23-KA***ARKS***AP-30

B

	ac
H3 K4ac	1-ART***K***Q***T***A-7
H3 K9ac	6-TA***RKST***G-12
H3 K27ac	24-AA***RKS***AP-30
H3 K56ac	53-RFQ***KST***E-59

C

	ac
H4 K5ac	3-R***GKG***G-7
H4 K8ac	6-G***GKG***L-10
H4 K12ac	10-L***GKG***G-14

Fig. 1 Histone PTMs that reside within similar sequence contexts. (**a**) H3K9 and K27 can be mono-, di-, or tri-methylated. (**b**) H3K4, K9, K27, and K56 can be acetylated. (**c**) H4K5, K8, and K12 can be acetylated. Sequence similarities are highlighted in *bold italic*

that occur in related sequence contexts and, therefore, even a small degree of cross-reactivity towards the high-abundance modification can lead to misleading results. This is exemplified here by antibodies against histone H3K56 acetylation (H3K56ac). Even in yeast, where H3K56ac is abundant, cross-reactivity against other modifications can be observed in strains that lack the enzyme that adds the modification (Rtt109) or a strain where the residue is mutated (*H3-K56R* mutant, Fig. 2). All the commercial antibodies against H3K56ac that we tested cross-react against H3K9ac and other PTMs. Under our experimental conditions, the entire signal arises from cross-reaction, rather than H3K56ac (Fig. 2, lanes 1 and 2).

Although it is not feasible for academic research laboratories to raise a large number of antibodies that recognize a diverse set of histone PTMs or combinatorial patterns of PTMs (e.g., H3-K9 methylation followed by H3-S10 phosphorylation [11, 12]), we feel that the numerous concerns about commercial antibody specificity fully justify the preparation of one or a few PTM-specific antibodies that can then serve as reliable tools for long-term studies. We present protocols and practical considerations for the preparation, purification, and usage of high-quality polyclonal antibodies against histone PTMs.

Our protocol is applicable to any chemical modification or combinatorial patterns of modifications that can be incorporated

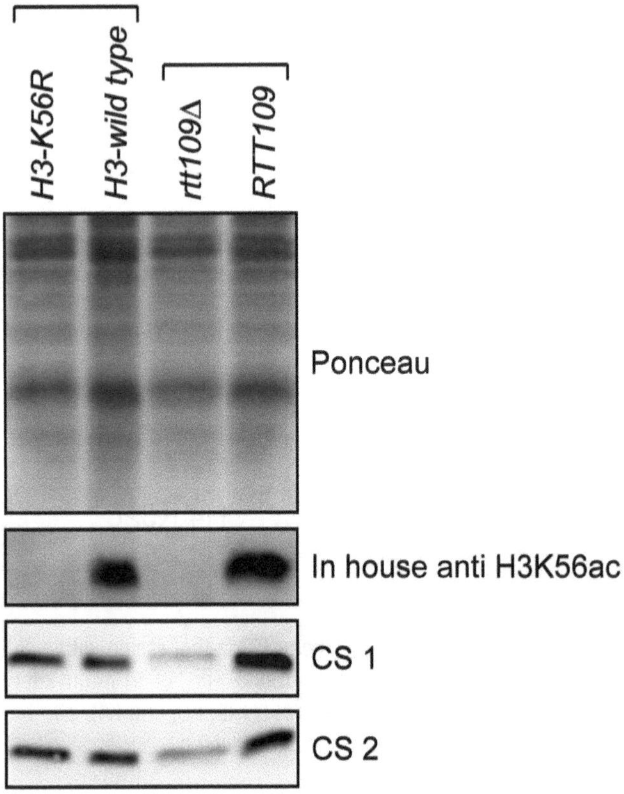

Fig. 2 Immunoblots to assess the specificity of antibodies directed against H3K56ac. *S. cerevisiae* whole-cell lysates from two types of mutant strains and their isogenic wild-type counterpart (indicated with *brackets*): a strain expressing an allele of H3 bearing the K56R mutation which cannot be acetylated (lane 1) and a strain in which *RTT109*, the gene encoding the enzyme that acetylates H3K56 in yeast, is deleted (lane 3). Immunoblots were performed with our in-house polyclonal antibody and two commercial sources of antibodies supposedly specific for H3K56ac. The experimental conditions for all three antibodies were identical

during chemical synthesis of short peptides. An important consideration is that several companies provide complete services for antibody production (peptide synthesis, animal immunization, and affinity purification of PTM-specific antibodies). However, the only assay for antibody sensitivity and specificity performed by these companies is an ELISA assay conducted with synthetic peptides. The companies often produce antibodies with very high titers in ELISA assays. Unfortunately, because ELISA assays are performed with amounts of synthetic peptides that are much higher than the in vivo abundance of the protein/PTM of interest, we often found that the commercial antibodies failed to work in our assays. Another limitation is that companies provide only 1–2 mg of antibody in the best of scenarios. In striking contrast, as much as 30 mg of specific antibodies can be affinity-purified from a single animal using our approach. This high yield of PTM-specific antibodies overcomes the extreme lot-to-lot variations in the specificity of commercial sources of antibodies. Hence, despite the fact that our procedure is more labor intensive than purchasing commercial antibodies, it is well worth investing in for histone PTMs that will be extensively studied over a long period of time.

2 Materials

1. Na phosphate-EDTA: 80 mM Na phosphate pH 8, 1 mM EDTA.
2. Ellman reagent solution: 4 mg/ml 5,5′-Dithio-bis-[2-nitrobenzoic acid] in 100 mM Na phosphate pH 8.0. Vortex hard and/or incubate at 37 °C to dissolve as much Ellman reagent as possible. Complete dissolution is hard to achieve and not necessary. However, it is important to clear the final Ellman reagent solution of non-dissolved chemical by low-speed centrifugation.
3. Spectrophotometer cuvettes.
4. Spectrophotometer.
5. Immobilized TCEP Disulfide Reducing Gel (ThermoFisher Scientific).
6. Lyophilized maleimide-activated mcKLH (*Megathura crenulata* keyhole limpet hemocyanin, ThermoFisher Scientific) (*Optional*).
7. Conjugation buffer: 10 mM Sodium phosphate pH 7.2, 150 mM NaCl, 10 mM EDTA, 50 mM sucrose.
8. Complete Freund's adjuvant.
9. Incomplete Freund's adjuvant.
10. New Zealand White rabbits.
11. Phosphate-buffered saline (PBS): 10 mM Na_2HPO_4, 2 mM KH_2PO_4, 137 mM NaCl, 2.7 mM KCl, pH 7.4.
12. SulfoLink Immobilization kit for peptides (ThermoFisher Scientific).
13. 50 ml Conical tubes.
14. 0.2-micron bottle-top filter.
15. 10 mM Tris–HCl pH 7.5.
16. 10 mM Tris–HCl pH 7.5, 0.02 % sodium azide (*see* **Note 1**).
17. PBS + 0.02 % sodium azide.
18. Gentle Ag/Ab Elution Buffer (ThermoFisher Scientific).
19. 0.1 M Glycine–HCl pH 2.5.
20. 1 M Tris–HCl pH 8.0.

21. Tris-buffered saline (TBS): 50 mM Tris–HCl pH 7.5, 150 mM NaCl.
22. 10 mg/ml BSA.
23. 6× SDS-PAGE loading buffer: 360 mM Tris–HCl pH 6.8, 47% v/v glycerol, 12% w/v SDS, 1.3 M 2-mercaptoethanol, 0.06% w/v bromophenol blue.
24. 15% SDS-PAGE resolving gel: 15% w/v Acrylamide:bis-acrylamide 29:1, 375 mM Tris–HCl pH 8.8, 0.1% w/v SDS, 0.1% w/v ammonium persulfate (APS), 0.1% v/v TEMED.
25. 6% SDS-PAGE stacking gel: 6% w/v Acrylamide:bis-acrylamide 29:1, 125 mM Tris–HCl pH 6.8, 0.1% w/v SDS, 0.1% w/v ammonium persulfate (APS), 0.1% v/v TEMED.
26. Running buffer: 25 mM Tris, 192 mM glycine, 0.1% w/v SDS.
27. SDS-PAGE apparatus.

3 Methods

3.1 Peptide Design

The peptide sequences that can be selected as antigen are restricted by the position of the posttranslational modification (PTM) against which one wishes to raise a specific antibody. In general, we order synthetic peptides that have the following structure: N-ter-Cys-X5-PTM-X5-C-ter, where X5 are the five residues on either side of the residue containing the target. For peptides that do not include the first residue of the target histone, the N-terminal amino group of the synthetic peptide is acetylated and the C-terminal carboxylic acid is amidated to "mimic" a peptide bond. Arguably, the N-terminus could be ordered unmodified to minimize the probability of obtaining antibodies against the N-terminal acetyl group, which may create a problem when the internal PTM is also an acetyl group. In spite of this, we were able to affinity purify antibodies against internal acetyl groups even when the synthetic peptide was also acetylated at the N-terminus.

In all eukaryotes where this question has been addressed, the most abundant/canonical forms of H2A, H2B, H3, and H4 all lack the N-terminal methionine and, with the notable exception of histone H3, their N-terminal amino groups are acetylated. Therefore, when designing peptides to raise antibodies against PTMs located near the N-termini of H2A, H2B, and H4 (e.g., H4S1 phosphorylation), but not H3, it is important to leave out the N-terminal methionine and order peptides in which the N-terminal amino groups are acetylated. For PTMs near the N-terminus of H3 (e.g., H3K4 methylation), our synthetic peptides lacked an N-terminal acetyl group and the cysteine was incorporated at the C-terminal end of the peptide. Although we have not investigated this possibility in a systematic manner, the cysteine can be inserted either at the N-terminal or the C-terminal end of the synthetic peptide for PTMs that are further away than six residues from the N-terminus of the peptide.

An important consideration is the number of amino acids on either side of the target PTM that should be included in the synthetic peptide(s) used for immunization and/or affinity purification of antibodies. This is important because epitopes recognized by antibodies can be very small. For instance, it has been possible to raise antibodies that recognized acetylated or methylated lysine residues, lysine residues linked via an isopeptide bond to a diglycine sequence (which occur at sites of ubiquitylation), or phosphorylated tyrosine, serine or threonine residues. In addition, as stated earlier, a number of PTMs that occur at different positions in the same core histone are flanked by related amino acid sequences (Fig. 1). In these cases, obtaining antibodies that are site specific requires that the immunogenic peptide contain residues beyond the regions that are identical at each position (*see* **Note 2**).

3.2 Ellman Assay to Determine the Molar Concentration of Free Thiol (-SH) Groups

The purpose of this protocol is to ensure that the fraction of peptide molecules containing a reduced cysteine is sufficient for cross-linking to maleimide-derivatized KLH (*see* **Note 3**). This is achieved by determining the molar concentration of peptides containing a reduced thiol group using the Ellman assay (*see* **Note 4**).

1. Dissolve the peptide at 5 or 10 mg/ml in H_2O (*see* **Note 5**).
2. After the peptide appears to be completely dissolved, centrifuge at 20,000 × *g* for 10 min at room temperature to get rid of small amounts of residual peptide that may not have been fully dissolved (*see* **Note 6**).
3. Use 5 μl of peptide stock solution. Add 963 μl of 80 mM Na phosphate pH 8.0–1 mM EDTA (*see* **Note 7**).
4. Add 32 μl of Ellman reagent solution to the peptide dilution and vortex. Incubate for 5 min at room temperature. Discard most of the supernatant.
5. Prepare a sample to "blank" the spectrophotometer at 410 nm by mixing 5 μl of water (instead of peptide stock solution), 963 μl of 80 mM Na phosphate pH 8.0–1 mM EDTA, and 32 μl of Ellman reagent solution. Read the optical density (OD) of your peptide dilution at 410 nm.
6. Based on the OD reading at 410 nm, you can calculate the molar concentration of your peptide–Cys-SH, using the Beer-Lambert equation:

$$\mathrm{OD} = \varepsilon \cdot c \cdot l$$

ε = molar extinction coefficient, in this case 13,600 M^{-1} cm^{-1}.

c = molar concentration (M).

l = length of the cuvette (cm).

The concentration of the peptide–Cys-SH stock solution is obtained by multiplying the molar concentration obtained above by the dilution factor of the peptide stock solution in the Ellman assay reaction, which is a 200-fold dilution when 5 μl of peptide stock solution is used in a 1 ml assay.

3.3 Conjugation of Peptide to KLH and Immunization

1. We typically immunize two rabbits simultaneously with the same peptide-KLH preparation. To prepare a single peptide-KLH antigen for multiple immunizations of two rabbits, dissolve 2 mg of lyophilized maleimide-activated KLH in 200 μl of conjugation buffer. The 10 mg/ml KLH solution should appear grey/bluish.
2. For peptides that contain 10–12 amino acid residues, dissolve 2 mg of peptide in 200–400 μl of H_2O (Subheading 3.2). Mix with 200 μl of 10 mg/ml KLH solution from **step 1** (*see* **Note 8**).
3. Incubate for 2 h at room temperature (*see* **Note 9**).
4. Dialyze the reaction mixture overnight at room temperature against 4 L of recently sterilized PBS (*see* **Note 10**).
5. Store the peptide-KLH conjugation at −20 °C or immediately proceed with the next step.
6. Collect an aliquot of pre-immune serum (1–10 ml per rabbit), which can serve as a valuable "negative" control when evaluating the production of antibodies that recognize the desired PTM.
7. Thaw out the peptide-KLH conjugate in preparation for rabbit immunization.
8. Immunize two New Zealand White rabbits (8-week-old females; *see* **Note 11**). For each animal, mix a volume of peptide-KLH equivalent to 250 μg KLH with 1 ml of "complete" Freund's adjuvant.
9. Inject this mixture at eight to ten subcutaneous locations.
10. Weeks 2 and 4: Prepare booster injections for each animal containing a volume of peptide-KLH equivalent to 150 μg KLH with 1 ml of "incomplete" Freund's adjuvant.
11. Week 5: Collect blood samples (1–10 ml test bleeds per rabbit).
12. Weeks 6, 8, and 10: Proceed with the third, fourth, and fifth booster injections as described in **step 10**.
13. Weeks 7, 9, and 11: Collect blood samples (1–10 ml test bleeds per rabbit) to monitor the production of antibodies (*see* **Note 12**).
14. Centrifuge the blood samples for 10 min at 2000 × *g* and 4 °C.
15. Recover the supernatant (serum) and add 0.02 % sodium azide to prevent microbial growth. Store some aliquots at 4 °C and

other aliquots at −20 °C to avoid exposing the antibodies to multiple cycles of freezing and thawing. We have kept large portions of our most precious antibodies at 4 °C for many years and they retain their activity. This is true for both crude antisera and affinity-purified antibodies.

3.4 Preparation of Affinity Purification Resin

This section contains only slight modifications of the protocols provided with the SulfoLink Immobilization Kit.

1. Dissolve peptides in water and ensure that most of the molecules contain a reduced cysteine (Subheading 3.2). In general, we prepare at least two peptide columns at the same time for each antibody that needs to be generated: one resin is covalently linked to the modified peptide and the other(s) is linked to the same peptide lacking the PTM of interest.
2. Resuspend the SulfoLink Resin by vortexing.
3. Remove the top cap followed by the bottom cap to avoid drawing air into the resin bed.
4. Place the column in a 15 ml conical tube and centrifuge the column at 1000 × *g* for 1 min at room temperature to remove the storage buffer.
5. Add 2 ml of coupling buffer and centrifuge as in the previous step. Repeat this step two more times (for a total of three washes).
6. Mix the peptide solution prepared in water (**step 1**) with an equal volume of conjugation buffer (*see* **Note 13**).
7. Insert the bottom cap and add the peptide solution to the SulfoLink column.
8. Insert the top cap and mix by rocking or end-over-end mixing at room temperature for 15 min.
9. Place the column upright and incubate at room temperature for 30 min without mixing.
10. Remove the top and bottom caps, place column into a fresh tube, and centrifuge at 1000 × *g* for 1 min at room temperature to collect the flow-through peptide fraction.
11. Save the flow through and determine the peptide coupling efficiency (*see* **Note 13**).
12. Add at least 2 ml of wash solution (supplied with the kit) to the column and centrifuge at 1000 × *g* for 1 min at room temperature. Repeat this wash three more times for a total of four total washes.
13. Add 2 ml of coupling buffer and centrifuge as above. Repeat this step once more for a total of two washes. Insert the bottom cap.
14. In order to block maleimide groups that did not react with peptide, prepare a solution 50 mM of cysteine–HCl (supplied in the kit) in 2 ml of coupling buffer. Apply the cysteine solution to the column and insert the top cap.

15. Mix by rocking or end-over-end mixing for 15 min at room temperature.
16. Incubate the column without mixing for 30 min.
17. Sequentially remove the top and bottom caps and allow the column to drain. At this point, the column is ready for affinity purification of antibodies (Subheading 3.5). Alternatively, the column can be stored for subsequent use by following **steps 18** and **19**.
18. For column storage, add 2 ml of PBS containing 0.02 % sodium azide (*see* **Note 1**) and centrifuge as above. Repeat this step three more times for a total of four washes.
19. Insert the bottom cap and add 2 ml of PBS containing 0.02 % sodium azide. Insert the top cap and store the column upright at 4 °C.

3.5 Affinity Purification

It is not always necessary or judicious to affinity purify antibodies because this may result in the loss or irreversible denaturation of high-affinity antibodies. We recommend testing the crude serum in the specific assays for which the antibody was developed to determine whether affinity purification is necessary.

1. Although this is not absolutely necessary, we often pool the crude sera derived from the positive test bleeds and the final bleed. This should amount to at least 30–50 ml of serum per rabbit.
2. Centrifuge the crude serum for 20 min at 4300 × *g* and 4 °C.
3. Recover the supernatant and transfer it to a fresh tube (*see* **Note 14**).
4. Filter through a 0.2 μm bottle-top filter in a sterile bottle.
5. Apply the filtered serum to the peptide affinity column (prepared in Subheading 3.4).
6. Let the serum flow through the column by gravity (*see* **Note 15**) and collect in a fresh tube.
7. Apply the flow through to the column a second time and let it flow by gravity.
8. Keep flow through at 4 °C in case your antibody did not bind to the peptide column.
9. Wash the column with 20 ml of PBS.
10. Wash the column with 20 ml of 10 mM Tris–HCl pH 7.5.
11. Elute with 10 ml of gentle Ag/Ab elution buffer (*see* **Note 16**).
12. Wash column with 20 ml of 10 mM Tris–HCl pH 7.5.
13. Elute with 10 ml of 0.1 M glycine–HCl pH 2.5 into a tube containing 1.5 ml of 1 M Tris–HCl pH 8.0 (*see* **Note 17**).
14. At this stage, it is often necessary to immunodeplete antibodies that bind to the SulfoLink resin covalently linked to the peptide lacking the PTM of interest (*see* **Note 18**).

15. Dialyze the antibodies eluted with gentle Ag/Ab elution buffer against two 4 L changes of either TBS or PBS at 4 °C.
16. Recover the resin covalently linked to the non-modified peptide by adding 10 mM Tris–HCl pH 7.5 to the SulfoLink column and transfer the beads and extra buffer to a conical tube.
17. Spin down the beads at 1000×*g* for 1 min at room temperature. Discard most of the supernatant.
18. To the beads, add the pool of antibody from which you wish to immunodeplete molecules that bind to the peptide lacking the PTM of interest (*see* **Note 19**).
19. In order to achieve extensive depletion of cross-reacting antibodies, incubate at room temperature for a long period of time (at least 12 h) by gentle rocking or using an end-over-end mixer.
20. Spin down the beads at 1000×*g* for 1 min at room temperature.
21. Resuspend the beads with the supernatant to form a slurry that can be readily transferred back into the empty SulfoLink column.
22. Remove the bottom cap and let the column flow by gravity to recover the eluate that contains your affinity-purified antibody pool from which antibodies that cross-react with the non-modified peptide have been depleted. You can proceed to **step 25**.
23. If necessary, the same SulfoLink resin can be immediately used (i.e., without further processing) for immunodepletion of cross-reacting antibodies from another pool of PTM-specific antibodies (*see* **Note 20**).
24. Storage of peptide columns: Wash the columns with 20 ml of 10 mM Tris–HCl pH 7.5 containing 0.02% sodium azide. Wrap the top and bottom caps in parafilm paper to prevent evaporation during long-term storage at 4 °C. Peptide columns can last for several years and can be recycled for multiple rounds of affinity purification or immunodepletion.
25. Add 0.02% sodium azide to prevent microbial growth. Store some aliquots at 4 °C and other aliquots at −20 °C to avoid having to expose the antibodies to multiple cycles of freezing and thawing. We have kept large portions of our most precious antibodies at 4 °C for many years and they retain their activity. This is true for both crude antisera and affinity-purified antibodies.
26. Run 30 μl aliquots through an SDS 15% polyacrylamide gel to assess the purity and concentration of your affinity-purified antibodies against a BSA standard curve (*see* **Note 21**).

4 Notes

1. Sodium azide is highly toxic; take precautions when handling it.
2. Because epitopes recognized by antibodies can be very small, obtaining antibodies with robust site specificity requires extremely careful immunodepletion of antibodies that cross-react against other residues that carry the same PTM in similar sequence contexts (Fig. 1). In those cases, obtaining site- and PTM-specific antibodies requires at least two rounds of immunodepletion (Subheading 3.5). The first is conducted with the non-modified peptide and the second with a peptide carrying the PTM present at the potentially cross-reacting position. For example, in order to obtain antibodies that are specific for H3K9ac, the first immunodepletion is carried out with a peptide containing non-modified H3K9 and the second is conducted with a peptide containing H3K27ac (Fig. 1b).
3. The N-terminal or C-terminal cysteine of the peptide must be reduced to form a covalent bond with the maleimide groups of KLH. Therefore the reduced state of the cysteine needs to be verified prior to incubation: this is best achieved using the Ellman assay (Subheading 3.2), which provides an estimate of the molar concentration of reduced thiol (-SH) groups. Although this is rarely necessary, thiol groups that were oxidized into disulfide bridges can be reactivated for conjugation to maleimide-activated KLH by incubation with a reducing gel composed of immobilized tris (2-chloroethyl) phosphine (TCEP) disulfide. Free TCEP, as opposed to immobilized TCEP, can also be used to reduce the peptide's cysteine residue; the TCEP needs to be removed by extensive dialysis against conjugation buffer prior to mixing the peptide with KLH. Because most peptides used to generate antigens are small (roughly 1–3 kDa), their retention during dialysis requires membranes with very low molecular weight cutoff (MWCO). Ultrafiltration would also require very low MWCO. Using dialysis, ultrafiltration or other related methods such as size exclusion (gel filtration), it is difficult to completely remove the TCEP under conditions where most of the reduced peptide can be recovered. Therefore, by far the simplest option is to reduce the peptide's cysteine with an immobilized TCEP disulfide reducing gel.
4. Small molecules with free thiol (-SH) groups that are components of numerous buffers, such as dithiothreitol or 2-mercaptoethanol, will interfere with the Ellman assay and must be removed before the peptide concentration can be determined using this assay. This is why we recommend dissolving the synthetic peptides in water.

5. Most histone peptides are sufficiently hydrophilic to be dissolved at high concentrations (>5 mg/ml) in pure water or aqueous buffers.
6. This step ensures that all the insoluble peptide molecules are precipitated. Otherwise, readings on the spectrophotometer are unreliable.
7. The total volume of the reaction must be 1 ml in order to obtain accurate readings with the spectrophotometer cuvettes that we use. If you suspect that your peptide solution is at a concentration considerably lower than 5 mg/ml (e.g., when the peptide cannot be readily dissolved in water), the volume of peptide stock can be increased beyond 5 μl. However, the total volume of the reagents (peptide, 80 mM Na phosphate-1 mM EDTA solution, and 32 μl Ellman reagent solution) must reach 1 ml.
8. This is sufficient for a first immunization of two rabbits (roughly 250 μg peptide-KLH per animal) and up to five booster injections (roughly 150 μg peptide-KLH per injection).
9. The turbid nature of the KLH solution is incompatible with monitoring the efficiency of peptide coupling to the KLH carrier protein using the Ellman assay. However, the molar ratio of maleimide groups per KLH monomer is >400 and the reaction efficiency of maleimides with reduced cysteines is very high. Therefore, as long as most of the peptide molecules initially contain a reduced cysteine, the final density of peptides covalently linked to KLH will be amply sufficient to serve as an efficient antigen for immunization.
10. The primary purpose of this dialysis step is to remove EDTA, an anticoagulant that should not be injected into animals. This step also removes peptide molecules that did not become covalently linked to KLH and would not be immunogenic. We prefer dialysis, rather than size-exclusion columns, for this step because, in a number of occasions, we observed that the peptide-KLH conjugates were semi-soluble. It would probably be impossible to remove EDTA without clogging up the size-exclusion columns.
11. Approval of any protocol using live animals is necessary. Enquire with your institution's animal care center to obtain an acceptable immunization protocol and all the necessary documents and authorizations. Alternatively, there are several companies that provide rabbit immunization and serum collection at a reasonable cost. There is a myriad of rabbit immunization protocols. Here we present a protocol that has worked well for production of antibodies against a number of different histone PTMs, but other immunization protocols may work just as well.

12. Blood samples (test bleeds) are collected every 2 weeks to monitor progression of the immune response. The immune response generally reaches its peak after two or three booster injection. Using our immunization protocol, we never observed the sudden emergence of a robust antigen-specific immune response after the third test bleed and, therefore, we never proceed beyond five booster injections. The terminal bleed (30–50 ml per animal) is generally collected after the fifth booster injection. However, it is unethical to pursue cycles of injection and test bleeds if the animals become ill.
13. Save 0.1 ml of peptide solution before applying it to the SulfoLink column. This is to determine the peptide coupling efficiency with the Ellman assay (Subheading 3.2). In general the coupling efficiency is very high (at least 80 %) provided that most of the peptide molecules contain a reduced cysteine. However, you can determine the coupling efficiency experimentally by comparing the concentration of the starting peptide solution with its concentration in the flow through the SulfoLink column.
14. You will probably notice a layer of lipids floating at the top of the serum after centrifugation. This is practically impossible to remove from the serum, but will not interfere with the protocol.
15. Alternatively, mix the resin with the serumin a 50 ml conical tube and incubate at room temperature for 30 min with gentle rocking. Turn the column upright and proceed with **step 6** (Subheading 3.5).
16. Although it is impossible to predict which conditions will result in elution of a significant portion of polyclonal antibodies against a specific antigen, several other elution buffers can be tested. These include non-denaturing high-salt buffers (5 M LiCl in 10 mM Na phosphate pH 7.2, 3.5 M $MgCl_2$ in 10 mM Na phosphate pH 7.2) or non-denaturing polyol buffers (50 % ethylene glycol pH 8.0 or 50 % propylene glycol pH 8.0). In addition, we occasionally included a denaturing high pH buffer (0.1 M triethylamine pH 11.5) after the acidic elution step (**step 13** in Subheading 3.5). As is the case for elution under acidic conditions, it is important to neutralize the pH of the basic eluate as soon as it emerges from the column by collecting the eluate into a tube containing 0.5 M Na phosphate pH 7.0. Prior to the purification, it is helpful to determine empirically the volume of 0.5 M Na phosphate pH 7.0 that is necessary to neutralize (pH 7.0–8.0) the volume of 0.1 M triethylamine–HCl pH 11.5 that will be used for elution. Using pH paper, make sure that the final solution of eluted antibodies is not too basic (pH 7.0–8.0). For the high-salt and polyol elution buffers, it is crucial to dialyze the antibody eluates

against a large volume (two 4 L changes) of either TBS or PBS to restore conditions where the antibodies can bind their cognate antigen.

17. Try to neutralize the antibody as soon as it emerges from the column by collecting the eluate into tubes that already contain 1 M Tris–HCl pH 8.0. Prior to the purification, it is helpful to determine empirically the volume of 1 M Tris–HCl pH 8.0 that is necessary to neutralize (pH 7.0–8.0) the volume of 0.1 M glycine–HCl pH 2.5 that will be used for elution. Using pH paper, make sure that the final solution of eluted antibodies is not too acidic (pH 7.0–8.0). Add more 1 M Tris–HCl pH 8.0 if necessary to reach a neutral pH.
18. It is important to monitor whether the immunodepletion step significantly enhances the specificity of the antibody for the PTM of interest. For this purpose, we save small aliquots of each antibody pool (e.g., the pool obtained with the gentle Ag/Ab buffer elution buffer and the pool obtained with the acidic elution buffer) prior to and following the immunodepletion step. These aliquots of antibody are then used to compare the sensitivity and specificity of the antibodies before and after the immunodepletion step using an assay for which a biologically relevant specificity control does exist. In yeast, this assay may for example consist in performing immunoblots or ChIP side by side with a lysate from a wild-type strain versus a lysate from a strain in which the PTM of interest cannot occur (*see* Fig. 2 for an example).
19. At this stage, a sensible course of action is to test the pools of antibodies obtained with the gentle Ag/Ab and the acidic elution buffers side by side to determine their respective sensitivity and specificity in assays such as those mentioned in **Note 18**. Based on the results, one may decide to perform the immunodepletion step on either pool of antibody or even possibly both. In the latter case, immunodepletion steps can be performed with two separate SulfoLink columns linked to the non-modified peptide. Alternatively, if the amount of non-modified peptide is limiting, it is possible to perform the two immunodepletions sequentially using a single SulfoLink column. In a few cases, the specificity assays performed prior to immunodepletion give rise to a high selectivity for the PTM of interest and, in those cases, immunodepletion of antibodies that recognize the non-modified residue is not even necessary.
20. This works because the number of non-modified peptide molecules covalently linked to the resin is in vast excess compared with the antibody molecules present in the pool that needs to be subjected to immunodepletion.

21. The final concentration of the affinity-purified antibodies is highly variable, but ranges from 10 ng/μl up to 300 ng/μl. Conditions for the use of the antibodies should be empirically optimized for each type of assay (e.g., immunoblotting, immunofluorescence) by preparing dilutions of the stock solution. Although there are considerable differences among antibodies, typically 1:500 to 1:10,000 dilutions are used for immunoblotting and 1:50 to 1:500 dilutions are used for immunofluorescence.

Acknowledgments

Research in A.V.'s laboratory is funded by the Canadian Institutes for Health Research (CIHR, grant MOP 125916). Research in H.W.'s laboratory is funded by grants from the Canadian Institutes of Health Research (MOP 123438), and Natural Sciences and Engineering Research Council of Canada (RGPIN 435636-2013).

References

1. Brumbaugh J, Phanstiel D, Coon JJ (2008) Unraveling the histone's potential: a proteomics perspective. Epigenetics 3:254–257
2. Egelhofer TA, Minoda A, Klugman S, Lee K, Kolasinska-Zwierz P, Alekseyenko AA et al (2010) An assessment of histone-modification antibody quality. Nat Struct Mol Biol 18:91–93
3. Fuchs SM, Strahl BD (2011) Antibody recognition of histone post-translational modifications: emerging issues and future prospects. Epigenomics 3:247–249
4. Nishikori S, Hattori T, Fuchs SM, Yasui N, Wojcik J, Koide A et al (2012) Broad ranges of affinity and specificity of anti-histone antibodies revealed by a quantitative peptide immunoprecipitation assay. J Mol Biol 424:391–399
5. Rothbart SB, Lin S, Britton LM, Krajewski K, Keogh MC, Garcia BA et al (2012) Polyacetylated chromatin signatures are preferred epitopes for site-specific histone H4 acetyl antibodies. Sci Rep 2:489
6. Tjeertes JV, Miller KM, Jackson SP (2009) Screen for DNA-damage-responsive histone modifications identifies H3K9Ac and H3K56Ac in human cells. EMBO J 28:1878–1889
7. Rothbart SB, Dickson BM, Raab JR, Grzybowski AT, Krajewski K, Guo AH et al (2015) An interactive database for the assessment of histone antibody specificity. Mol Cell 59:502–511
8. Guillemette B, Drogaris P, Lin HH, Armstrong H, Hiragami-Hamada K, Imhof A et al (2011) H3 lysine 4 is acetylated at active gene promoters and is regulated by H3 lysine 4 methylation. PLoS Genet 7, e1001354
9. Drogaris P, Villeneuve V, Pomies C, Lee EH, Bourdeau V, Bonneil E et al (2012) Histone deacetylase inhibitors globally enhance h3/h4 tail acetylation without affecting h3 lysine 56 acetylation. Sci Rep 2:220
10. Baker M (2015) Reproducibility crisis: blame it on the antibodies. Nature 521:274–276
11. Hirota T, Lipp JJ, Toh BH, Peters JM (2005) Histone H3 serine 10 phosphorylation by Aurora B causes HP1 dissociation from heterochromatin. Nature 438:1176–1180
12. Fischle W, Tseng BS, Dormann HL, Ueberheide BM, Garcia BA, Shabanowitz J et al (2005) Regulation of HP1-chromatin binding by histone H3 methylation and phosphorylation. Nature 438:1116–1122

Chapter 10

Immunofluorescence of Histone Proteins

Hilmar Strickfaden and Michael J. Hendzel

Abstract

The identification of a vast array of posttranslational modifications of histone proteins during cell cycle, repair, replication, and transcription has created the challenge of determining structure-function relationships for individual modifications and combinations of modifications. Some of this information can be gathered from indirect immunofluorescence, where the location and cell cycle relationships can be readily identified. Here we present an immunofluorescence protocol that is adapted for the use in histone modifications.

Key words Immunofluorescence, Histone modifications

1 Introduction

Immunofluorescence is a light microscopy technique that is used to specifically image populations of biomolecules in situ by combining the epitope specificity of mono- and polyclonal antibodies with fluorescent dyes that emit light of a defined range of wavelengths when excited with light within a specific range of shorter wavelength. Using different antibody/dye combinations for staining different epitopes allows mapping the distribution of different structures simultaneously. This technique can assist in the identification of potential functions for proteins or modified proteins. For example, histone H2A.X is phosphorylated in response to DNA damage and histone H2A.X that is phosphorylated on serine 139 in response to DNA double-strand breaks has a characteristic staining pattern. In contrast to the very successful and popular use of fluorescent proteins, which are suitable for imaging epigenetic machinery and histone proteins, most posttranslational modifications of histones must be visualized by antibodies that specifically bind to the modification site on the histone proteins. Hence, immunofluorescence remains an important technique for the study of histone posttranslational modifications. Moreover, with the advent of "super-resolution microscopy" [1], immunofluorescent

Benoit Guillemette and Luc Gaudreau (eds.), *Histones: Methods and Protocols*, Methods in Molecular Biology, vol. 1528,
DOI 10.1007/978-1-4939-6630-1_10, © Springer Science+Business Media New York 2017

staining of modified histone proteins may begin to reveal the relationship between the presence or absence of a histone posttranslational modification and the fine structure of the chromatin that contains/does not contain the modification. A separate advantage of imaging histone posttranslational modifications by immunofluorescence is that quantification of the fluorescence can reveal information about heterogeneity on the single-cell level.

2 Materials

1. 1× PBS, pH 7.4: NaCl 137 mM, KCl 2.7 mM, Na_2HPO_4 10 mM, KH_2PO_4 1.8 mM. Adjust pH with HCl to 7.4.
2. Cells (this protocol is optimized for adherent mammalian cells) (*see* **Note 1**).
3. 4% Paraformaldehyde in 1× PBS (*see* **Notes 2, 3** and **11**).
4. 20 mM Glycine in 1× PBS.
5. 0.4% Triton X-100 in 1× PBS.
6. PBST: 1× PBS, 0.05% Tween 20.
7. 4% Bovine serum albumin (BSA) dissolved in 1× PBS.
8. Antibodies (primary, secondary) (e.g., Millipore 05-636|anti-phospho-histone H2A.X (Ser139) antibody, clone JBW30 (dilution 1:500)).
9. Counterstain solution: 1 mg/ml DAPI in H_2O.
10. Antifade: Vectashield or equivalent (Vectashield, Vectorlabs).
11. Lint-free tissue paper: Kimwipes or equivalent.
12. Coverslips: #1.5 for use with immersion objectives.
13. Microscopy glass slides.
14. Nail polish.
15. Humid chamber.
16. Tweezers.
17. Parafilm.

3 Methods

1. Grow cells on sterile coverslips in 35 mm petri dishes or 6-well plates in the recommended culturing medium for the respective cell line (*see* **Note 1**). Make sure that the coverslips have the appropriate thickness for the use of oil immersion objective lenses (170 μm or thickness #1.5). For mammalian cells culturing conditions are in general 37 °C in a humidity-saturated atmosphere containing 5% CO_2.

2. When the cells have reached the desired density (e.g., 80%), fix for 10 min in 1× PBS containing 4% paraformaldehyde. Make sure that you use freshly prepared paraformaldehyde (*see* **Notes 2** and **3**). In some cases, the primary antibody will only recognize its epitope when alcohol or acetone fixation is used. In this case, the fine structure of the chromatin will not be preserved but the coarse structure and the cell cycle relationships will be maintained. It is preferable to fix with paraformaldehyde when staining for chromatin-associated proteins or modified proteins.
3. After fixation, wash the cells three times in 1× PBST at room temperature on a rotating platform and then let them sit for 5–10 min in the same buffer containing 20 mM glycine. The glycine will react with aldehyde groups that have not reacted. Reactive aldehyde groups have the potential to cross-link the antibody to the cell independent of the presence of the epitope.
4. Permeabilize the cells by incubating in 1× PBS with 0.4% Triton X-100 for 15 min at room temperature on a rotating platform in order to wash away the membranes and thus make the cell and cell nucleus accessible to the antibodies.
5. Wash three times in 1× PBST at room temperature on a rotating platform.
6. Block the cells for 10–20 min in 4% BSA dissolved in 1× PBST to prevent nonspecific binding of the antibodies. Blocking is not necessary for many antibody preparations; however it is recommended if high-quality samples have to be prepared for quantitative analysis or for super-resolution light microscopy (*see* **Notes 4** and **5**).
7. Prepare the antibody solution for the primary antibody or antibodies by diluting in 1× PBST 4% BSA according to the manufacturer's recommendation. When antibodies for different species are being used to stain the same cells, they can be mixed together and stained simultaneously. Usually working concentrations for antibodies range between 0.5 and 5 μg/ml. If you are not sure what concentration to use, a 1:200 dilution of a 1 mg/ml antibody solution or directly from immune sera is a good starting point (*see* **Notes 6** and 7). If the antibody produces a good signal at 1:200, increase the dilution until the signal begins to decrease. The objective is to use the antibody at the highest dilution possible that still provides good signal. The higher the dilution used, the less likely cross-reactivity to related epitopes contributes to the signal. Make enough of the primary antibody solution to make sure that it can completely cover the cells and will not dry out over the planned incubation period. Using our technique of staining the coverslip inverted onto a drop of antibody solution placed on Parafilm, approximately 20 μl of antibody for an 18 mm^2 coverslip is sufficient.

8. Place a 20 μl drop of antibody solution onto a piece of Parafilm and place the coverslip with the cells upside down onto the drop so that the primary antibody solution completely covers the cells. Avoid air bubbles by placing the coverslip down on one edge first and then gradually lowering the coverslip while avoiding trapping any air bubbles (*see* **Note 8**).
9. Incubate with the primary antibody(s) for 1 h at room temperature. Alternatively 0.5 h at 37 °C or overnight at 4 °C may also be used.
10. Wash away the primary antibody away by washing three times for approximately 5 min with 1× PBST at room temperature on a rotating platform.
11. Prepare the secondary antibody solution by diluting the secondary antibody according to the recommended or experimentally determined ratio in 1× PBST with 4% BSA. Make sure that the chosen secondary antibody exclusively detects the species of the primary antibody (*see* **Notes 9** and **10**). Spin down the stock solution at high speed in a table centrifuge to avoid precipitates of antibodies in your working solution.
12. Spot about 20 μl onto a Parafilm and cover the drop with the coverslip facing the cells. Avoid air bubbles and avoid exposure to bright light to protect the fluorochromes. Incubate the secondary antibody for 1 h at room temperature (alternatively, 0.5 h at 37 °C or even overnight at 4 °C could be used) (*see* **Note 12**).
13. Wash away the secondary antibody three times for approximately 5 min with 1× PBST at room temperature on a rotating platform.
14. Spot about 10 μl of 1 μg/ml DAPI onto a Parafilm and cover the drop with the coverslip facing the cells. Incubate for 5 min (wear gloves!). Wash away excess of the DAPI three times for approximately 5 min with 1× PBST. Alternatively, DAPI can be added directly to the embedding medium. Since DAPI only fluoresces when bound to nucleic acids, this method has the advantage that you save one step and can provide more consistent staining with DAPI.
15. Rinse with water to prevent salt crystal formation on the upper surface of the coverslip.
16. Spot approximately 20 μl of the embedding medium or antifade onto a microscopic slide and carefully place the coverslip with the cells facing down onto the slide to prevent bubbles.
17. Seal with nail polish to fix the position of the coverslip and prevent the coverslip from sliding across the glass slide, which can damage the cells on the coverslip and destroy the morphology. If nail polish is not permitted for use with the microscopes that you are using, press the coverslip down firmly, removing as much excess mounting media as possible, and move the slide with care when on the microscope. It is possible to use glycerol-based

antifade solutions without sealing the coverslips while not damaging the specimen during visualization in the microscope as long as care is taken when moving the microscope stage.

18. If necessary carefully wipe away salt crystals that form at the surface with a lint-free tissue paper that was dipped in 70% ethanol, avoiding the areas that were sealed with nail polish.
19. Store the slides in the dark at 4 °C.

4 Notes

1. This protocol focuses mainly on cells that grow adherently on coverslips (*see* Subheading 2, **item 1**). In order to immunostain cells that grow in suspension, you can perform all preparation steps in a centrifuge tube, spinning down the cells after every step using the same speed necessary to pellet intact cells while preserving their viability and keeping the pellet until you place them between a coverslip and a microscopic slide [2]. Other ways of using immunofluorescence with cells in suspension are to immobilize and grow cells on coverslips that are coated with poly-L-lysine [3]. A third method, that is not recommended if the 3D structure of a nucleus is of interest, is to stick the cells to a coverslip using centrifugation with a centrifuge designed for this purpose (Cytospin) [4].
2. If the antigen is denatured by aldehydes switch to a different fixing method (e.g., methanol or acetone) (*see* **step 2**). Organic fixatives do not preserve the ultrastructure of chromatin and organic fixation is therefore not suitable for super-resolution fluorescence microscopy.
3. Some antibodies need a methanol treatment of the cells to expose the epitopes that are recognized by the antibody. If this is the case treat the cells with 100% methanol for 5–10 min following paraformaldehyde fixation in place of the permeabilization (*see* **step 2**).
4. For many sample preparations the use of blocking buffer like 1× PBS with 4% BSA is not required and can be omitted for the sake of speed and simplicity. Initially, a control experiment should be performed comparing blocking and non-blocking conditions. If the results are identical, the blocking step is unnecessary (*see* **step 6**).
5. If you want to detect epitopes that include a phosphorylated site do not use dry milk powder as a blocking agent, since that contains casein, which is a phosphoprotein and thus can produce undesired background (*see* **step 6**).
6. To test for antibody specificity towards specific modified histone proteins, experiments with blocking peptides mimicking the epitope of interest should be carried out. The primary

antibodies should be incubated with a 100-fold molar excess of the respective peptide and then be added to the cells [5]. If the primary antibody in question is specific and only detects the epitope that is also present in the peptide, there should be no staining present in the nucleus (*see* **step** 7).

7. If cell lines are available that are null for the enzyme(s) responsible for the modification or the modification is known to be absent from a cell line, this is the optimal control for verifying the specificity of the antibody (*see* **step** 7).
8. It is essential that the cells do not dry out at any point of the protocol as this leads to artifacts in both staining and cell morphology. This is easiest to prevent by staining the coverslips inverted on Parafilm. It is possible that some mitotic cells are displaced using this approach. If the objective is to quantify the cells in mitosis, the coverslips can be stained with the cell surface facing upwards but it is essential to maintain a fully humidified staining environment (*see* **step 8**).
9. Select the fluorophores that are attached to the secondary antibodies carefully to optimize for the filter sets you have at your microscope. This will enable proper spectral separation of the signals representing one epitope and to avoid "bleed through". For multicolor fluorescence microscopy, this typically requires the use of band-pass filters rather than long-pass filters, which transmit any wavelength of light greater than the cutoff point of the filter (*see* **step 11**).
10. If you consider using super-resolution microscopy, carefully select antibodies carrying appropriate fluorochromes [6] (*see* **step 11**).
11. It is recommended to produce the buffered paraformaldehyde solution yourself, since the commercially available formaldehyde that can be bought contains ca. 10% methanol, which could precipitate proteins and thus change the structure of the biomolecules to be detected (*see* Subheading 2).
12. It is possible to use isotype-specific secondary antibodies for monoclonal antibodies that differ by subtype. In each case, a secondary antibody control containing no primary antibody should be included as a control until the reactivity of the secondary antibody is firmly established.

Acknowledgements

Hilmar Strickfaden holds a postdoctoral fellowship by the Alberta Cancer foundation and was supported by the Bayerische Forschungsallianz. This work was supported by grants from the Canadian Institutes of Health Research and Alberta Cancer Foundation.

References

1. Markaki Y, Gunkel M, Schermelleh L, Beichmanis S, Neumann J, Heidemann M, Leonhardt H, Eick D, Cremer C, Cremer T (2011) Functional nuclear organization of transcription and DNA replication: a topographical marriage between chromatin domains and the interchromatin compartment. Cold Spring Harb Symp Quant Biol 75:475–492. doi:10.1101/sqb.2010.75.042
2. Wang CC, Bajikar SS, Jamal L, Atkins KA, Janes KA (2014) A time- and matrix-dependent TGFBR3-JUND-KRT5 regulatory circuit in single breast epithelial cells and basal-like premalignancies. Nat Cell Biol 16:345–356
3. Mazia D, Schatten G, Sale W (1975) Adhesion of cells to surfaces coated with polylysine. Applications to electron microscopy. J Cell Biol 66:198–200
4. Koh CM (2013) Preparation of cells for microscopy using cytospin. Methods Enzymol 533:235–240
5. Zinner R, Teller K, Versteeg R, Cremer T, Cremer M (2007) Biochemistry meets nuclear architecture: multicolor immuno-FISH for colocalization analysis of chromosome segments and differentially expressed gene loci with various histone methylations. Adv Enzyme Regul 47:223–241
6. Chozinski TJ, Gagnon LA, Vaughan JC (2014) Twinkle, twinkle little star: photoswitchable fluorophores for super-resolution imaging. FEBS Lett 588:3603–3612

Chapter 11

Acid-Urea Gel Electrophoresis and Western Blotting of Histones

Catherine A. Hazzalin and Louis C. Mahadevan

Abstract

Acid-urea gel electrophoresis offers significant advantages over SDS-PAGE for analysis of post-translational protein modifications, being capable of resolving proteins of similar size but varying in charge. Hence, it can be used to separate protein variants with small charge-altering differences in primary sequence, and is particularly useful in the analysis of histones whose charge variation arises from post-translational modification, such as phosphorylation or acetylation. On acid-urea gels, histones that carry multiple modifications, each with a characteristic charge, are resolved into distinct bands, the so-called "histone ladder." Thus, the extent and distribution of different modification states of histones can be visualized. Here, we describe the analysis of histone H3 by acid-urea gel electrophoresis and western blotting.

Key words Acid-urea (AU) gel electrophoresis, Histone extraction, Western blotting, Histone ladder, Acetylation, Phosphorylation, Histone H3

1 Introduction

AU gel electrophoresis of histones was originally described by Panyim and Chalkley in ref. [1]. This method has been widely used to analyze the post-translational modification of histones in their amino-terminal tail ([1, 2] and references therein). The acetic acid in the gel system reduces the pH to approximately 3.0, at which pH most proteins will be positively charged and migrate towards the cathode (−). Urea acts as a protein denaturant and is required for enhanced resolution. The net charge of a protein is determined by amino acid sequence and modifications carried on specific residues. Acetylation of lysine neutralizes its positive charge [3] and references therein (Fig. 1a). Phosphorylation of serine/threonine or tyrosine residues adds negative charge [4] and references therein (Fig. 1b). Each such modification causes an incremental reduction in the net positive charge of histones, which reduces the mobility of the protein in AU gels. The technique is widely applicable to the analysis of any protein that is modified by acetylation or

Benoit Guillemette and Luc Gaudreau (eds.), *Histones: Methods and Protocols*, Methods in Molecular Biology, vol. 1528, DOI 10.1007/978-1-4939-6630-1_11, © Springer Science+Business Media New York 2017

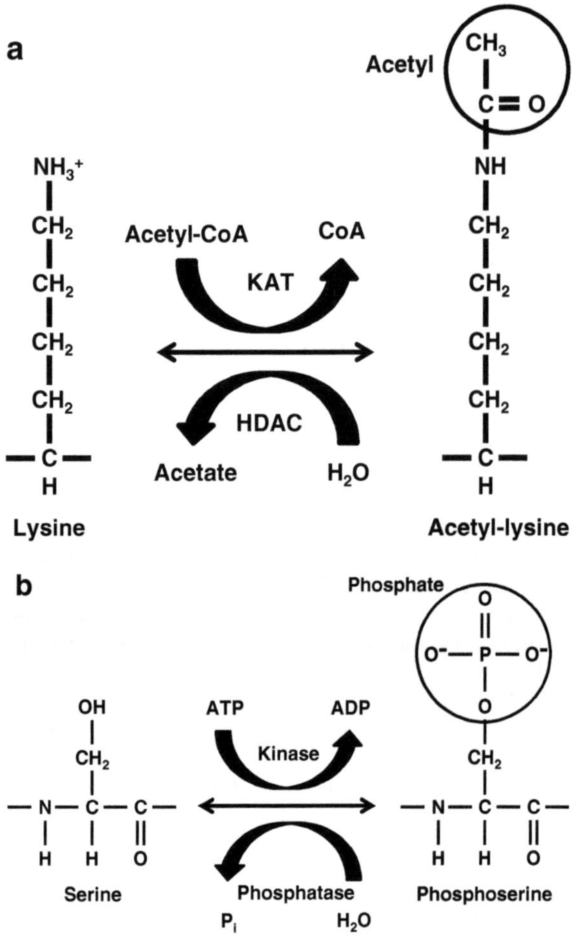

Fig. 1 Effect of charge on protein mobility: acetylation and phosphorylation reactions. (**a**) Lysine acetylation reaction. Acetylation of lysine residues on histone H3 tails is a dynamic process resulting from the continuous action of lysine acetyl transferases (KATs) and histone deacetylases (HDACs). HDAC inhibitors can be used to investigate this response. (**b**) Serine phosphorylation reaction. Phosphorylation of serine residues is regulated by activating kinases and the opposing action of phosphatases. Phosphorylation at serine residues in histone H3 is stimulated by activators of MAP kinase cascades such as anisomycin, 12-0-Tetradecanoylphorbol 13-acetate (TPA), and Epidermal Growth Factor (EGF)

phosphorylation. Here, we describe our laboratory techniques optimised for electrophoretic separation of specific modification states of histone H3 carrying characteristic post-translational modifications. Histone H3 tails contain a number of lysine and serine residues that undergo dynamic acetylation and phosphorylation respectively (Fig. 2a). Each additional acetyl or phosphate modification causes an incremental change in mobility producing up to six distinct H3 bands ("histone ladder"; Fig. 2b). Using the AU gels described here, acetylated histone H4 is also resolved into a

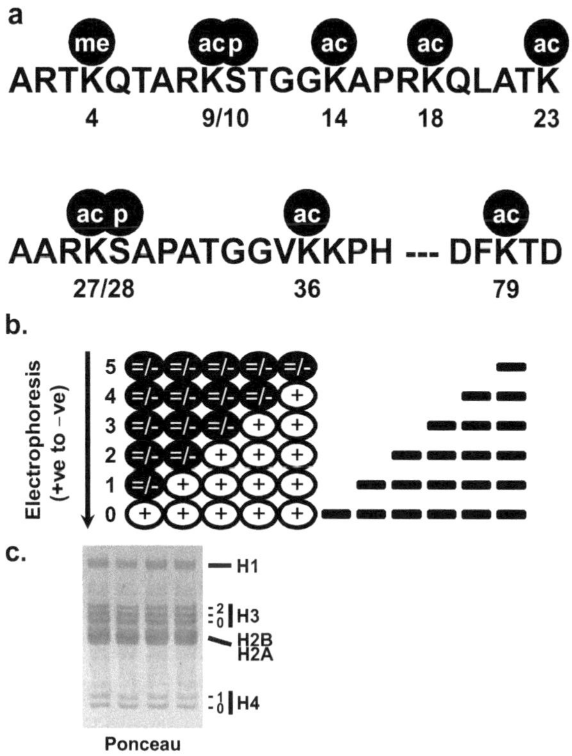

Fig. 2 Histone H3 residues modified by phosphorylation and acetylation: the "histone ladder". (**a**). Many residues in histone H3 tails are specifically targeted for post-translational modification. Shown here are the main sites in histone H3 tails associated with dynamic acetylation and phosphorylation. The lysine 4 methylation site is also shown, as histone H3 tails carrying the tri-methyl modification at this position become highly acetylated at other lysines on treatment with HDAC inhibitors (*see* Fig. 3a and ref. 2). Note that modification of lysine residues by methylation does not affect net charge. Some sites can be modified by either acetylation or methylation (not discussed here). For reviews on histone H3 post-translational modifications *see* ref. [3]. (**b**). Schematic showing how the histone H3 ladder is produced during AU gel electrophoresis, where proteins are separated by both size and charge. At low pH unmodified histone H3 (band 0) is positively charged (+++++). Modification of lysine residues by acetylation neutralizes the positive charge (=), while phosphorylation of serine residues adds a negative charge (–). Modification of a single residue causes a net reduction in the overall charge of the protein, resulting in reduced mobility of histone H3. As additional residues become modified, and histone H3 becomes less positively charged, this causes incremental reductions in mobility. For example: histone H3 in band 1 carries a single modification (acetylation or phosphorylation), in band 2, H3 carries two modifications (both acetylation, both phosphorylation, or one of each) and so on. (**c**). Acid-soluble proteins were extracted from nuclear pellets for analysis on AU gels. Ponceau staining of the membrane shows the resolution of histones in this gel system. Bands 0, 1, and 2 (Histone H3) and bands 0 and 1 (H4) forming the histone ladder for each protein are clearly resolved. Histone H1, H2B, and H2A are also indicated

ladder (bands 0 and 1 are visible following Ponceau staining of membrane) and H1 into a discrete band (Fig. 2c). To improve the separation of histones H2B and H2A, the gel running conditions can be modified (*see* Subheading 2.2, **item 3** and [5]).

2 Materials

All solutions are prepared using high quality demineralized or ultrapure water (water purified to a resistivity of 18 MΩ.cm). We routinely use demineralized water for all solutions described here. All reagents are stored at room temperature unless indicated. Dispose of acrylamide waste according to local regulations. Appropriate safety measures should be taken when handling toxic protease and phosphatase inhibitors, particularly when in powder form. All reagents should be analytical grade or equivalent.

2.1 Sample Preparation

1. Protein samples were prepared from C3H10T½ mouse embryo cells grown in culture. Cell culture dishes: 90 mm (*see* **Note 1**).
2. Cell culture growth medium: Dulbecco's Modified Eagle Medium (DMEM, high glucose, GlutaMAX™, pyruvate).
3. Fetal bovine serum (FBS).
4. 10% FBS: Cells are grown to confluence (~3 days) in DMEM containing 10% FBS (v/v).
5. 0.5% FBS: Cells are quiesced/serum-starved overnight (~16–18 h) in DMEM containing 0.5% FBS (v/v).
6. 1.5 ml microfuge tubes.
7. Sodium hydroxide (NaOH): NaOH pellets.
8. Magnetic stirrer.
9. 1 M Hepes pH 8.0: Mix 23.83 g of Hepes in 80 ml of water. Stir on a magnetic stirrer. Adjust pH to 8.0 with NaOH pellets. Add water to a final volume of 100 ml. Sterilize by autoclaving or filtration using a 0.2 μm filter.
10. 0.5 M EDTA pH 8.0: Mix 18.61 g of ethylenediaminetetraacetic acid disodium salt in 80 ml of water. Stir on a magnetic stirrer. Adjust pH to 8.0 with NaOH pellets. Add water to a final volume of 100 ml. Sterilize by autoclaving or filtration using a 0.2 μm filter (*see* **Note 2**).
11. 0.5 M EGTA pH 8.0: Mix 19.02 g of EGTA in 80 ml of water. Stir on a magnetic stirrer. Adjust pH to 8.0 with NaOH pellets. Add water to a final volume of 100 ml. Sterilize by autoclaving or filtration using a 0.2 μm filter (*see* **Note 3**).
12. Tube rotator (*see* **Note 4**).
13. 200 mM sodium fluoride (NaF): Weigh 0.42 g of NaF into a 50 ml tube. Add 40 ml of water and mix on a tube rotator until

dissolved. Adjust volume to 50 ml with water. Store at 4 °C (*see* **Note 5**).

14. Glycerol: 50% solution in water. Pour 25 ml of glycerol into a 50 ml tube. Adjust volume to 50 ml with water and mix on a tube rotator until solution is clear (*see* **Note 6**).
15. 1 M dithiothreitol (DTT): Weigh 2.31 g of DTT into a 15 ml tube. Add 14 ml of water and mix on a tube rotator until dissolved. Adjust final volume to 15 ml with water. Dispense 1 ml aliquots into microfuge tubes and store at −20 °C in the dark (*see* **Note 7**).
16. 5 M sodium chloride (NaCl): Dissolve 29.22 g NaCl in 80 ml of water. Adjust volume to 100 ml with water. Sterilize by autoclaving or filtration using a 0.2 μm filter.
17. Triton X-100: 10% (v/v) solution in water. Pour 5 ml of Triton X-100 into a 50 ml tube using gradations on the side of the tube. Adjust volume to 50 ml with water and mix on a tube rotator until solution is clear (*see* **Note 8**).
18. Protease Inhibitor Cocktail A: 100× stock—Aprotinin (200 μg/ml); Leupeptin (500 μg/ml); Trypsin inhibitor from *Glycine max* (soybean), (5 mg/ml); N_α-Tosyl-L-lysine-chloromethyl ketone hydrochloride (TLCK, 5 mg/ml). Weigh 250 mg of Trypsin inhibitor and 250 mg of TLCK into a 50 ml tube and add water to 20 ml. Add water to a 10 mg bottle of Aprotinin and a 25 mg bottle of Leupeptin, mix and leave to dissolve for 3–5 min. Carefully transfer dissolved Aprotinin and Leupeptin to the 50 ml tube containing Trypsin inhibitor and TLCK. Rinse chemical bottles with water and add to inhibitor cocktail in 50 ml tube. Adjust volume to 50 ml with water and mix on a tube rotator until dissolved. Dispense 1 ml aliquots into microfuge tubes. Store at −20 °C (*see* **Note 9**).
19. Ethanol: 100% ethanol.
20. Protease Inhibitor Cocktail B: 100× stock—N-*p*-Tosyl-L-phenylalanine chloromethyl ketone (TPCK, 10 mg/ml); Phenylmethanesulfonyl fluoride (PMSF, 10 mg/ml). Weigh 500 mg of TPCK and 500 mg of PMSF into a 50 ml tube. Adjust volume to 50 ml with ethanol and mix on a tube rotator until dissolved. Dispense 1 ml aliquots into microfuge tubes. Store at −20 °C (*see* **Notes 9** and **10**).
21. 1 M NaOH: Weigh 2 g of NaOH into a 50 ml tube. Add water to 40 ml and mix on tube rotator until dissolved. Adjust volume to 50 ml with water.
22. 1 M HCl: Prepare 50 ml of HCl from concentrated HCl (*see* **Note 11**).
23. 200 mM sodium orthovanadate (Na_3VO_4): Dissolve 3.68 g of Na_3VO_4 in 90 ml of water. Adjust volume to 100 ml with

water. (1) Depending on the pH of the solution, adjust to pH 10.0 using either 1 M NaOH or 1 M HCl (*see* **Note 12**). At pH 10.0 the solution is yellow. (2) Boil the solution until it turns clear, then cool to room temperature. (3) Readjust to pH 10.0 and repeat boiling step until the solution turns clear again. Repeat **steps (1)—(3)** until the pH stabilizes at ~pH 10.0 and the solution remains clear. Usually two to three repetitions are required (*see* **Note 13**).

24. 1 M β-glycerophosphate: Weigh 10.80 g of β-glycerophosphate disodium salt hydrate into a 50 ml tube. Add 40 ml of water and mix on a tube rotator until dissolved. Adjust volume to 50 ml with water. Sterilize by filtration using a 0.2 μm filter and store at 4 °C (*see* **Note 14**).
25. Dimethyl sulfoxide (DMSO).
26. 1 mM microcystin: Add 1.0 ml of DMSO to 1 mg of microcystin-LR. Recap vial and leave for 3–5 min at room temperature to dissolve, then vortex. Dispense 50 μl aliquots into microfuge tubes and store at −20 °C (*see* **Note 15**).
27. 1 M sodium butyrate; Weigh 5.50 g of sodium butyrate into a 50 ml tube. Add 40 ml of water and mix on a tube rotator until dissolved. Adjust volume to 50 ml with water. Store at 4 °C (*see* **Note 16**).
28. Lysis buffer: 20 mM Hepes pH 8.0, 5 mM EDTA pH 8.0, 10 mM EGTA pH 8.0, 5 mM NaF, 10% Glycerol, 1 mM DTT, 400 mM NaCl, 0.4% Triton X-100, 1× Protease Inhibitor Cocktail A, 1× Protease Inhibitor Cocktail B, 1 mM sodium orthovanadate, 20 mM β-glycerophosphate, 1 μM microcystin, 10 mM sodium butyrate (*see* **Note 17**).
29. Acid-extraction buffer: 0.4 M HCl. Prepare 50 ml of 0.4 M HCl from concentrated HCl (*see* **Note 11**).
30. Acetone: ice-cold 100% acetone (*see* **Note 18**).
31. Urea: Make solutions containing urea just before use or freeze in small aliquots (*see* **Note 19**).
32. 1 M Tris–HCl pH 7.5: Mix 12.11 g of Tris base in 80 ml of water. Stir on a magnetic stirrer. Adjust pH to 7.5 with concentrated HCl. Add water to a final volume of 100 ml. Sterilize by autoclaving or filtration using a 0.2 μm filter.
33. 2-Mercaptoethanol.
34. Glacial acetic acid: Anhydrous (100%) acetic acid.
35. AU sample loading buffer: 8 M Urea, 10 mM Tris–HCl pH 7.5, 5% 2-Mercaptoethanol, 5% glacial acetic acid. Weigh 0.48 g of urea into a 1.5 ml microfuge tube, add water to 800 μl, 10 μl of 1 M Tris–HCl pH 7.5 and mix on tube rotator for 5–10 min. Add 50 μl of 2-mercaptoethanol, 50 μl of glacial

acetic acid. Mix solution and adjust volume to 1 ml with water. Make fresh sample buffer just before use (*see* **Note 20**).

36. Methyl Green loading dye: 8 M Urea, 10 mM Tris–HCl pH 7.5, 5% glacial acetic acid solution. Weigh 0.48 g of urea into a 1.5 ml microfuge tube, add water to 800 μl, 10 μl of 1 M Tris–HCl pH 7.5 and mix on tube rotator for 5–10 min. Add 50 μl of glacial acetic acid. Mix solution and adjust volume to 1 ml with water. Dissolve a small amount of methyl green powder in this solution, until desired green color is achieved. Store small aliquots at −20 °C to prevent cyanate production (*see* **Note 21**).

2.2 Acid-Urea Polyacrylamide Gel Electrophoresis

1. Tube rotator.
2. Disposable 15/50 ml polypropylene tubes (*see* **Note 22**).
3. 4 M Urea: Weigh 4.8 g of urea per 20 ml final volume of polyacrylamide resolving gel solution required (*see* Subheading 3.2, **step 6** and *see* **Note 23**).
4. 8 M Urea: Weigh 2.4 g of urea per 5 ml final volume of polyacrylamide stacking gel solution required (*see* Subheading 3.2, **step 15**).
5. Acrylamide: 40% acrylamide solution in water (*see* **Note 24**).
6. *N*,*N*′-methylenebisacrylamide: 2% bis-acrylamide solution in water (*see* **Note 24**).
7. Glacial Acetic Acid: Anhydrous (100%) acetic acid.
8. Ammonium Persulphate (APS): 10% (w/v) solution in water. Weigh 1 g APS into a 15 ml tube and add 8 ml of water. Mix on a tube rotator for 5–10 min until dissolved. Adjust volume to 10 ml with water and store at 4 °C (*see* **Note 25**).
9. *N*,*N*,*N*′,*N*′-tetramethylethylenediamine (TEMED; *see* **Note 26**).
10. Isopropanol: 100% isopropanol.
11. Gel casting apparatus (*see* **Note 27**).
12. Glass plates (20 × 20 cm) for gel electrophoresis (*see* **Note 28**).
13. 0.75 mm spacers and 0.75 mm 20 well comb (*see* **Note 29**).
14. Ethanol: 100% ethanol.
15. Wash bottle (*see* **Note 30**).
16. Filter paper: Whatman 3MM paper (or equivalent).
17. Vertical gel electrophoresis apparatus for 20 × 20 cm gels (*see* **Note 31**).
18. Running buffer: 5% (~0.9 M) Acetic Acid. Add 50 ml of glacial acetic acid to 950 ml of water.
19. Plastic syringe: Disposable 20 ml (*see* **Note 32**).
20. Fine needles: 23G (*see* **Note 33**).

21. Glass syringe: 25–50 μl (Hamilton Microliter Syringe or similar; *see* **Note 34**).
22. Polyethylene Glycol average MW 8000 (PEG): 20% (w/v) solution in water. Weigh 2 g of polyethylene glycol, add water to 8 ml and mix on tube rotator until PEG goes into solution. Adjust volume to 10 ml with water.
23. 1 M Tris–HCl pH 7.5.
24. Buffered 8 M Urea solution: 8 M urea, 10 mM Tris–HCl pH 7.5. Weigh 0.48 g of urea into a 1.5 ml microfuge tube, add water to 800 μl, 10 μl 1 M Tris–HCl pH 7.5 and mix on tube rotator for 5–10 min. Adjust volume to 1 ml with water. Make fresh buffer just before use.
25. 4 M urea/10% PEG solution: Mix equal volumes of buffered 8 M urea solution and 20% PEG. Prepare solution just before use.

2.3 Western Blotting

1. Polyvinylidene fluoride (PVDF) membranes (Merck Millipore Immobilon-P membrane 0.45 μm; *see* **Note 35**).
2. Acid-urea gel equilibration buffer: 5% acetic acid, 10% methanol. Add 10 ml of glacial acetic acid and 20 ml of methanol to 170 ml of water (*see* **Note 36**).
3. Plastic containers: Lunch box containers large enough to hold fully flat (~18 × 15 cm) gels and membranes are used.
4. Acid-urea gel transfer buffer: 0.1% acetic acid, 10% methanol. Add 1 ml of glacial acetic acid to 800 ml of water, add 100 ml of methanol and adjust volume to 1 L (*see* **Note 36**).
5. Filter paper: Whatman 3MM or equivalent.
6. Electroblotting apparatus for tank transfer (*see* **Note 37**).
7. Ponceau S stain: 0.4% Ponceau S, 5% acetic acid. Weigh 1 g Ponceau S stain and add directly to a 250 ml bottle, add 247.5 ml of water and 12.5 ml of glacial acetic acid (*see* **Note 38**).
8. Methanol destain: 100% methanol to destain Ponceau S stained membranes (*see* **Note 39**).
9. Magnetic stirrer.
10. 10× phosphate buffered saline (10× PBS): Dissolve 80 g NaCl, 2 g KCl, 11.5 g Na_2HPO_4, 2 g KH_2PO_4 in 800 ml of water in a 1 L bottle containing a magnetic stir bar (flea). Mix on magnetic stirrer until dissolved, adjust volume to 1 L. There is no need to pH this solution.
11. Tube rotator.
12. 10% Tween 20: 10% (v/v) solution in water. Pour 5 ml of Tween 20 into a 50 ml tube using gradations on the side of the tube. Adjust volume to 50 ml with water and mix on a tube rotator until solution is clear (*see* **Note 40**).

13. PBS containing 0.1% Tween 20 (PBST). Add 100 ml of 10× PBS and 10 ml of 10% Tween 20 to 890 ml of water.
14. Antibody diluent solution: 5% milk powder in PBST. Weigh 0.5 g of dried milk powder directly into a 10 ml tube, add PBST up to 9 ml and mix on tube rotator. Adjust volume to 10 ml with PBST.
15. Glass hybridization tubes: Large hybridization tubes (Techne; *see* **Note 41**).
16. Hybridization oven or roller mixer: Hybridization oven (Techne; *see* **Note 42**).
17. Plastic containers: All membrane washes during western blotting are carried out in plastic containers.

2.4 Antibodies to Histone H3 Modifications

1. Primary antibodies: anti-histone H3 (K4 tri-methyl) rabbit polyclonal (H3K4me3), anti-histone H3 (K4 di-methyl) rabbit polyclonal (H3K4me2), anti-histone H3 (K4 mono-methyl) rabbit polyclonal (H3K4me1), anti-histone H3 (K9 acetyl) rabbit polyclonal (H3K9ac), anti-histone H3 (S10 phospho, K9 acetyl) sheep polyclonal (H3KS10phK9ac). We use both antibodies raised in-house and from commercially available sources.
2. Secondary antibodies: Anti-rabbit IgG HRP-linked, Anti-goat (sheep) IgG HRP-linked.

2.5 Enhanced Chemiluminescence (ECL) Detection

1. DMSO.
2. Luminol (3-aminophthalhydrazine): 250 mM in dimethyl sulphoxide (DMSO). Weigh 442.9 mg luminol into a 10 ml tube, add 8 ml of DMSO, vortex briefly and mix on tube rotator until dissolved. Adjust volume to 10 ml with DMSO. Store 500 μl aliquots at 4 °C. This solution is a solid at 4 °C. Thaw in a room temperature water bath before use.
3. *p*-Coumaric acid: 90 mM in DMSO. Weigh 147.7 mg of *p*-Coumaric acid into a 10 ml tube, add 8 ml of DMSO, vortex briefly and mix on tube rotator until dissolved. Adjust volume to 10 ml with DMSO. Store 500 μl aliquots at 4 °C. This solution is a solid at 4 °C. Thaw in a room temperature water bath before use.
4. Hydrogen peroxide: 30% solution. Store bottle at 4 °C.
5. 1 M Tris–HCl pH 8.0; Mix 121.1 g of Tris base in 800 ml of water. Stir on a magnetic stirrer. Adjust pH to 8.0 with concentrated HCl. Add water to a final volume of 1 L. Sterilize by autoclaving or filtration using a 0.2 μm filter.
6. ECL Solution A: 2.5 mM Luminol, 0.4 mM *p*-Coumaric acid, 100 mM Tris–HCl pH 8.0. Add 1 ml of 1 M Tris–HCl pH 8.0, 100 μl of 250 mM Luminol and 44.4 μl of 90 mM *p*-Coumaric

acid to 8 ml of water, mix and adjust volume to 10 ml with water (*see* **Note 43**).

7. ECL Solution B: 5.4 mM H_2O_2, 100 mM Tris–HCl pH 8.0. Add 6.12 μl of 30% hydrogen peroxide and 1 ml of 1 M Tris–HCl pH 8.0 to 8 ml of water, mix and adjust volume to 10 ml with water (*see* **Note 43**).
8. Alternatively, a commercial enhanced chemiluminescent detection kit may be used (e.g., Immobilon Western Chemiluminescent HRP Substrate: Merck Millipore; *see* **Note 44**).
9. Saran Wrap or equivalent.
10. X-ray film.
11. X-ray film processor or chemiluminescent imaging system (*see* **Note 45**).

3 Methods

As AU gel electrophoresis of proteins is particularly sensitive to alterations in charge, artifactual modification of proteins during sample preparation and electrophoresis should be avoided. The two main causes of artifactual protein modification are: carbamylation of proteins by urea during sample preparation (*see* **Note 19**) and oxidation during electrophoresis (by residual APS/TEMED in the gel; *see* **Note 46**). All procedures are carried out at room temperature unless indicated otherwise.

3.1 Preparation of Samples for Acid-Urea Gel Electrophoresis

Protein samples were prepared from quiescent C3H10T½ cells grown to confluence on 90 mm dishes.

1. Prior to harvest, cells are left unstimulated or stimulated as required.
2. To harvest cells, aspirate culture medium from dishes, standing each dish on the bottom edge to drain for ~30 s.
3. Aspirate any remaining medium (*see* **Note 47**).
4. Add 200 μl of lysis buffer directly to dishes (*see* **Note 48**).
5. Use a clean cell scraper for each dish to spread the lysis buffer over the entire surface to ensure all cells are lysed immediately.
6. Clean cell scraper with paper towel wet with ethanol and dry between dishes.
7. Use the edge of the cell scraper to collect lysate at the bottom of the dish.
8. Transfer the cell lysate to a 1.5 ml microfuge tube and place on ice.
9. When all dishes have been harvested incubate samples for 15 min on ice.

10. Centrifuge at 17,000 ×*g* for 10 min at 4 °C to pellet nuclei (*see* **Note 49**).
11. Transfer the soluble lysate to a fresh 1.5 ml microfuge tube if analyzing lysis soluble proteins. Discard if not required.
12. Add 1 volume of 4× SDS loading buffer [6] to three volumes of soluble lysate.
13. Vortex and heat at 95 °C for 5 min. Store samples at −20 °C.
14. Acid-extract histones from nuclei by adding 100 μl of 0.4 M HCl to the pellet.
15. Vortex to mix, then incubate for 30 min to 1 h on ice (*see* **Note 50**).
16. Centrifuge at 17,000 ×*g* for 5 min at 4 °C.
17. Transfer supernatant to a fresh microfuge tube.
18. Discard pellet containing DNA and acid-insoluble proteins.
19. Add 1 ml of ice-cold acetone to supernatant, mix by inverting tubes.
20. Incubate for 2 h to overnight at −20 °C.
21. Centrifuge at 17,000 ×*g* for 30 min at 4 °C.
22. Remove supernatant (*see* **Note 51**).
23. Wash pellet with 1 ml of ice-cold acetone.
24. Centrifuge at 17,000 ×*g* for 20–30 min at 4 °C.
25. Remove supernatant.
26. Repeat acetone wash and centrifugation.
27. Remove supernatant.
28. Air-dry pellet for 10–15 min (*see* **Note 52**).
29. Solubilize pellet by adding 120 μl of AU sample loading buffer (*see* **Note 53**).
30. Allow samples to stand for 2–3 min, then vortex briefly.
31. Add ~10 μl methyl green loading dye to the sample. Store samples at −20 °C until required (*see* **Note 54**).
32. Alternatively, solubilized acid-extracts can be split into separate aliquots to enable analysis of the same protein preparation by AU- and SDS-PAGE. Dried pellets (from **step 28**) are resuspended in 120 μl of 8 M urea, 10 mM Tris–HCl pH 7.5.
33. Allow samples to stand for 2–3 min, then vortex.
34. Centrifuge briefly to pellet solution and remove an aliquot to a fresh tube (e.g., 60 μl).
35. For AU gel analysis add 2-mercaptoethanol (5 % final concentration) and glacial acetic acid (5 % final concentration). Vortex and store samples at −20 °C until required (*see* **Note 55**).

36. For SDS gel analysis add an equal volume of 2× SDS loading buffer [6].

3.2 15% Acid-Urea Gel Electrophoresis

This AU gel electrophoresis protocol is modified from Lennox and Cohen [7]. Optimal resolution of some histone proteins may require higher (or lower) percentage gels. If separation of histone primary structure variants is required, Triton acid-urea gel electrophoresis should be used (not discussed here; *see* refs. 7, 8).

Analysis of histone H3 modified by acetylation and methylation on AU (Fig. 3a i) and SDS gels (Fig. 3a ii) and phosphoacetylation on AU gels (Fig. 3b) are shown. On AU gels, histone H3 carrying different modifications are resolved into characteristic histone ladders. By contrast, analysis of the same samples by SDS-PAGE produces a single band. The resolution of proteins on AU gels is dramatically affected by a number of factors: well distortion, protein sample concentration and loading volume. Figure 3c clearly shows how these factors can affect resolution of phosphoacetylated H3.

Fig. 3 (continued) (TSA; 1, 10 ng/ml for 5 min to 2 h). "C" indicates control unstimulated. The most abundant histone H3 forms detected by Ponceau staining are found in bands 0, 1, and 2. Using antibodies against specific modifications, additional bands (up to band 5) are detected. Histone H3 tails become acetylated on K9 (H3K9ac) and these proteins are also acetylated on other lysine residues after TSA treatment (main bands detected at 10 ng/ml, 2 h are bands 3, 4, and 5). Using antibodies to mono-, di-, and tri-methyl K4 (H3K4me1, H3K4me2, H3K4me3), variations in the distribution of proteins carrying distinct modifications are clearly demonstrated. TSA treatment shows that H3 tri-methylated at K4 is subject to dynamic acetylation. Positions of histone isoforms are shown on the right of each panel, with 0 being unmodified histone H3. (ii) Analysis of histone H3 samples by 15% SDS-PAGE. Ponceau staining shows resolution of four histone bands: H3, H2B, H2A, and H4. Antibodies to histone H3 K9 acetylation (H3K9ac) or K4 tri-methylation (H3K4me3) detect a single band on mini-gels (shown here) or large gels (data not shown). Increased acetylation is indicated by an increase in the intensity of the bands. (**b**). Acid-soluble proteins were extracted from nuclear pellets for analysis on AU gels. Analysis of histone H3 phosphorylation and phosphoacetylation by AU gel electrophoresis. Quiescent C3H 10 T½ cells were untreated (–) or pretreated with increasing concentrations of TSA (1, 10 ng/ml; 15 min). Cells were left unstimulated (C) or stimulated with TPA (15–60 min). Using antibodies against histone H3 phosphorylated at serine 10 (H3S10ph) or both phosphorylated at serine 10 and acetylated at lysine 9 (anti-phosphoacetyl H3; H3S10phK9ac), the targeting of phosphorylation and acetylation to the same H3 tails is clearly demonstrated. Histone H3 carrying up to six modified residues is detected in cells treated with TSA and the HDAC inhibitor TSA. An equivalent gel was stained with Coomassie to control for protein loading. Positions of histone isoforms are shown on the right of each panel, with 0 being unmodified histone H3. Figure adapted from ref. [2]. (**c**). Effect of well distortion, loading volume and protein concentration on resolution of histone H3 bands. Quiescent C3H10T½ cells were stimulated with sub-inhibitory anisomycin (sAn; 25 ng/ml, 30 min) or epidermal growth factor (EGF; 50 ng/ml, 30 min). Nuclear pellets were acid-extracted to analyze histone H3. Samples were loaded in twice the standard volume (2×V), standard volume (Std) or twice the standard amount of protein was loaded in the wells (2×P). One of the main factors affecting resolution of proteins is distortion of wells. This can occur during removal of the comb and unlike for SDS-PAGE, straightening the wells does not solve the problem. Loading volume as well as total protein concentration also affects the resolution of bands. Note that a shorter autoradiograph exposure is shown for lanes containing 2×P

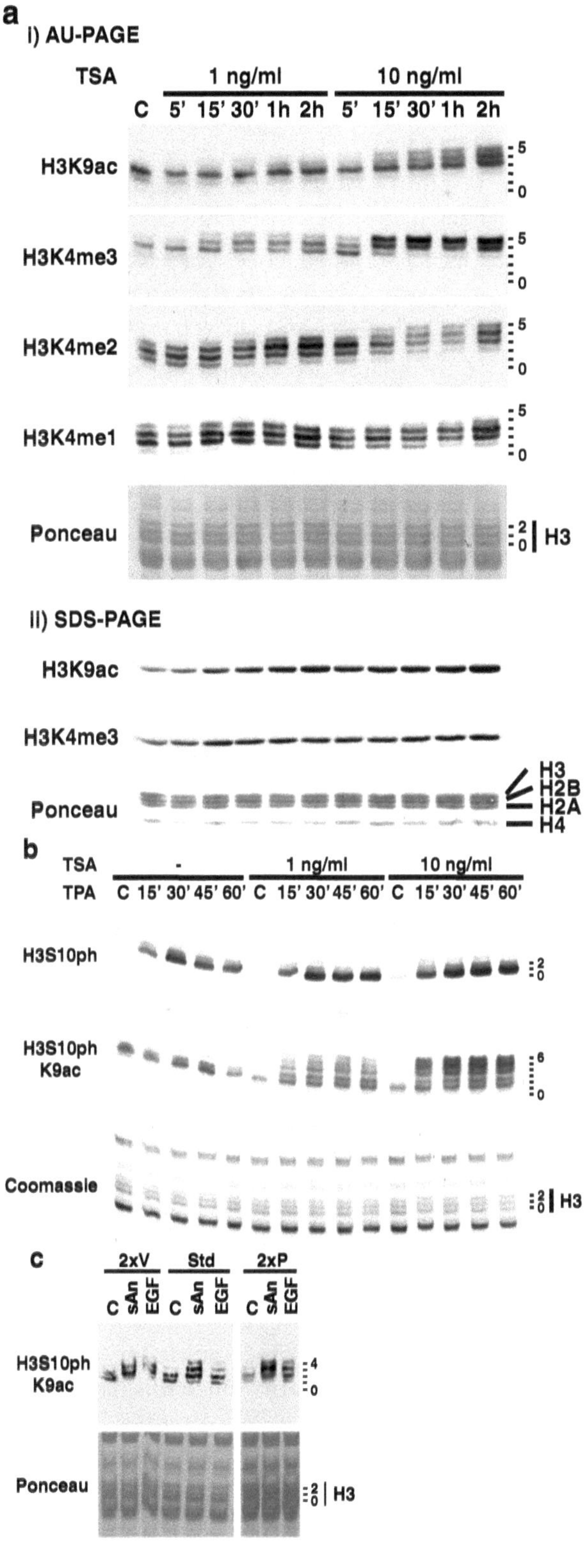

Fig. 3 AU gel electrophoresis of histone H3. (**a**). Acid-soluble proteins were extracted from nuclear pellets and following solubilization of histones in urea, samples were split (*see* Subheading 3.1, **steps 32–36**), for analysis on both AU and SDS gels. (i) Analysis of histone H3 acetylation and methylation by AU gel electrophoresis. Quiescent C3H10T½ cells were treated with increasing concentrations of the HDAC inhibitor trichostatin A

1. Lay glass plates on bench and clean the sides of the glass plates that will be in contact with the gel with 100% ethanol and paper towel.
2. Position spacers at the edge of the glass plates, align front & back plates then insert in gel casting apparatus (*see* **Note 56**).
3. Once assembled, check for leaks by filling the area between the glass plates with water using a wash bottle. Fill up to the top of the back plates and allow to stand for 15 min.
4. If the plates are completely sealed, pour off the water and remove any remaining water by sliding filter paper between the glass plates (*see* **Note 57**).
5. If leaks are detected, pour off the remaining water, dismantle the apparatus, check the alignment of spacers and plates then reassemble.
6. To prepare a 4 M urea/15% resolving gel, weigh 4.8 g of urea into a 50 ml tube and add 7.5 ml of 40% acrylamide solution, 2 ml of 2% bis-acrylamide solution, 1 ml of glacial acetic acid, and water to 15 ml.
7. Mix on a tube rotator until the urea is completely dissolved.
8. Equilibrate the solution in a room temperature water bath for 10–15 min (*see* **Note 58**).
9. Add 400 μl of 10% APS and 80 μl of TEMED, mixing after each addition (*see* **Note 59**).
10. Adjust the final volume to 20 ml with room temperature water and mix gently.
11. Pour the resolving gel between 20 cm × 20 cm × 0.75 mm gel plates using a 10 ml pipette. At the start of pouring, holding the gel plates at a 45° angle helps prevent bubbles forming. Allow for 2 cm of stacking gel from the bottom of the wells to the top of the resolving gel. If any bubbles appear they can be removed by tilting the gels in the casting apparatus or by gentle tapping of gel plates.
12. Gently overlay the gel with 300 μl of isopropanol (*see* **Note 60**). Do not move gels until completely polymerized (this will take longer than for SDS gels).
13. Monitor gel polymerization by checking the remaining gel solution in the tube. Once this has polymerized, wait a further 15–20 min before removing the isopropanol from the gel surface by rinsing the space between the gels plates with two to three changes of water, using a wash bottle.
14. Remove all water before pouring the stacking gel by sliding filter paper between the glass plates, being careful not to touch the gel surface. The gel can be tilted so that the water collects in one corner for easy removal.

15. To prepare an 8 M urea/7.6% stacking gel, weigh 2.4 g of urea into a 15 ml tube and add 0.95 ml of 40% acrylamide solution, 0.25 ml of 2% bis-acrylamide solution, 0.25 ml of glacial acetic acid, and water to 4.5 ml (*see* **Note 61**).
16. Mix on a tube rotator until the urea is completely dissolved (*see* **Note 62**).
17. Equilibrate the solution in a room temperature water bath for 10–15 min.
18. Add 200 μl of 10% APS and 40 μl of TEMED, mixing after each addition. Adjust the volume to 5 ml with water and mix gently.
19. Pour the stacking gel between the plates using a 10 ml pipette and insert a 20 well comb. If any bubbles are present, remove by gently tapping the glass plates (*see* **Note 63**).
20. Monitor gel polymerization by checking the remaining gel solution in the tube. Once this has polymerized wait a further 20–30 min before removing the comb (*see* **Note 64**).
21. Remove the comb before positioning the plates in the gel tank. Pour water along the top of the comb and gently remove the comb trying to avoid vacuum formation which will result in well distortion.
22. Immediately rinse wells with water to remove any unpolymerized acrylamide using a plastic syringe and needle. If necessary straighten wells using Hamilton syringe.
23. Assemble gel plates in the electrophoresis apparatus and fill upper and lower reservoirs with gel running buffer (5% acetic acid).
24. Rinse wells with gel running buffer using a plastic syringe and needle.
25. Load 20 μl of 4 M urea/10% PEG solution into every well using a Hamilton syringe (*see* **Note 65**).
26. Pre-electrophorese the gel overnight before loading samples (*see* **Note 46**).
27. Connect the electrodes to a power supply with the cathode (−) at the bottom of the gel, the anode (+) at the top and pre-electrophorese overnight at 150 V (constant voltage). In AU gels proteins migrate from the anode (+) towards the cathode (−). Take care to ensure the electrodes are connected the opposite way round compared to conventional SDS-PAGE.
28. At this voltage, when electrophoresis begins, the current for each gel is normally 20–25 mA. As the charged ions move out of the gel during electrophoresis this causes a progressive decrease in current. If pre-electrophoresis is omitted, this results in poor resolution of bands [1].

29. Once all the ions have been removed from the gel by electrophoresis, the current for each gel will drop to approximately 5 mA (*see* **Note 66**).
30. Following pre-electrophoresis, remove the 4 M urea/10% PEG in wells by rinsing with gel running buffer using a syringe and needle.
31. Discard running buffer in the upper and lower reservoirs and replace with fresh running buffer.
32. Immediately before loading samples rinse wells once more in running buffer using a syringe and needle.
33. Load 15 μl of sample (~1/9th plate) per well using a Hamilton syringe.
34. Rinse the Hamilton syringe with several changes of running buffer before loading next sample, to prevent cross-contamination. As the stacking effect of the 7.6% gel is not as great as for SDS-PAGE, ensure each sample is loaded as compactly as possible (*see* **Note 67**).
35. Empty wells should be loaded with AU sample loading buffer alone.
36. Change the settings on power pack to 15 mA (constant current) per gel and commence run. As for pre-electrophoresis, ensure electrodes are connected the opposite way round compared to conventional SDS-PAGE as proteins migrate from the anode (+) towards the cathode (−).
37. Monitor progression of the blue dye component of methyl green dye in samples. In acetic acid-urea gels the green and yellow dyes run faster than the blue dye. For the analysis of histone H3, gels are electrophoresed until 1 h after the blue dye front runs off the bottom of the gel.
38. Using our apparatus, 15% AU gels are electrophoresed for approximately 5 h 30 min. For analysis of histone H4 (not discussed here) gels may be run for a shorter time as H4 migrates ahead of other histones (Fig. 2c).

3.3 Western Blotting

1. Following electrophoresis, separate the gel plates and remove the stacking gel (*see* **Note 68**).
2. Cut one small corner off the gel so that gel and membrane can be aligned during assembly.
3. Wash the resolving gel/gels in 200–300 ml of equilibration buffer for 30 min with gentle agitation (*see* **Note 69**).
4. Carefully pour off the equilibration buffer and rinse the gel in transfer buffer before proceeding to assemble the gel and PVDF membrane in an electroblotting cassette.
5. Wet PVDF in methanol for at least 15 s.

6. Pour off methanol then rinse the membrane in transfer buffer until completely wet and no "beading" is visible on the membrane surface.
7. Wet foam pads and filter paper in transfer buffer before use.
8. Assemble transfer components in gel holder cassette on the anode (+) side, in the same order as for standard electroblotting: foam pad, filter paper, PVDF membrane, gel, filter paper and foam pad (*see* **Note 70**).
9. To ensure good gel-membrane contact and to remove air bubbles, roll a 10 ml pipette gently across the surface after each layer is added.
10. Close cassette carefully without moving the assembled stack and lock in closed position using latch or rubber bands depending on cassette design.
11. For AU gel transfer, resolved proteins migrate towards the cathode (−), the opposite direction to standard transfer. Turn cassette the opposite way around in the tank (*see* **Note 71**).
12. Add sufficient transfer buffer to cover the gels completely. Transfer proteins to PVDF membrane at 200 mA (constant current) overnight in a cold room at 4 °C.
13. Use a magnetic stirrer in the transfer tank to maintain an even temperature and ion distribution (*see* **Note 72**).
14. Following transfer of proteins to PVDF membrane, remove cassette from the tank, disassemble transfer stack and carefully remove membrane with forceps.
15. Visualize proteins by incubating in Ponceau S stain. Incubate the wet membrane directly in Ponceau S, for 1–2 min with gentle agitation, or if PVDF membrane is allowed to dry before staining, the membrane should be pre-wet with methanol before incubation in Ponceau S.
16. Destain with several changes of 100% methanol or water (*see* **Notes 39** and **73**). When bands are clearly visible and background staining has been removed, dry membranes on filter paper (*see* **Note 74**).
17. If water has been used to destain the membrane, rinse briefly in 100% methanol to drive the water out and dry the membrane more quickly (*see* **Note 75**).
18. If no protein staining is observed check that protein samples were solubilized (in 4–8 M urea) before electrophoresis. Insoluble precipitated protein will not enter the gel.
19. Once dry, cut the membrane to size leaving an extra ~5–10 mm at each edge. Scan membranes on flatbed document scanner (*see* **Note 76**). Stained membranes from AU gels are shown in Fig. 2c, 3a i, c, and an SDS gel in Fig. 3a ii.

3.4 Rapid Immunodetection Method for Western Blotting

1. The PVDF membrane must be completely dry before western blotting when using the rapid immunoblotting method. Membranes are semi-transparent when wet and as the blot dries, become more opaque.
2. Once the methanol-destained membrane is completely opaque, allow to air-dry for a further 30 min before proceeding with rapid immunodetection.
3. No blocking step is required for rapid immunodetection (*see* **Note** 77). Incubate the dry membrane for 1 h at room temperature in 10 ml of primary antibody diluted in 5% milk in PBST in a rotating hybridization bottle.
4. Wash the membrane 3 times in fresh PBST for 5 min with gentle agitation (*see* **Note 78**). Insufficient washing may result in high background and conversely, over-washing may remove bound antibody and reduce signal detected.
5. Incubate the membrane for 45 min at room temperature in 10 ml of secondary antibody solution diluted in 5% milk in PBST in a rotating hybridization bottle. During incubation in secondary antibody the PVDF membrane will begin to wet, changing from opaque to semi-transparent (*see* **Note 79**).
6. Wash the membrane 3 times in fresh PBST for 5–10 min with gentle agitation.

3.5 Protein Detection by Enhanced Chemiluminescence

1. Freshly prepare ECL solutions A and B and mix just before use. Allow 0.1 ml of ECL solution (A + B mixed) per square cm of membrane (*see* **Note 80**).
2. After the final wash in PBST drain excess buffer by touching a corner of the membrane on paper towel.
3. Lay the membrane protein side up on a sheet of Saran Wrap (surface smoothed to remove bubbles and wrinkles).
4. Pour or pipette the ECL detection solution to cover the membrane surface evenly.
5. Incubate for 1 min.
6. Drain excess ECL solution from the membrane.
7. Place the membrane face down on fresh sheet of Saran Wrap, (smoothed to remove bubbles and wrinkles using paper towel) fold Saran Wrap over the back of the membrane.
8. Remove excess ECL substrate by flattening and smoothing out the Saran Wrap surface using paper towel.
9. Cut the Saran Wrap to size, allowing ~1 cm around edges, turn over edges to seal.
10. Ensure the outside of the wrapped membrane is dry and all bubbles/wrinkles are removed (*see* **Note 81**).

11. Place the membrane in an X-ray film cassette and proceed to a darkroom for exposure to X-ray film or detection using an imaging system.

4 Notes

1. We mainly use 90 mm dishes, but lysis volumes can be adjusted for smaller (60 mm) or larger dishes (145 mm). We culture cells in dishes rather than flasks when preparing cell lysates to enable rapid lysis. Using cell scrapers, lysed cells are quickly harvested with very good and reproducible recovery of lysates.
2. The disodium salt of EDTA does not go into solution until the pH is close to 8.0.
3. EGTA does not go into solution until the pH is close to 8.0.
4. Tube rotators can be adjusted to hold a variety of different sized tubes (microfuge to 50 ml tubes) and gently mix buffers/samples using an end over end motion.
5. NaF is highly toxic, appropriate safety measures should be taken, especially when handling the chemical in powder form. NaF is an inhibitor of serine/threonine phosphatases.
6. As glycerol is viscous, pour 25 ml directly into a 50 ml tube using gradations on the side of the tube.
7. Do not autoclave DTT solutions.
8. As Triton X-100 is viscous it is easier to pour into the tube than pipette.
9. All these protease inhibitors are highly toxic and appropriate safety measures should be taken when handling these compounds. We purchase Aprotinin and Leupeptin in 10 mg and 25 mg quantities to avoid weighing these chemicals. To completely avoid handling toxic proteases in powder form, commercially available inhibitor cocktails may be used.
10. The protease inhibitors in this 100× solution do not dissolve completely, but do go into solution at 1× concentration. When aliquoting or adding to lysis buffer ensure the suspension is evenly distributed.
11. Always add acid to water (not the other way round) and add slowly to avoid violent reaction.
12. This is due to variation between batches of the supplied chemical.
13. This process to stabilize the pH activates Na_3VO_4, converting it to a more potent inhibitor of protein tyrosine phosphatases.
14. β-glycerophosphate is a serine/threonine phosphatase inhibitor.

15. Make microcystin-LR solution directly in the supplied vial. This compound is highly toxic and may be omitted if not looking at phosphorylation. Microcystin is a potent inhibitor of protein phosphatase 1 (PP1) and 2A (PP2A).
16. Sodium butyrate may be omitted if not looking at acetylation. Sodium butyrate is an inhibitor of HDACs.
17. We routinely use the cell lysis buffer described here to study signaling proteins/transcription factors as well as histones. This allows analysis of a wide variety of proteins from the same dish of cells and produces a nuclear pellet which can be acid-extracted to analyze histones by AU gel electrophoresis. Any lysis buffer which keeps nuclei intact may be used to prepare acid-extracts containing histones suitable for AU gel electrophoresis [4, 9]. Lysis buffer contains protease inhibitors to prevent protein degradation and phosphatase/HDAC inhibitors to preserve in vivo post-translational modification of proteins. Note that high salt lysis buffers extract proteins peripherally associated with chromatin including histone H1 and high mobility group (HMG) proteins. For the analysis of these proteins by AU gel electrophoresis prepare nuclei using a low salt buffer.
18. For convenience store acetone in 50 ml tubes at −20 °C in laboratory (flammable) safe freezer.
19. In acid-urea gel electrophoresis, urea is used to denature and solubilize proteins in the absence of SDS. However, care should be taken when using urea as it can modify proteins by carbamylation, affecting the charge and hence mobility in AU gels. Urea in solution is unstable, resulting in the production of cyanate ions. The isocyanic acid (cyanate) reacts with amino groups of proteins, causing carbamylation which is pH-dependent and irreversible. If this affects the amino terminus of proteins, it blocks N-terminal sequencing. It also reacts with lysine and arginine residues of proteins. Carbamylated residues are neutral, so affected proteins carry less charge. Solutions containing urea should be freshly prepared and acidified to change the equilibrium of reaction to favor urea. Protein solutions containing urea should never be heated, as heating increases the carbamylation of proteins.
20. 8 M urea in loading buffer solubilizes and denatures proteins (acid-extracted histones require at least 4 M urea to solubilize) and also increases the density of samples for loading.
21. Methyl green dye contains blue, yellow and green dye components which have different electrophoretic mobilities in AU gels.
22. For convenience acrylamide gel solutions are prepared in plastic tubes. After gels are cast, the sealed tubes containing any remaining polymerized acrylamide are disposed of via appropriate waste channels.

23. The urea concentration in the resolving gel can be varied depending on the protein of interest. For example, we have observed that at a low urea concentration (1 M) histone H2B co-migrates with H3 (data not shown) and at 4–8 M urea, H2B is poorly resolved from H2A ([10], *see* Fig. 2c].
24. Commercially available aqueous solutions of acrylamide/bis-acrylamide in deionized water offer a convenient and safe alternative to the preparation of acrylamide solutions from powder. Solutions are supplied pre-filtered, have a long shelf-life (up to 2 years if stored correctly) and have a high level of reproducibility between different batches. Acrylamide, a potent neurotoxin, is absorbed through contact with skin or inhalation of powder. Wear protective gloves when handling acrylamide and *N,N'*-methylenebisacrylamide solutions. Once polymerized, the gels are generally considered non-toxic, but the gels may still contain some unpolymerized acrylamide.
25. When stored at 4 °C, the solution is stable for several weeks.
26. To avoid the pungent smell, store and add to acrylamide solutions in a fume cupboard.
27. We use a commercial gel casting apparatus, consisting of a casting base, side clamps and silicone mats to seal the gel plates and spacers. When aligned correctly in the casting unit, the gel spacers and plates are pulled down onto the silicon mat using cams to create a tight seal.
28. Each pair of plates should comprise one blank plate (front) and one notched plate (back).
29. Each well on the combs that we use is ~ 5 mm wide. Combs with smaller or larger width wells may be used, but the amount of protein loaded should be adjusted to take this into account.
30. Using a wash bottle is the easiest way to add water between the glass plates, when checking for leaks and to remove isopropanol overlaid on the resolving gel.
31. We use double vertical gel systems to run two gels at the same time.
32. It is quicker to fill the syringe before fitting the needle.
33. Needles of this gauge are thin enough to fit easily between the glass plates (0.75 mm spacers). Carefully recap the needle and remove from the syringe before refilling to avoid needle stick injuries. Needles sometimes get blocked with acrylamide washed from wells and may need replacing.
34. Glass syringes with fine blunt needles fit easily between glass plates (0.75 mm spacers) and can be rinsed after loading each sample to prevent cross-contamination. The volumes dispensed are accurate and we find it much quicker and easier to load samples using a Hamilton syringe than gel loading tips.

35. Use flat edged forceps to handle PVDF to avoid damaging the membrane. Cut PVDF membrane according to dimensions of the gel (allow an extra ~10–15 mm on each side to allow for gel swelling during incubation in equilibration buffer). Cut corner off membrane to align with gel. PVDF is hydrophobic and must be wet in methanol before use. Ethanol or isopropanol may also be used to wet PVDF membrane.
36. Methanol in this buffer and transfer buffer reduces swelling of gels during these incubations and helps stabilize gel size.
37. The transfer tank we use can hold three cassettes, allowing the transfer of up to three large gels at the same time.
38. Ponceau S stain is used to detect proteins bound to PVDF before western blotting, verifying protein transfer and whether lanes are loaded evenly. Ponceau S is a negative stain that binds reversibly to positively charged amino groups on proteins. Ponceau S stain can be reused many times. After membrane staining, pour back into the bottle. Make fresh Ponceau S stain when band intensity of stained membranes appears reduced.
39. Water may also be used to destain membranes. However, more Ponceau S dye bound to proteins is removed under these conditions, resulting in a reduced signal to noise ratio.
40. Tween 20 is a viscous solution that is difficult to pipette accurately. We make 10% (v/v) Tween 20 and use this solution to make up buffers containing lower concentrations of Tween 20.
41. Large tubes can hold multiple blots, depending on size. This allows more membranes to be analyzed in parallel under identical conditions and also reduces the amount of antibody required. A minimal volume of 10 ml of antibody solution is required for large tubes (or 5 ml for small tubes).
42. The oven holds up to six large tubes and all antibody incubations are carried out at room temperature.
43. Make the ECL solution fresh just before use.
44. If using commercial kits, we have found that reducing the substrate incubation time (from 5 min to less than 1 min) produces good signal to noise ratio, but with less nonspecific background.
45. When using an X-ray film processor ensure the machine is well maintained and cleaned regularly. The water reservoir should be drained at the end of each day and refilled with fresh water every time the machine is switched on.
46. AU gels are pre-electrophoresed to remove APS/TEMED to prevent oxidation of proteins in the gel, which affects the net charge of proteins. Methionine and cysteine residues are particularly sensitive.

47. As cells are lysed in small volumes it is important to remove as much culture medium as possible, as if dishes are not drained sufficiently and reproducibly, the volume of soluble lysate recovered will vary between samples. This is not as critical if only analyzing the pellet.
48. Work in our lab is focused on dynamic post-translational modification of MAP kinase signaling proteins and histone H3/H4. Many of these changes occur over very short time periods and therefore we do not wash dishes with PBS before adding lysis buffer.
49. The soluble fraction contains cytoplasmic proteins and nuclear proteins solubilized by 400 mM NaCl (including many transcription factors, HMGs, and some histone H1). The pellet fraction contains mainly core histones and DNA.
50. Depending on the lysis buffer used to prepare the nuclear pellet, this pellet may break up completely when the sample is vortexed, or not at all. Acid-extraction works efficiently even when the nuclear pellet appears intact.
51. To remove acetone following precipitation/washes, carefully pour the supernatant into a small beaker, closely observing the position of the pellet. If the pellet starts to move and is in danger of being lost, remove remaining acetone using a pipette (pouring off the supernatant is quicker than removing by pipetting). Briefly place microfuge tubes upside down on paper towel, until all samples are processed and then turn upright for further washing/drying the pellet. Collect all acetone waste for safe disposal according to local regulations.
52. Following the final wash, dry samples in the upright position to prevent loss of the pellet. If microfuge tubes are left upside down to dry completely, the pellets may detach and fall out of tubes.
53. The concentration of proteins should be as high as possible (without overloading the wells), so that the volume loaded is as small as possible. This improves protein resolution.
54. Methyl green loading dye may be added to the whole sample or an aliquot just before loading. In AU sample loading buffer this dye appears green when it is first added, but if samples are frozen and then thawed, the color changes to a lighter blue color. This does not affect the integrity of the sample.
55. If samples for AU gel electrophoresis are solubilized in 8 M urea in the absence of acetic acid, acidify the solution as soon as possible as the proteins may become carbamylated under these conditions (*see* **Note 19**).
56. Apply a very small amount of vacuum grease evenly along both sides of the spacers, as this helps prevent leakage.

57. As the filter paper becomes saturated, cut off wet areas and use dry edges to remove water. If any wet filter paper breaks off, push to the bottom of the gel and compact as much as possible. As long as the filter paper is removed before electrophoresis, it will not affect protein resolution.
58. As the solubilization of urea is an endothermic reaction and the rate of acrylamide polymerization is reduced at lower temperatures, equilibrating the gel solution to room temperature speeds up polymerization of the gel.
59. APS and TEMED initiate the gel polymerization reaction through the production of radicals that interact with acrylamide. Urea increases the rate of acrylamide polymerization, but acetic acid strongly inhibits this reaction. Relatively larger amounts of APS and TEMED are added here to counteract this effect.
60. Isopropanol is used to form a flat smooth surface at the top of the gel and also forms a barrier against oxygen that inhibits polymerization.
61. The purpose of lower percentage acrylamide in the AU stacking gel is to produce a "stacking" effect of proteins, similar to SDS gels, to improve resolution of proteins and to enable easy removal of the comb. Good separation/resolution of proteins is also obtained with a lower acetic acid concentration in the stack [5]. If the acetic acid concentration in the stack is reduced, the APS and TEMED volumes should be halved.
62. This will take longer than for the resolving gel as the concentration of urea is higher. This solution may need to be vortexed to solubilize the urea completely.
63. During polymerization the resolving gel sometimes shrinks away from the spacers slightly. This causes bubbles to appear at the edges of the gel as the stack is poured. The bubbles can be removed by gentle tapping, but should not interfere with electrophoresis of samples. The comb can be put in position before or after adding the stack solution.
64. We find it difficult to see the wells clearly in our electrophoresis apparatus. Before removing the comb, we mark the position of the bottom of each well using a fine marker pen on the front glass plate.
65. This helps to maintain the size and shape of the wells during pre-electrophoresis [11].
66. We find it most convenient to pre-electrophorese gels overnight, but it can be done at higher voltage (300 V) for shorter times, until the current has dropped to ~5 mA. Note that if pre-electrophoresis is carried out by mistake with the electrodes connected as for SDS-PAGE, this does not prevent use

of the gel for sample analysis. Provided the electrodes are reconnected in the correct way for AU gel electrophoresis before loading samples (cathode (−) at the bottom of the gel, the anode (+) at the top), sample electrophoresis is not affected.

67. Positioning the tip of the Hamilton syringe at the very bottom of the well, in one corner before expelling the sample produces the best result. The needle is then carefully removed to avoid disturbing the sample.
68. Use a spatula or flat edged forceps to carefully prise the plates apart. Insert spatula/forceps away from the top of the gel/notched part of plates as the notches are more susceptible to breakage.
69. This step is required to remove urea and stabilize the gel size before transfer. If the gel swells during transfer, resolution of bands may be lost. During this incubation the gel will swell slightly.
70. We find that layering the gel onto the membrane (rather than the opposite way round) results in better gel-membrane contact.
71. Do not plug the electrodes in the opposite way round (as for gel electrophoresis) as this will damage the cathode in the transfer tank.
72. To eliminate the possibility of SDS contamination, keep transfer apparatus, stocks of foam pads and cassettes for use only in AU gels.
73. After the first methanol wash, the solution is very pink/red and this solution is discarded and disposed of according to local regulations. Subsequent washes contain less Ponceau S stain and can be collected and reused for the first wash in later experiments.
74. If nonspecific staining is visible on the membrane during destaining, rinse the membrane in several changes of water until removed and then repeat incubation in Ponceau S before destaining in methanol.
75. Drying the membrane enhances protein binding and reduces background.
76. For histone H3 analysis, after scanning, we mark the position of the three visible histone H3 bands with a pencil to enable assignment of detected bands to correct position on the histone ladder.
77. This protocol for western blotting takes advantage of the hydrophobicity of the PVDF membrane. During incubation, primary antibodies are unable to bind to the (non-wetted) hydrophobic membrane, but are able to bind to proteins on the membrane. This prevents nonspecific background on blots.

78. It is not necessary to wash individual membranes separately; we place multiple membranes together in one box, in relatively large volumes (100–150 ml) of buffer per wash.
79. Membrane wetting at this stage does not result in increased nonspecific background.
80. We make up just enough ECL solution to cover the membrane/s.
81. If bubbles or wrinkles in the Saran Wrap are present on the protein side of the membrane, they will interfere with signal detection. Any wrinkles in Saran Wrap running across detected bands will be visible on X-ray film.

References

1. Panyim S, Chalkley R (1969) High resolution acrylamide gel electrophoresis of histones. Arch Biochem Biophys 130:337–346
2. Hazzalin CA, Mahadevan LC (2005) Dynamic acetylation of all lysine 4-methylated histone H3 in the mouse nucleus: analysis at c-fos and c-jun. PLoS Biol 3, e393
3. Thomson S, Clayton AL, Mahadevan LC (2001) Independent dynamic regulation of histone phosphorylation and acetylation during immediate-early gene induction. Mol Cell 8:1231–1241
4. Mahadevan LC, Willis AC, Barratt MJ (1991) Rapid histone H3 phosphorylation in response to growth factors, phorbol esters, okadaic acid, and protein synthesis inhibitors. Cell 65:775–783
5. Smith BJ (1984) Acetic acid-urea polyacrylamide gel electrophoresis of proteins. Methods Mol Biol 1:63–73
6. Laemmli UK (1970) Cleavage of structural proteins during the assembly of the head of bacteriophage T4. Nature 227:680–685
7. Lennox RW, Cohen LH (1989) Analysis of histone subtypes and their modified forms by polyacrylamide gel electrophoresis. Methods Enzymol 170:532–549
8. Cano E, Barratt MJ, Mahadevan LC (1992) Which histone kinase? Nature 360:116
9. Barratt MJ, Hazzalin CA, Zhelev N et al (1994) A mitogen- and anisomycin-stimulated kinase phosphorylates HMG-14 in its basic amino-terminal domain in vivo and on isolated mononucleosomes. EMBO J 13:4524–4535
10. Hardison R, Chalkley R (1978) Polyacrylamide gel electrophoretic fractionation of histones. Methods Cell Biol 17:235–251
11. Paulson JR, Higley LL (1999) Acid-urea polyacrylamide slab gel electrophoresis of proteins: preventing distortion of gel wells during preelectrophoresis. Anal Biochem 268:157–159

Chapter 12

Chromatin Immunoprecipitation of Histone Modifications in Fission Yeast

Jean Mbogning and Jason C. Tanny

Abstract

Chromatin immunoprecipitation (ChIP) is a sensitive, accurate, and reliable technique widely used to analyze protein-DNA interactions at specific binding sites in vivo. It has been a particularly powerful technique for mapping of histone modification patterns both at individual loci and genome-wide. Here we provide a detailed protocol for ChIP of histone modifications associated with active transcription in fission yeast (*Schizosaccharomyces pombe*).

Key words Chromatin immunoprecipitation, ChIP, Histone modification, *S. pombe*, Fission yeast

1 Introduction

Posttranslational modification of histones regulates numerous aspects of chromatin structure and function [1]. Chromatin immunoprecipitation (ChIP), a method in which chromosomal proteins are isolated in complex with the genomic DNA they associate with in vivo, is currently the most powerful method to study patterns of histone modifications [2]. Genome-wide ChIP studies have found strong correlations between most histone modifications and transcriptional activity. Indeed, a stereotypic pattern of histone modifications is associated with transcribed genes throughout eukaryotes [3, 4]. Major features of this pattern include histone acetylation (at multiple sites on histones H3 and H4) proximal to gene 5′ ends, methylation at histone H3 lysine 4 proximal to gene 5′ ends, methylation of histone H3 lysine 36 proximal to gene 3′ ends, and mono-ubiquitylation of histone H2B throughout gene-coding regions. How this pattern is established during transcription and how it impacts expression of individual genes are important unresolved issues.

The fission yeast *S. pombe* has emerged as a useful model system for the study of various aspects of chromatin structure and function.

Benoit Guillemette and Luc Gaudreau (eds.), *Histones: Methods and Protocols*, Methods in Molecular Biology, vol. 1528,
DOI 10.1007/978-1-4939-6630-1_12, © Springer Science+Business Media New York 2017

Like its distant relative *S. cerevisiae*, fission yeast is easy to grow and to manipulate genetically. However, its genome architecture more closely resembles that of mammalian cells. For example, *S. pombe* centromeres are flanked by heterochromatin, and intron-containing genes are common [5]. Work in *S. pombe* has led to elucidation of pathways linking the RNA interference pathway to transcriptional regulation and chromatin structure [6]. Study of transcription-associated histone modifications in this system is also likely to lead to valuable insights into their functions in mammalian cells.

ChIP involves formaldehyde cross-linking of live cells followed by chromatin solubilization and immunoprecipitation with antibodies against a protein of interest. Here we outline our method for ChIP of histone modifications in *S. pombe*. Genomic DNA isolated using this procedure can be processed for genome-wide or locus-specific analyses; here we describe analysis by quantitative PCR (qPCR).

2 Materials

2.1 Yeast Culture and Cross-Linking

1. YES media: 5 g/L Yeast extract, 30 g/L glucose, 1 g/L amino acid supplement. The amino acid supplement is typically added as a premixed powder containing equal parts adenine, histidine, leucine, uracil, and lysine (*see* **Note 1**).
2. 37 % Formaldehyde solution.
3. 2.5 M Glycine (autoclaved).
4. TBS: 20 mM Tris–HCl pH 7.5, 150 mM NaCl.
5. 50 mL Conical tubes.
6. 2 mL Screw-cap tubes.

2.2 Yeast Extract Preparation and Immunoprecipitation

1. Lysis buffer: 50 mM Hepes pH 7.5, 150 mM NaCl, 1 mM EDTA, 1 % Triton X-100, 0.1 % Na deoxycholate, 1 mM PMSF (add right before using), protease inhibitor tablet (Roche) (add one complete mini tablet per 10 mL right before using).
2. Acid-washed glass bead diameter 425–600 μm.
3. 5 mL Snap-cap tube.
4. 15 mL Polycarbonate tubes.
5. Dynabeads protein G (Life Technologies).
6. Lysis buffer + 0.1 % SDS.
7. Lysis buffer + 0.1 % SDS + 500 mM NaCl.
8. LiCl buffer: 10 mM Tris 7.5, 1 mM EDTA, 250 mM LiCl, 0.5 % Na deoxycholate, 0.5 % NP-40 (Nonidet P 40 substitute, Fluka #74385).
9. TE: 10 mM Tris pH 7.5, 1 mM EDTA.

10. Elution buffer: 50 mM Tris pH 7.5, 10 mM EDTA, 1% SDS.
11. TE+0.67% SDS.
12. Antibodies: Anti-ubiquityl-histone H2B, anti-histone H3, anti-trimethyl histone H3K4.

2.3 DNA Purification

1. 10 mg/mL DNase-free RNase A.
2. 20 mg/mL Glycogen.
3. 10 mg/mL Proteinase K.
4. 4 M LiCl (filter-sterilized).
5. Phenol:chloroform:isoamyl alcohol (25:24:1).
6. Chloroform.
7. 95% Ethanol.
8. 70% Ethanol.

2.4 Real-Time Quantitative PCR

1. 2× qPCR reaction mix containing SYBR® Green (*see* **Note 2**).
2. 100 μM Gene-specific primers.
3. PCR plates 96-well, clear.
4. Sealing film for 96-well plates, clear.

3 Methods

3.1 Cell Culture

1. Inoculate a single yeast colony in 5 mL YES medium. Grow cells under vigorous shaking (200–250 rpm) to saturation (1–2 days). Inoculate a 1 L sterile Erlenmeyer flask containing 50 mL YES medium with an appropriate amount of the saturated culture and grow overnight to an OD_{600} between 0.3 and 0.5 (*see* **Note 3**).
2. Add 37% formaldehyde to a final concentration of 1% (1.35 mL) to the cell culture and incubate for 30 min at room temperature with gentle shaking in the fume hood (*see* **Note 4**).
3. Stop the cross-linking by adding 2.5 M glycine to 125 mM final concentration (2.5 mL) and continue shaking for 5 min.
4. Mix the culture thoroughly and transfer approximately 1.5×10^7 cells to a 50 mL Falcon tube in the fume hood. Use a hemocytometer to determine the amount of cells per milliliter of culture (*see* **Note 5**).
5. Collect the cells by centrifugation at 2650×*g* for 3 min at +4 °C in a benchtop centrifuge with a swinging-bucket rotor. Pour off the supernatant and wash the pellet twice with 10 mL cold 1×TBS (resuspend the pellet with 10 mL cold 1×TBS, spin down, pipette off supernatant, and repeat the washing one more time).

6. Resuspend the cell pellet with 1 mL of cold 1×TBS and transfer to 2 mL screw-cap tube.
7. Spin down for 1 min at 16,100×*g* in a microcentrifuge and snap-freeze the pellet in dry ice or liquid nitrogen. Store the sample at −80 °C.

3.2 Extract Preparation

1. Resuspend the cell pellet with 0.4 mL of ice-cold lysis buffer, and add 0.4 mL of glass beads (*see* **Note 6**).
2. Bead-beat in a minibeadbeater (Biospec) in the cold room (+4 °C). Use four pulses of 30 s each (at full power) with a 1-min rest on ice in between (*see* **Note 7**).
3. Pierce the bottom of the tube with a red-hot 22-G needle and insert the tube into a 5 mL snap-cap plastic tube. Collect the lysate by centrifugation at 168×*g* in a swinging-bucket rotor for 1 min at +4 °C (*see* **Note 8**).
4. Transfer the lysate to a new 1.5 mL Eppendorf tube (transfer all the contents of the tube, including the pellet, by gently pipetting up and down before transfer). Spin down in a microcentrifuge at +4 °C for 15 min at 16,100×*g*.
5. Gently remove the supernatant with a pipette. Wash the pellet containing chromatin (without resuspending) (*see* **Note 9**) with 0.4 mL of ice-cold lysis buffer and spin down at 16,100×*g* for 15 min in a microfuge at +4 °C. Remove the supernatant by pipetting.
6. Resuspend the pellet with 1 mL lysis buffer. Since the pellet is very sticky at this stage, we often use a sterile toothpick or a sealed pipette tip for the resuspension. Transfer the lysate to 15 mL polycarbonate (*see* **Note 10**) Falcon tube.
7. Sonicate the lysate to shear the chromatin. We use a water bath sonicator that accommodates six samples at a time (Bioruptor; Diagenode Model #UCD-200). The Bioruptor should be set up in a cold room, and the water should be supplemented with a handful of ice just before starting sonication to minimize sample heating. Use output setting of HIGH, and set on time to 30 s and off time to 30 s. Set total sonication time to 10 min (corresponding to 5 min of sonication).
8. Keep sample on ice for 5 min and repeat **step** 7 once more (*see* **Note 11**).
9. Transfer sonicated extracts to 1.5 mL Eppendorf tube. Spin down at 16,100×*g* in a microcentrifuge for 5 min at +4 °C. Transfer the supernatant to a new 1.5 mL Eppendorf tube and proceed to immunoprecipitation or alternatively snap-freeze in dry ice or liquid nitrogen and store at −80 °C.

3.3 Immunoprecipitation (IP)

1. Before starting the IP, keep 100 μL aliquots of each extract at +4 °C to prepare the input sample. Add an appropriate amount of antibody (*see* **Note 12**) to the remaining extract (~0.9 mL) and rock in the cold room for at least 3 h. If the extracts were stored at −80 °C, thaw on ice and spin down at 16,100 × *g* in a microcentrifuge for 5 min at +4 °C before using for IP.
2. Add 15 μL of pre-equilibrated Dynabeads protein G to each IP and rock for 1 h in the cold room. To equilibrate, add 1 mL of lysis buffer to the beads, mix briefly by inverting the tube, use the magnet to collect the beads on the wall of the tube (~1 min), and remove the liquid. Sepharose or agarose beads may also be used.
3. Use the magnet to collect the beads and pipette off the supernatant.
4. Wash beads as follows: once with 1 mL of lysis + 0.1 % SDS, once with 1 mL of lysis +0.1 % SDS + 500 mM NaCl, once with 1 mL of LiCl buffer, and once with 1 mL of TE pH 7.5. Carefully remove the supernatant as in **step 3** during all washes.
5. Add 100 μL of the elution buffer and mix by pipetting up and down. Incubate at 65 °C for 20 min. Use the magnet to collect the beads and then transfer the supernatant to a new 1.5 mL Eppendorf tube. Wash beads once with 150 μL of TE + 0.67 % SDS, and combine the supernatant with the eluted fraction (the combined eluted material should be 250 μL for each IP).
6. Add 150 μL of TE + 0.67 % SDS to the input samples. Incubate all IP and input samples overnight at 65 °C (in a heat block) to reverse the cross-linking.

3.4 DNA Purification

1. Add 250 μL of TE, 2 μL of glycogen, and 1 μL of RNase A, to each sample. Incubate at 37 °C for 1 h.
2. Add 5 μL of Proteinase K to each tube, and incubate at 37 °C for 2–3 h.
3. Add 55 μL of 4 M LiCl to each sample, and mix by gently vortexing. Add 500 μL of phenol:chloroform:isoamyl alcohol, vortex vigorously for 10 s, and spin down for 1 min in microcentrifuge at 16,100 × *g*. Pipette the aqueous phase (upper phase) to a new Eppendorf tube.
4. To the aqueous phases from **step 3**, add 500 μL of chloroform, vortex briefly, and spin down at 16,100 × *g* for 1 min. Pipette the aqueous phase (upper phase) to a new Eppendorf tube.
5. To the aqueous phase from **step 4**, add 1 mL of −20 °C 95 % ethanol. Mix gently and incubate at −20 °C for 1 h.
6. Spin down the sample at 16,100 × *g* for 15 min at 4 °C, and carefully remove the supernatant by aspiration.

7. Wash the DNA pellets by adding 500 μL of ice-cold 70% ethanol to each tube. Centrifuge at 16,100×*g* for 5 min at 4 °C. Carefully remove 70% ethanol by pipetting (*see* **Note 13**). Air-dry the pellets (no more than 5 min on the benchtop) and resuspend in 50 μL TE pH 7.5. Store the ChIP samples at −20 °C until analysis.

3.5 Quantitative Real-Time PCR (qPCR)

After purification, the immunoprecipitated DNA is suitable for analysis by quantitative PCR or genome-wide methods. Here we describe our qPCR analysis methodology.

We perform qPCR analysis using the SYBR® Green method using an instrument that accommodates 96-well plates. DNA amounts in each sample are interpolated from a standard curve generated using serial dilutions of a DNA sample known to contain the amplicons of interest (usually input DNA). The sample values should fall within the linear range of the standard curve. PCR primer pairs for qPCR are typically 20–25 nucleotides in length with a T_m of 58–63 °C, and the amplicon size is 50–150 base pairs. We regularly use Primer3 (http://bioinfo.ut.ee/primer3/) for primer design. Sequences for primers that we have used successfully (located in three constitutive genes and in mitochondrial DNA as a negative control) are provided in Table 1. We set up the qPCR reaction as follows:

1. Dilute the input DNA sample by a factor of 100 (1/100) with dH_2O. From this initial diluted sample (dilution 1), set up a series of fivefold dilutions to obtain a total of six dilutions (undiluted, 1/5, 1/25, 1/125, 1/625, 1/3125).
2. Dilute primers (*see* **Note 14** and Table 1) by a factor of 100 to make a primer mix. Pipette 5 μL of each of the two primers and combine to 490 μL of dH_2O.
3. Dilute input DNA samples 1/100 and ChIP DNA samples 1/5 with dH_2O. Add 3.6 μL of each sample into separate wells of a 96-well optical plate. Each sample should be assayed in triplicate. Add 2.4 μL of the primer mix and 6 μL of the 2× SYBR® Green qPCR master mix.
4. Cover the 96-well plate with a sealing film and ensure that all wells are tightly sealed. Briefly centrifuge the plate for 1 min at 42×*g* (in a benchtop centrifuge fitted with an appropriate swinging-bucket rotor) and place the plate into the real-time thermocycler.
5. Run qPCR. We use the following protocol, which has been optimized for the primer pairs listed in Table 1. Before using new primer pairs we typically test a range of annealing temperatures (cycle 2, **step 2**) on the standards to optimize the amplification (*see* Subheading 3.5). We also use melt-curve determination (included in this protocol) to ensure that a single amplicon is produced (i.e., a single peak is observed on the melt curve) and that it has the expected T_m.

Table 1
Sequences of PCR primers in *act1*⁺, *adh1*⁺, and *cdc2*⁺. F + R primer pairs are given for the promoter, 5′ end, and 3′ end of each gene

Primer name	Sequence (5′–3′)
act1-1F	GGTTGCTCAATGTTATCCGTTTC
act1-1R	TGATAAAGCCACACACAGCGTTA
act1-2F	CTCAAAGCAAGCGTGGTATTT
act1-2R	TCTTTTCCATATCATCCCAGTTG
act1-3F	CCACTATGTATCCCGGTATTGC
act1-3R	CAATCTTGACCTTCATGGAGCT
adh1-1F	TTTTGCTTTGCCGATGTTACTT
adh1-1R	CACCTGTCCACCACCCAC
adh1-2F	ATTGGTGACCGTGTTGGTG
adh1-2R	GAAAGTTGAATGTGAGGGCAG
adh1-3F	GAAGGAAGCCGACATGATTG
adh1-3R	CAGCTTGCTCGTAAGACTTGG
cdc2-1F	ATCATTCTCGCATCTCTATTA
cdc2-1R	ATTCTCCATTGCAAACCACTA
cdc2-2F	CCAGCTAGTGAACGGTGTA
cdc2-2R	AATTGCCTTCTTTGTCTATCA
cdc2-3F	GAAGTATGGCCTGGAGTCACG
cdc2-3R	ATGGTAATGCAAAATCTT
mit-F	GCAAAAGCCAAAAGAGAAAA
mit-R	AACTATCCAATCATCAGCGTATCG

Cycle 1: (1×)
Step 1: 95.0 °C for 3 min

Cycle 2: (40×)
Step 1: 95.0 °C for 10 s
Step 2: 56.0 °C for 30 s (plate is read after each cycle)

Cycle 3: (1×)
Step 1: 95.0 °C for 10 s

Step 2: Cool to 65 °C

Cycle 4 (melt curve determination): (1×)

Step 1: 65.0–95.0 °C (0.5° intervals for 5 s each; plate is read at each interval)

Cool to 4 °C

3.6 Interpretation of ChIP-qPCR Results

Quantification relies on the C_T value, which corresponds to the number of cycles required to reach the threshold SYBR® Green fluorescence in each well. The C_T values for each of the standards are used to generate a standard curve. We typically perform these calculations after export of the raw C_T results into a spreadsheet program. Ideally, a plot of the standard C_T values against a –log transformation of the relative concentrations should yield a line with slope −3.32. Significant deviations from this slope indicate inefficient or off-target amplification. The standard curve is then used to interpolate the relative DNA amounts for each sample based on the C_T values. The IP efficiency for each IP is then expressed as a % of input calculated as follows:

$$\% = (\text{SA} - \text{IP} \times 5 \,/\, \text{SA} - \text{input} \times 100) \times 100$$

% is the ChIP signal expressed as a percentage of input sample.

SA-IP is the starting amount of DNA in the IP sample obtained during the amplification and 5 is the dilution factor of the IP sample.

SA-input is the starting amount of the input sample obtained during the amplification and 100 is the dilution factor of the input sample.

An appropriate negative control for histone ChIP in experiments is the use of a primer pair in the mitochondrial DNA (*see* Table 1). It is also important to note that ChIP signals for histone modifications can be affected by the levels of the modified histone of interest that is associated with a given genomic region, or by the level of total histone present. Thus, the % for histone modifications is typically normalized to the % measured for total histone. We typically use histone H3 for this purpose. A typical set of ChIP data for histone H2B monoubiquitylation and histone H3 lysine 4 methylation over the constitutive *act1+* gene is shown in Fig. 1.

The method we describe allows for comparison of ChIP efficiency across various genomic regions in a given strain. Comparisons across strains can be confounded by strain-to-strain variation in ChIP efficiency. To make these comparisons it is important to define a reference genomic region where ChIP efficiency is known not to change in the strains being compared.

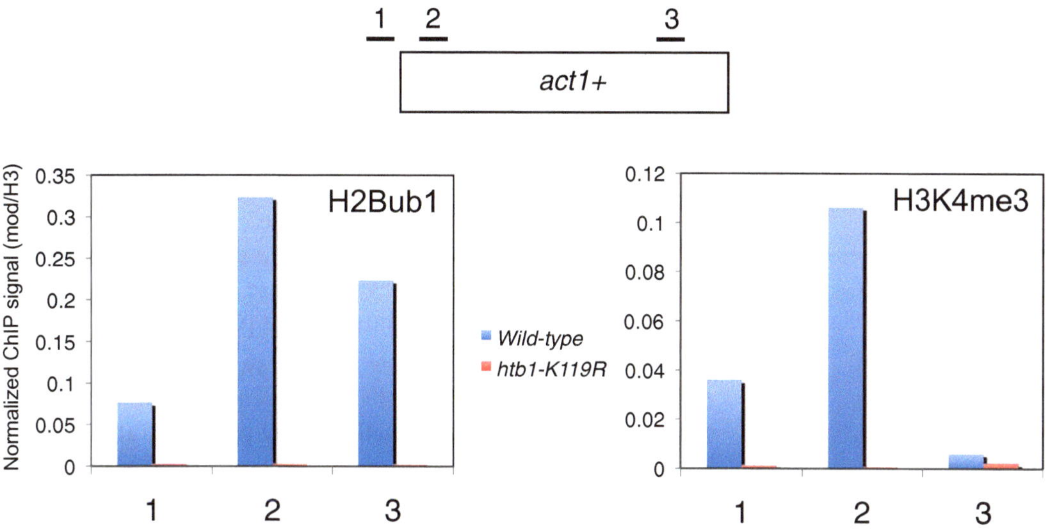

Fig. 1 ChIP of histone modifications at *act1*⁺. ChIP signals for the indicated histone modifications (H2Bub1 denotes histone H2B monoubiquitylation; H3K4me3 denotes trimethyl histone H3 at lysine 4). Primer pairs used for quantification by qPCR are shown (*see* Table 1). ChIP signals are reported as % input after normalization to total histone H3 signals at each position. The *htb1-K119R* strain, in which both modifications are absent, is shown for comparison

4 Notes

1. To make a premixed powder, 10 g each of adenine, histidine, leucine, uracil, and lysine are combined as solid powder into a beaker and mixed thoroughly using a spoon or spatula to break up any clumps. For larger mixtures a standard blender may be used. The powder is added to YES at 1 g/L.
2. We use the iQ SYBR® Green Supermix from Bio-Rad but other equivalent products may be used as well.
3. The amount of the saturated culture to be inoculated depends on the growth rate of the strain. The generation time or doubling time of the wild-type strain grown in YES medium at 30 °C is approximately 2.5 h. Cells in a saturated culture are in stationary phase. After inoculation in fresh media they will require a certain time (lag phase) to reenter the cell cycle. The lag phase is usually equal to one generation time. Use the following formula to calculate the volume of the saturated culture to be inoculated:

$$\mathrm{Vi} = (\mathrm{Vf} \times \mathrm{ODf}) / (2^{n} \times \mathrm{ODi})$$

Vi is the volume of the saturated culture to be inoculated.

Vf is the desired final volume of the exponential culture (to be used for the experiment).

ODf is the desired optical density of the exponential culture (OD_{600} of 0.4).

n is the generation time or doubling time.

ODi is the OD_{600} of the saturated culture (which should be measured after a 1/100 dilution of the culture to ensure that the OD is within the linear range of the spectrophotometer).

If using cells from stationary phase, use ($n-1$) instead of n in the formula to take into account the lag phase.

4. During cross-linking cultures may be swirled gently by hand every 5 min (such that the cells are prevented from settling), or swirled continuously on a small rotating platform (at a speed of less than 100 rpm).

5. We usually use 1.5×10^7 cells per ChIP, but if working with a very slow-growing strain, 4×10^6 cells also give a similar result. Before adding formaldehyde, mix the cell culture vigorously and remove 1 mL for counting. Make a 1/10 dilution of the culture with distilled water. Clean the hemocytometer and the coverslip with water and a Kimwipe. Place the coverslip on the hemocytometer, on top of the coverslip support chamber. Mix the diluted sample and pipette 15 μL into the chamber. There are 25 smaller squares in the central square of the hemocytometer, each of which is approximately 0.2 mm. Count cells in all the 25 smaller squares. The minimum number of counted cells should be 100. If there are too many cells on the central square after a 1/10 dilution, try 1/100 dilution. Use the following formula to calculate the number of cells per milliliter of your culture:

$$N = (\text{counted cells} / 25) \times 10{,}000 \times \text{dilution factor}$$

N is the number of cells per mL culture.

Counted cells is the number of cells counted on the 25 smaller squares of the hemocytometer.

Dilution factor is the dilution factor of the 1 mL culture (10 or 100).

An OD_{600} of 0.4 corresponds to approximately $4–6 \times 10^5$ cells/mL in a wild-type strain. This value can differ significantly in mutant strains, particularly those that show a defect in cell size.

6. The glass beads should reach just below the meniscus of the liquid in the tube. To conveniently measure a consistent amount of beads, we typically use an Eppendorf tube previously marked at 0.4 mL as a "scoop" for the glass beads. Glass beads can be stored in a small beaker at 4 °C.

7. The efficiency of cell lysis by bead beating can be estimated by light microscopy. Intact cells appear "phase bright" with a

defined interior, whereas lysed cells appear "phase dark." Lysis efficiency should be at least 90%. Critical parameters affecting lysis efficiency include the volume of beads, the cell density, and bead-beating time. Additional cycles of bead-beating may be required for efficient lysis of some mutant strains.

8. The lysate should appear cloudy and there should be a small pellet of cell debris at the bottom of the 5 mL tube.
9. In **steps 5** and **6**, Subheading 3.2, the chromatin should be visible as a glassy layer on top of a white pellet containing cell debris. In **step 9**, Subheading 3.2 (after sonication and chromatin shearing), this pellet is no longer visible.
10. Polycarbonate tubes are used in order to optimize sonication using the Bioruptor (**step 7**, Subheading 3.2).
11. Sonication time should be determined to achieve the optimal chromatin shear size. We have optimized our sonication to yield chromatin fragments in the range of ~300 base pairs in length. Approximate shear size can be determined by subjecting DNA purified from the sonicated extracts (i.e., after **step 7**, Subheading 3.2) to electrophoresis on a 2% agarose gel followed by staining (using ethidium or SYBR gold stains; *see* Fig. 2). A small sample of extract (100 μL) should typically be analyzed this way before commencing with the IP steps.

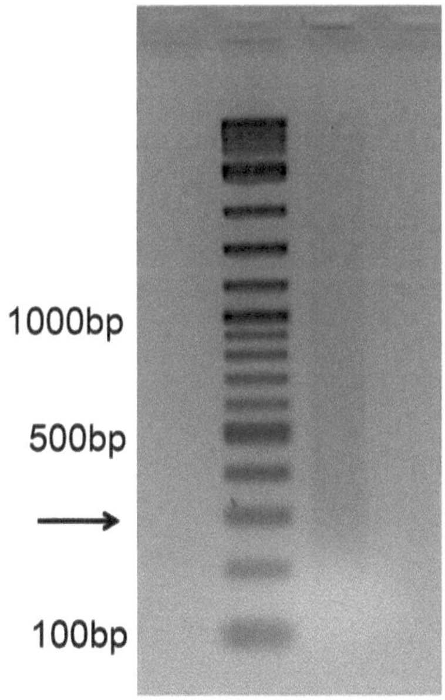

Fig. 2 Sonicated ChIP DNA. Input DNA from a typical ChIP experiment was subjected to 2% agarose gel electrophoresis and stained with ethidium bromide. Shown are size markers (*left*) and the estimated average DNA fragment size (*arrow*)

12. The amount of antibody is assay dependent and will need to be determined empirically. We have had success using 4 μg of the indicated antibodies per 1 mL of lysate for IPs (antibodies are typically supplied at a concentration of 1 mg/mL).
13. Extreme care must be taken in removing the supernatants at this stage as the DNA pellets are small and easily dislodged from the wall of the tube.
14. We resuspend primers in 1× TE at a final concentration of 100 μM. For qPCR, each primer is diluted to 1 μM and 1.2 μL of the diluted primer is used per well.

References

1. Bannister AJ, Kouzarides T (2011) Regulation of chromatin by histone modifications. Cell Res 21:381–395
2. Aparicio O, Geisberg JV, Sekinger E, Yang A, Moqtaderi Z et al (2005) Chromatin immunoprecipitation for determining the association of proteins with specific genomic sequences in vivo. Curr Protoc Mol Biol Chapter 21:Unit 21.23
3. Li B, Carey M, Workman JL (2007) The role of chromatin during transcription. Cell 128:707–719
4. Smolle M, Workman JL (2013) Transcription-associated histone modifications and cryptic transcription. Biochim Biophys Acta 1829:84–97
5. Rhind N, Chen Z, Yassour M, Thompson DA, Haas BJ et al (2011) Comparative functional genomics of the fission yeasts. Science 332:930–936
6. Moazed D (2009) Small RNAs in transcriptional gene silencing and genome defence. Nature 457:413–420

Chapter 13

A Spiking Strategy for ChIP-chip Data Normalization in *S. cerevisiae*

Célia Jeronimo and François Robert

Abstract

Chromatin immunoprecipitation coupled to DNA microarrays (ChIP-chip) is widely used in the chromatin field, notably to map the position of histone variants or histone modifications along the genome. Often, the position and the occupancy of these epigenetic marks are to be compared between different experiments. It is now increasingly recognized that such cross-sample comparison is better done using externally added exogenous controls for normalization but no such method has been described for ChIP-chip. Here we describe a spiking normalization strategy that makes use of phiX174 phage DNA as a spiked control for normalization of ChIP-chip signals across different experiments.

Key words Chromatin immunoprecipitation, ChIP-chip, DNA tiling microarrays, Microarray normalization, Spike-in control, Chromatin, Histone variant, Histone modifications, *S. cerevisiae*

1 Introduction

Chromatin immunoprecipitation coupled to DNA microarrays (ChIP-chip) is a widely used technology for the determination of the genome-wide location pattern of nuclear proteins in vivo [1]. In recent years, ChIP-chip has been progressively supplemented by ChIP-Seq, a related technique where high-throughput sequencing is used as a readout of the immunoprecipitated DNA [2]. For organisms with a small genome like yeast, however, ChIP-chip remains widely used, in part due to its low cost relative to sequencing.

Most often, ChIP-chip and ChIP-Seq experiments are normalized using whole data points (all the spots on the microarray or all the reads from the sequencing), generally using nonlinear models such as Limma Loess or quantile normalization [3]. While these approaches are adequate in most cases, alternatives are sometimes necessary in order to get a more faithful representation of protein localization, especially when different experiments are to be compared. For example, if a protein binds to a large fraction of the genome, normalizing using all data points would dampen the

Benoit Guillemette and Luc Gaudreau (eds.), Histones: Methods and Protocols, Methods in Molecular Biology, vol. 1528,
DOI 10.1007/978-1-4939-6630-1_13,

enrichment observed for this protein. On itself, this is without strong consequences as the ranked order of binding sites is independent of the normalization. In other words, top hits will remain on top regardless of the normalization scheme. When comparing different samples (i.e., different mutants, different growth conditions), however, using global normalization may lead to erroneous interpretations if the studied protein drastically changes its localization, from binding a few discrete sites to promiscuously binding to a large fraction of the genome for example (Fig. 1). For instance, in yeast, if a protein binds to all promoters (~14% of the genome) in condition A but also populates ORFs (~80% of the genome) in condition B, global normalization will not allow for a faithful representation of condition B. Indeed, the data will rather look as if binding decreases at promoters while in fact it rather increases at other sites (ORFs). This happens because global normalization forces the average to one, which leads to a collapse of the enrichment

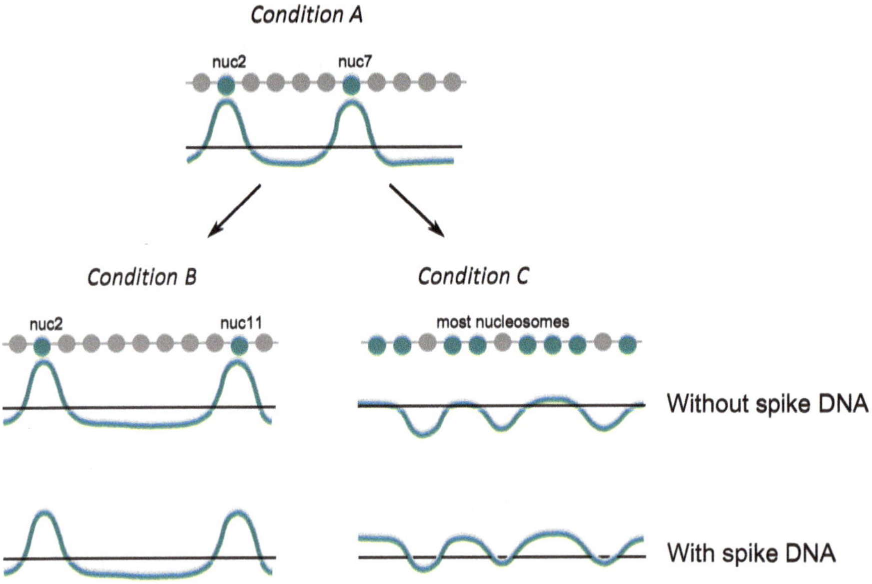

Fig. 1 A cartoon representation illustrating how the use of spike DNA is, in some cases, required in order to accurately interpret ChIP-chip data. A stretch of chromatin is shown where two nucleosomes (nuc2 and nuc7) carry the modification of interest in *Condition A*. The green trace below the cartoon illustrates the ChIP-chip pattern that would be obtained by ChIP-chip in such a scenario. In *Condition B*, the modification disappears from nuc7 and is gained on nuc11. In such a scenario where few changes occur on the number of binding or their locations, the spiking is not necessary as standard normalization techniques can easily capture these changes. In *Condition C*, however, the modification spreads on a large number of nucleosomes such that the pattern is radically different from the one observed in *Condition A*. In such a scenario, standard normalization techniques fail to capture the widespread presence of the modification. Instead, the enriched regions are dampened to baseline and the few regions unoccupied by the modification appear as depleted. This happens because these standard normalization techniques set the genome average to baseline, which, in cases where most of the genome is occupied, constitutes a flawed premise. Because the spiked DNA is added exogenously, its enrichment can be used to scale data from *Conditions B* and *C* so that they can be compared to *Condition A*

when a large fraction of the genome is actually bound by the immunoprecipitated protein. In other words, and taking an extreme scenario, global normalization makes "binding everywhere" looking like "binding nowhere."

A normalization method that prevents such problem in ChIP-Seq experiments has been described [4]. This method relies on spiking (adding) in a fixed amount of exogenous (foreign) chromatin in the extract prior to immunoprecipitation. Alternatively, a fixed amount of foreign cells could be added prior to the extract preparation. In both cases, reads corresponding to the foreign species are used to normalize the data.

To our knowledge, spike-in strategies have not been described for ChIP-chip experiments. Because ChIP-chip relies on hybridization to tiling microarrays, exogenous DNA from a foreign species might cross-hybridize and jeopardize the experiment. Here, we describe a spike-in strategy for ChIP-chip experiments in *Saccharomyces cerevisiae* (Fig. 2). In these experiments, sonicated phiX174 phage DNA is added to the chromatin extract prior to immunoprecipitation. The sequence from the phage DNA being radically different from that of *S. cerevisiae*, signal from control spots (corresponding to phiX174 sequences) on the microarray can be used to normalize the data (Fig. 3). This approach was

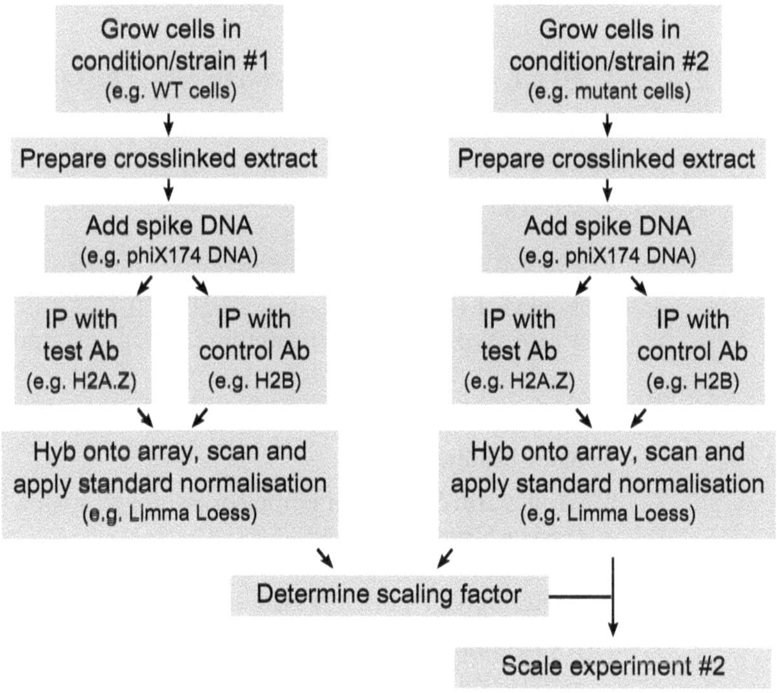

Fig. 2 A flowchart of the experimental setup used in this protocol. The different steps are shown in *grey boxes* and examples matching the specific case of the H2A.Z ChIP experiment described in our protocol are stated in *parentheses*

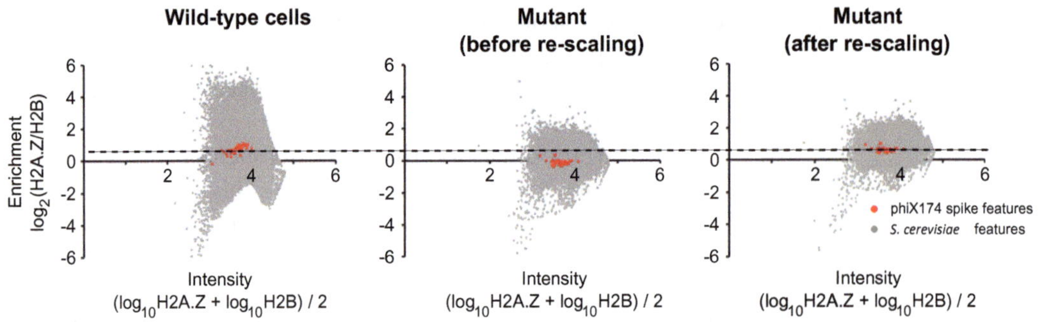

Fig. 3 MA plots of ChIP-chip experiments measuring H2A.Z enrichment relative to H2B in wild-type cells (*left*) and in a mutant for a histone chaperone (*middle* and *right*) are shown. The data from the mutant cells are shown before (*middle*) and after (*right*) rescaling using the data collected on the phiX174 microarray features (*red*). Note how the average enrichment of the phiX174 microarray features in the rescaled dataset matches that of the wild-type experiment, as shown by the *dashed line*

instrumental in our ability to detect pervasive localization of the histone variant H2A.Z in mutants for factors involved in restricting the histone variant to promoter regions [5]. In principle, this approach should be applicable to other histone variants or histone modifications.

2 Materials

2.1 Growing and Cross-Linking Cells

1. 50 mL of exponentially growing cells (i.e., OD_{600} of 0.6–1.0).
2. 37% Formaldehyde solution.
3. 2.5 M Glycine.
4. Ice-cold TBS.

2.2 Chromatin Preparation

1. Lysis buffer: 50 mM HEPES–KOH pH 7.5, 140 mM NaCl, 1 mM EDTA, 1% Triton X-100, 0.1% Na-deoxycholate, 1 mM PMSF, 1 mM benzamidine, 10 μg/mL aprotinin, 1 μg/mL leupeptin, 1 μg/mL pepstatin.
2. Acid-washed glass beads (425–600 μm).
3. Mini-beadbeater.
4. Model 100 Sonic dismembrator equipped with a micro probe (Fisher Scientific).

2.3 Spiking Samples

1. Bacteriophage phiX174 RF1 DNA (50 μg).
2. 6×16 mm Glass microtubes with AFA fiber and pre-slit snap caps (Covaris).
3. Covaris E220 ultrasonicator (Covaris).

2.4 Immunoprecipitation and Washes

1. Magnetic beads coated with protein G.
2. Rabbit anti-H2A.Z (Active Motif, 39647).
3. Rabbit anti-H2B (Abcam, ab1790).
4. Magnet-rack for 1.5 mL tubes.
5. PBS/BSA: PBS supplemented with 0.5% bovine serum albumin.
6. Lysis buffer (same as for chromatin preparation).
7. Lysis buffer 500: Lysis buffer supplemented with 360 mM NaCl (500 mM NaCl final).
8. Wash buffer: 10 mM Tris–HCl pH 8.0, 250 mM LiCl, 0.5% NP40, 0.5% Na-deoxycholate, 1 mM EDTA.
9. TE buffer: 10 mM Tris pH 8.0, 1 mM EDTA

2.5 DNA Isolation

1. TE/SDS: 10 mM Tris pH 8.0, 1 mM EDTA, 1% SDS.
2. TE buffer: 10 mM Tris pH 8.0, 1 mM EDTA.
3. 10 mg/mL DNase-free RNase A solution in ddH_2O.
4. 20 mg/mL Glycogen.
5. 10% SDS solution in ddH_2O.
6. 20 mg/mL Proteinase K solution.
7. Phenol:chloroform:isoamyl alcohol (25:24:1) saturated with 10 mM Tris, pH 8.0, 1 mM EDTA.
8. 5 M NaCl.
9. 100% Ethanol (−20 °C).
10. 70% Ethanol (−20 °C).

2.6 Amplification of ChIP DNA by Ligation-Mediated PCR (LM-PCR)

1. 3 U/μL T4 DNA polymerase (New England Biolabs).
2. 10× NEBuffer 2 (New England Biolabs).
3. 10 mg/mL Ultrapure BSA.
4. dNTP mix: 20 mM dATP, 20 mM dCTP, 20 mM dGTP, and 20 mM dTTP.
5. Blunting mix: 11 μL 10× NEBuffer 2, 0.5 μL BSA, 0.5 μL dNTPs, 0.2 μL T4 DNA polymerase, and 57.8 μL ice-cold water.
6. 3 M NaOAc solution pH 5.2.
7. 20 mg/mL Glycogen.
8. Phenol:chloroform:isoamyl alcohol (25:24:1) saturated with 10 mM Tris, pH 8.0, 1 mM EDTA.
9. 100% Ethanol (−20 °C).
10. 70% Ethanol (−20 °C).
11. 5 U/μL T4 DNA ligase (Life Technologies).
12. 5× T4 DNA ligase buffer (Life Technologies).

13. Annealed linkers (*see* protocol below).
14. 40 μM Oligo 1 (GCGGTGACCCGGGAGATCTGAATTC) stock.
15. 40 μM Oligo 2 (GAATTCAGATC) stock.
16. Ligase mix: 8 μL ice-cold water, 10 μL 5× T4 DNA ligase buffer, 6.7 μL annealed linkers, 0.5 μL T4 DNA ligase.
17. 10× ThermoPol buffer (New England Biolabs).
18. 5-(3-Aminoallyl)-dUTP (aa-dUTP).
19. aa-dUTP/dNTP mix: 2 mM aa-dUTP, 3 mM dTTP, 5 mM each dATP, dCTP, dGTP.
20. Labeling mix: 7.75 μL ice-cold water, 4 μL 10× ThermoPol buffer, 2 μL aa-dUTP/dNTP mix, 1.25 μL oligo 1.
21. 5 U/μL Taq DNA polymerase.
22. 2.5 U/μL Pfu DNA polymerase.
23. Enzyme mix: 8 μL Ice-cold water, 1 μL 10× ThermoPol buffer, 1 μL Taq DNA polymerase, 0.01 μL Pfu DNA polymerase.

2.7 Fluorescent Labeling of Amplified ChIP DNA

1. QIAquick PCR purification kit (Qiagen).
2. Phosphate wash buffer: 5 mM KPO_4 pH 8.5 buffer, 80% ethanol.
3. Phosphate elution buffer: 4 mM KPO_4 pH 8.5 buffer.
4. 1 M Na_2CO_3 pH 9.0 (make fresh every few weeks to a month).
5. 0.1 M Na_2CO_3 pH 9.0 (diluted from fresh 1 M stock solution).
6. Cy5 Mono-Reactive NHS Ester (GE Healthcare).
7. Cy3 Mono-Reactive NHS Ester (GE Healthcare).
8. 0.1 M NaOAc pH 5.2.

2.8 Hybridization

1. Ozone-free enclosure.
2. Microarray hybridization chamber (Agilent).
3. Hybridization oven with rotator slide rack (Agilent).
4. *S. cerevisiae* DNA tiling microarray containing several hundreds of probes covering the phiX174 genome (Agilent, custom-designed arrays).
5. Microarray gasket slide (Agilent).
6. DIG Easy Hyb buffer (Roche Life Science).
7. 10 mg/mL Salmon sperm DNA (50 μg per microarray).
8. 8 mg/mL Yeast tRNA per microarray (40 μg per microarray).
9. Hybridization mix: 100 μL DIG Easy Hyb buffer, 5 μL salmon sperm DNA, and 5 μL yeast tRNA.

10. 20× SSPE buffer: 3 M NaCl, 200 mM NaH_2PO_4, and 20 mM EDTA (pH 7.4).
11. Wash solution I: 6× SSPE, 0.005 % N-lauroylsarcosine.
12. Wash solution II: 0.06× SSPE.
13. Staining dishes.
14. Slide rack.
15. Orbital shaker.

3 Methods

3.1 Growing and Cross-Linking Cells

Nearly any growth condition can be used depending on the experimental design needed. Generally, we use 50 mL of a fresh logarithmically growing yeast culture (i.e., OD_{600} of 0.6–1.0) per ChIP. Once cells reach the desired OD, protein-DNA cross-linking is performed by treating the cells with formaldehyde. The cross-linking time and the formaldehyde concentration should be optimized if initial attempts to immunoprecipitate a protein fail. The cross-linking reaction is terminated by adding glycine to the culture medium, which quenches formaldehyde. When comparing different samples [for example, wild-type (WT) and mutant cells], strains should be grown and treated in tandem and at least in duplicates. In this protocol, WT and temperature-sensitive cells were grown in yeast extract-peptone-dextrose medium to an OD_{600} of 0.6 at 30 °C and transferred to 37 °C for 80 min before cross-linking. The genotypes for the two yeast strains used here are as follows: WT (MATα leu2Δ1) and *spt16-197* mutant (MATα leu2Δ1 spt16-197).

1. For each strain, grow 50 mL of yeast in a flask containing the appropriate medium until OD_{600} reaches 0.6–1.0.
2. Transfer the content of the flask into a 50 mL conical tube containing 1.4 mL of 37 % formaldehyde.
3. Incubate for 30 min at room temperature with agitation.
4. Add 2.5 mL of 2.5 M glycine to quench the formaldehyde.
5. Centrifuge the tubes for 5 min at 2000 ×*g* in a refrigerated (4 °C) table-top centrifuge.
6. Pour off the supernatant and vigorously wash the cells with 40 mL of ice-cold TBS. Centrifuge again as in **step 5**.
7. Repeat **step 6** for a total of two washes.
8. Using the remaining liquid, resuspend the cell pellet and transfer into a 1.5 mL screw-cap tube.
9. Centrifuge for 1 min at maximum speed in a microcentrifuge. Remove the supernatant with a pipette and snap freeze in liquid nitrogen.
10. At this point, samples can either be used directly or stored at −80 °C.

3.2 Chromatin Preparation

Yeast cells, notoriously difficult to break due to their cell wall, can be broken by vigorous shaking in the presence of glass beads using a mini-beadbeater (often referred to as bead beating). After obtaining the cell lysate, sonication conditions should be determined empirically. This can be done by looking at the size of the sheared DNA on an agarose gel. The DNA fragments should have an average size of ~400 bp. The sonication protocol described in this section was optimized for a Fisher sonic dismembrator 100 equipped with a micro probe. All the steps indicated here must be performed with cold buffers, on ice or at 4 °C.

1. Thaw cell pellet on ice and resuspend it in 700 μL of lysis buffer.
2. Add the equivalent of a 0.5 mL PCR tube of glass beads.
3. Screw cap tightly and secure tubes to the beadbeater.
4. Beat for 5 min and then put the tubes on ice for 5 min.
5. Repeat **step 4** three more times (*see* **Note 1**).
6. Punch a hole at the bottom of the tube using an 18G1 ½ needle (stop as soon as liquid or air is going through) and set up over a 2 mL screw-cap tube.
7. Loosen the cap and spin for 3–4 s. This should allow the material to transfer into the 2 mL tube while the beads stay trapped into the top tube.
8. Resuspend the chromatin pellet with its own supernatant and transfer into a new 1.5 mL tube (*see* **Note 2**). Proceed to sonication.
9. Sonicate the sample four times for 20 s with an output of around 7 W. Between each sonication cycle, keep the tube on ice for at least 1 min.
10. Centrifuge for 5 min at maximum speed in a microcentrifuge and transfer 600 μL of the supernatant containing the solubilized chromatin fragments to a new 1.5 mL tube. From now on, this will be referred to as the chromatin extract.

3.3 Spiking Samples

In order to correct for microarray normalization artifacts caused by global occupancy changes, ChIP-chip experiments are done here using a spiking strategy. In these experiments, a fixed amount of foreign DNA (in our case, sonicated bacteriophage phiX174 genomic DNA) is spiked in (added) to chromatin extracts prior to immunoprecipitation. The sonication conditions described here were optimized for a Covaris E220 ultrasonicator, well known for its reproducibility, precision, and rapidity. The amount of sonicated phiX174 genomic DNA to be added to each chromatin extract must be determined empirically and it depends on the experimental design used.

1. Dissolve 50 μg of bacteriophage phiX174 RF1 genomic DNA (commercial stock at 0.5 μg/μL) in 1990 μL of MilliQ water for a final concentration of 25 ng/μL.
2. Transfer 130 μL of diluted phiX174 DNA to a 6 × 16 mm glass microtube with AFA fiber and pre-slit snap cap. Make six of these 130 μL aliquots in total. Store the remaining diluted phiX174 DNA at −20 °C for future experiments.
3. Cool down and degas the Covaris sonicator with water level at "6" and ensure that the amplifier is in.
4. Shear phiX174 DNA for 105 s using the following Covaris sonicator settings: duty factor: 10%, PIP: 140, cycles per burst: 200, volume: 130 μL, amplifier: yes.
5. Run 10 μL on 1.5–2% agarose gel next to a 100 bp DNA ladder (expect a smear between 100 and 600 bp with an average size of ~300 bp).
6. Make aliquots (50 μL) of the sonicated phiX174 DNA (25 ng/μL) and store them at −20 °C. The sonicated phiX174 DNA is ready to be used in spike-in experiments.
7. Add 300 ng of sonicated phiX174 DNA to each soluble chromatin extract (from Subheading 3.2, **step 10**) (*see* **Note 3**).

3.4 Immunoprecipitation and Washes

To immunoprecipitate the protein–DNA complexes, a great variety of systems can be used, including both monoclonal and polyclonal antibodies as well as different types of resins (agarose, sepharose, or magnetic beads). We favor the use of magnetic beads since they reduce nonspecific binding, limit the loss of resin, and save time. The protocol described here uses anti-H2A.Z (test) and anti-H2B (control) antibodies bound to protein G-coupled magnetic beads.

1. Soluble chromatin supplemented with sonicated phiX174 DNA (from Subheading 3.3, **step** 7) is further diluted with 600 μL of lysis buffer allowing incubating the same chromatin extract with two antibodies: anti-H2A.Z and anti-H2B. Each immunoprecipitation is thus done in the presence of 150 ng of sonicated phiX174 DNA.
2. The beads should be prepared the day prior to use in the following way. Take enough beads to get 50 μL per immunoprecipitation. Put into a conical plastic tube, and centrifuge for 3 min at low speed (2000 × *g*) to pellet the beads. Wash twice with PBS+BSA and incubate them overnight at 4 °C with agitation in the presence of anti-H2A.Z antibody (2 μL per ChIP) or anti-H2B antibody (5 μg per ChIP) and enough PBS+BSA to ensure a good agitation. The next day (prior to use) wash twice in lysis buffer to remove non-bound antibodies and resuspend beads in 30 μL of lysis buffer per immunoprecipitation.

3. Set up a tube with 600 μL of diluted chromatin extract containing the spike DNA and 30 μL of anti-H2A.Z-coupled magnetic beads and a second tube with the remaining 600 μL of diluted chromatin extract and 30 μL of anti-H2B-coupled magnetic beads.
4. Incubate with rotation overnight at 4 °C.
5. Using a magnet, wash the magnetic beads with 1 mL of lysis buffer as follows. Put the tubes to the magnetic system and wait for a few seconds for the beads to get captured by the magnet. By keeping the tubes in the system, remove the liquid by inverting the tubes. Add the washing solution, close the tubes, remove the magnet, and agitate for a few seconds to resuspend the beads. Put the magnet back on and remove the liquid as above. Repeat the wash step with 1 mL of lysis buffer, twice with 1 mL of lysis buffer 500, twice with 1 mL of wash buffer, and once with 1 mL of TE.
6. After the last wash (TE), centrifuge briefly and remove the last bit of liquid using a P200 pipette. This step should leave a dry bead pellet.

3.5 DNA Isolation

After reversing formaldehyde cross-linking in the presence of heat, the immunoprecipitated DNA is purified by a succession of RNase A and protease K treatments, followed by phenol extraction and ethanol precipitation.

1. To the bead pellets (from Subheading 3.4, **step 5**), add 50 μL of TE/SDS, vortex, and incubate overnight at 65 °C (use an oven instead of a water bath in order to minimize evaporation and condensation).
2. Centrifuge for 1 min at maximum speed and transfer the supernatant to a new tube.
3. Add 350 μL of a mixture of TE, RNase A, and glycogen (345 μL TE, 3 μL RNase A, 2 μL glycogen). Mix by vortexing and incubate for 2 h at 37 °C.
4. Add 15 μL of 10% SDS and 7.5 μL of proteinase K solution. Mix by vortexing and incubate for 2 additional hours at 37 °C.
5. Extract twice with 400 μL of phenol/chloroform/isoamyl alcohol solution.
6. Add 14 μL of 5 M NaCl (200 mM final concentration) and vortex. Add 1 mL of cold 100% ethanol, vortex, and centrifuge for 20 min at maximum speed at 4 °C.
7. Pour off liquid and wash pellet with cold 70% ethanol.
8. Centrifuge for 5 min and pour off liquid.
9. Centrifuge briefly, remove the last bit of liquid with a pipette, and let dry for 1 min.

10. Resuspend the pellet (which should be very small but still visible) in 50 μL of TE.
11. Store at −20 °C. The ChIP DNA is now ready to be amplified and labeled before being hybridized on microarrays.

3.6 Amplification of ChIP DNA by Ligation-Mediated PCR (LM-PCR)

The immunoprecipitated DNA fragments are first blunted, and then ligated to short unidirectional double-stranded oligonucleotide linkers, and finally amplified via PCR where aminoallyl-modified dUTP is incorporated in the product.

1. Transfer 40 μL of ChIP DNA to a 0.5 mL PCR tube. Keep the sample on ice.
2. Add 70 μL of blunting mix.
3. Mix by pipetting and incubate for 20 min at 12 °C (a thermocycler block is used here).
4. Place on ice and add 12 μL of the following mix: 11.5 μL 3 M NaOAc and 0.5 μL glycogen.
5. Vortex and add 120 μL of phenol/chloroform/isoamyl alcohol solution.
6. Vortex hard and centrifuge for 5 min at maximum speed.
7. Transfer 110 μL to a new 1.5 mL tube and add 230 μL of cold 100% ethanol.
8. Vortex and centrifuge for 20 min at 4 °C.
9. Pour off supernatant and wash pellet with 500 μL of cold 70% ethanol.
10. Vortex and centrifuge for 5 min at 4 °C.
11. Pour off liquid. Centrifuge briefly and remove the remaining liquid with a pipette.
12. Let air-dry for 1 min. Add 25 μL of ice-cold water and put on ice for about 30 min.
13. Vortex, centrifuge briefly, and put on ice.
14. Add 25 μL of ligase mix (*see* **Note 4** for linker preparation).
15. Mix by pipetting and incubate overnight at 16 °C (a water bath is fine for this step).
16. Add 6 μL of 3 M NaOAc, vortex, and add 130 μL of cold 100% ethanol.
17. Precipitate samples for at least 30 min at −20 °C.
18. Centrifuge for 20 min at 4 °C, maximum speed.
19. Pour off supernatant and wash pellet with 500 μL of cold 70% ethanol.
20. Vortex and centrifuge for 5 min at 4 °C, maximum speed.

21. Pour off supernatant, centrifuge briefly, and remove remaining liquid with a pipette.
22. Let air-dry for a minute and add 25 μL of ice-cold water. Vortex at low speed for about 5 min to rinse the tube (the pellet at this stage is usually diffused on the side of the tube).
23. Leave on ice for about 30 min, vortex, centrifuge briefly, and put on ice.
24. Add 15 μL of labeling mix.
25. Mix by pipetting, rinsing the side of the tube, and transfer to a new 0.5 mL PCR tube.
26. Put in thermocycler and start the following PCR program for amplification: step 1: 4 min at 55 °C, step 2: 5 min at 72 °C, step 3: 2 min at 95 °C, step 4: 30 s at 95 °C, step 5: 30 s at 55 °C, step 6: 1 min at 72 °C, step 7: go back to step 4 31 more times, step 8: 4 min at 72 °C, step 9: 4 °C forever.
27. When temperature reaches 55 °C, add 10 μL of enzyme mix and mix by pipetting. Proceed with the PCR program.

3.7 Fluorescent Labeling of Amplified ChIP DNA

The labeling is done using mono-reactive NHS esters of fluorescent Cy dyes that will react specifically with the aminoallyl-modified dUTP incorporated in the ChIP DNA during the LM-PCR reaction (*see* **Note 5**).

1. Purify the LM-PCR reactions using the QIAquick PCR purification kit following the manufacturer's instructions, but replace buffer PE with the phosphate wash buffer and buffer EB with the phosphate elution buffer. Do the wash step twice. Elute with 50 μL of phosphate elution buffer.
2. Dry the eluates in a speed-vac.
3. Resuspend the pellets in 4.5 μL of fresh 0.1 M Na_2CO_3 pH 9.0 buffer.
4. Add 4.5 μL of the appropriate Cy dye [we use Cy5 for the test (H2A.Z) and Cy3 for the control (H2B)] (*see* **Note 6**).
5. Incubate for 1 h in the dark at room temperature.
6. Add 35 μL of 0.1 M NaOAC pH 5.2.
7. Purify the samples using the QIAquick PCR purification kit following the manufacturer's instructions. Do the wash step twice. Elute with 50 μL of buffer EB. Color should be visible on the filter and the eluate should have a faint but visible taint of blue/pink (*see* **Note 7**).
8. Combine the corresponding Cy5- and Cy3-labeled samples (test and control ChIPs).
9. Dry the samples in a speed-vac until they reach ~20 μL.
10. Store the combined labeled samples at −20 °C in the dark or proceed with hybridization.

3.8 Hybridization

The combined fluorescently labeled samples are co-hybridized on DNA microarrays, competitively binding to the probes, and the resulting fluorescence intensity ratios reflect the relative abundance of both samples. In this protocol, to allow for external normalization, we hybridize the samples on DNA tiling microarrays containing a total of about 180,000 Tm-adjusted 60-mer probes covering the entire yeast genome as well as several hundred probes covering the phiX174 genome custom designed by us and manufactured by Agilent (*see* **Note 8**). Hybridization is done in Agilent hybridization chambers, ensuring a homogenous hybridization of the slide surface. All the process and especially the washes and scanning are done in an ozone-free chamber to minimize the degradation of the Cy dyes (*see* **Note 9**).

3.8.1 Hybridization

1. Add 110 μL of fresh hybridization mix to each combined sample. Mix by pipetting up and down.
2. Denature samples for 5 min in a 95 °C heat block.
3. Vortex, spin briefly, and transfer to a 45 °C heat block for another 5 min.
4. Assemble the gasket slide and the base of the hybridization chamber.
5. Dispense the sample on the gasket slide (avoid bubbles).
6. Carefully cover with the microarray slide and assemble the hybridization chamber upper part.
7. Slip the clamp assembly onto the chamber and tighten the thumbscrew. Vertically rotate the slide to ensure that the bubbles are moving freely.
8. Incubate in the hybridization oven with rotation at 42 °C for at least 18 h.

3.8.2 Microarray Washing

1. Pour wash solution I (~250 mL) into a standard staining dish with a slide rack and into a second dish that will be used as "disassembly dish."
2. Remove the hybridization chamber from the oven, disassemble all metal parts, submerge the microarray slide/gasket slide sandwich into the "disassembly dish," and separate the slides using plastic tweezers.
3. Transfer the microarray slide to the slide rack in the second dish containing wash solution I.
4. Proceed to the next slide.
5. Wash the microarray slides with wash solution I on an orbital shaker for 5 min.
6. Transfer the slide rack to a dish filled with wash solution II and wash on the orbital shaker for another 5 min. Slowly remove the rack from the liquid.

7. Store the microarray slides in the dark and proceed to scanning rapidly.

3.9 Data Analysis

After data acquisition (using the scanner of your choice), the data is analyzed using your favorite global normalization technique (we use Limma Loess) and replicates are combined. Next, the average of $\log_2$ ratios for the control (phiX174) spots is calculated and the data for each experiment is scaled by setting this ratio to a common number, which is usually the ratio obtained in the experiment considered to be the reference. For example, when comparing ChIP signals for a given protein in a WT cell with ChIP for the same protein in different mutants, the data from the mutants can be rescaled by adjusting the average ratio of their control spots to the corresponding value obtained in the WT experiment. Analogously, if different growth conditions or different treatments are compared, each condition/treatment can be rescaled by setting the average ratio of their control spots to the corresponding value in the reference experiment (no treatment or mock treated). Note that the average ratio of the phiX174 spots is not set to one (or zero in $\log_2$) since it is not necessarily expected to be. Indeed, the amount of phiX174 DNA that is pulled down is likely to be different between different antibodies. However, this amount should be independent of the strain background, the presence of a mutation, or the growth condition used since the DNA is spiked in the sonicated chromatin extract. Setting the ratio of the spiked DNA to a common number therefore allows for a more accurate comparison between experiments performed with the same antibodies but in different strains/conditions (*see* **Note 10**).

4 Notes

1. The amount of beating that is required may vary depending on the equipment, the yeast strain, the age of the culture, and the sample/bead ratio used. Cell breakage can be easily monitored by looking at cells under a light microscope.
2. Alternatively, the supernatant can be discarded and the chromatin pellet resuspended in 700 μL of fresh lysis buffer. For proteins with a prominent soluble form (not associated with chromatin), this may prevent unbound proteins from competing for antibody during the subsequent immunoprecipitation step.
3. 300 ng of sonicated phiX174 DNA represents a 100-fold molar excess relative to the yeast chromatin present in the extract. This amount was tested empirically and was shown to lead to an appropriate amount of phiX174 DNA in the immunoprecipitated sample when using our H2A.Z and H2B anti-

bodies. This step may have to be optimized when using other antibodies.

4. The annealed linkers are prepared by annealing oligos 1 and 2 the following way. Mix 250 μL 1 M Tris pH 7.9, 375 μL oligo 1, and 375 μL oligo 2. Make 100 μL aliquots in 1.5 mL tubes. Place in a 95 °C heat block for 5 min. Transfer tubes to a 70 °C heat block filled with water. Place the block at room temperature and let it cool down to 25 °C. Transfer the block to 4 °C and allow it to cool down overnight. Store annealed linkers at −20 °C. The Tm of the linkers being very low, always keep the linkers on ice and use ice-cold buffers when manipulating it to avoid de-annealing.
5. This method is less expensive and yields better results than the direct incorporation of Cy-dye-modified dNTPs during PCR. We found that dye swap experiments are not necessary with this protocol.
6. Resuspend new tubes of the Cy3 and Cy5 dyes in 73 μL of DMSO (kept in the dark, under desiccant and vacuum). Store at −80 °C.
7. Optionally, DNA amount and dye incorporation can be estimated by spectrometry prior to speed-vac using 50 μL of the undiluted elution (which is recovered from the cuvette after measurement). For this, we measured the absorbance at 260 nm, 320 nm, 550 nm, and 650 nm for Cy3 and 260 nm, 320 nm, 650 nm, and 750 nm for Cy5 and apply the following formulas:

$$\text{DNA}(\mu\text{g}) = (A_{260} - A_{320}) \times 50\,\mu\text{g/mL} \times \text{volume}(\text{mL}) \times \text{dilution factor}$$

$$\text{Base}/\text{Cy3} = \left(\left((A_{260} - A_{320}) - \left((A_{550} - A_{650}) \times 0.08\right)\right) \times 150{,}000\right) / \left((A_{550} - A_{650}) \times 6600\right)$$

$$\text{Base}/\text{Cy5} = \left(\left((A_{260} - A_{320}) - \left((A_{650} - A_{750}) \times 0.05\right)\right) \times 250{,}000\right) / \left((A_{650} - A_{750}) \times 6600\right)$$

$$\text{Dye molecules}/100\,\text{bp} = 100/(\text{base}/\text{dye})$$

Expect about 5 μg of DNA (typically between 4 and 6 μg) and an incorporation of 1.5 dyes per 100 bp (typically between 1.2 and 1.8 dyes per 100 bp).

8. These arrays have been deposited in the Gene Expression Omnibus (GEO) repository from NCBI under the platform

number GPL18340. More details concerning the genomic sequences present in these arrays can be provided upon request.

9. Cyanide dyes (Cy5 particularly) are sensitive to even low levels of atmospheric ozone. Working under an ozone-free enclosure is especially important when the dyes are dry and thus directly exposed to the atmospheric ozone (after the last wash). Alternatively, Agilent sells a solution saturated with ozone scavengers that will help protect the slide from ozone during the scanning.

10. The method described here only works for ChIP-chip experiments that are designed such that two different immunoprecipitations using distinct antibodies are competitively hybridized to get enrichment ratios. In the case described here (*see* Fig. 2), the histone variant H2A.Z localization is assessed by performing ChIP experiments with an antibody against H2A.Z (test), which is hybridized against a ChIP performed with an antibody against H2B (control). The latter serves as baseline for nucleosome occupancy. Analogously, one could look at a histone modification such as H3K4me3, by hybridizing an H3K4me3 ChIP against a total H3 ChIP. If this experiment was repeated in different conditions/strains, the spiking method described here could allow for a better comparison of the effect of these conditions/strains on H3K4me3 occupancy/distribution. While the method should, in principle, function for any antibody, it does not work for ChIP-chip experiments designed to be hybridized against input DNA or DNA from mock (IgG or naked beads) ChIP samples. Indeed, the input DNA contains too much spiked DNA to be accurately measured on the microarray. The fluorescence at these spots becomes saturated, preventing the calculation of any ratio. We noticed a similar problem when using a mock IP performed with IgG-coupled beads as the control. In these experiments, the amount of spiked DNA also exceeds the dynamic range of the scanner. While we do not fully understand why, we speculate that the absence of target in the extract causes the spiked DNA (which is in 100 molar excess compared to genomic DNA) to bind more efficiently to the bead/antibody complex. Inversely, when using naked beads, too little amount of spiked DNA is recovered. In sum, this design performs best when using a pair of antibodies that recover similar amounts of spiked DNA in the immunoprecipitation.

Acknowledgments

This work was supported by a grant from the Canadian Institutes of Health Research (MOP-133648).

References

1. Buck MJ, Lieb JD (2004) ChIP-chip: considerations for the design, analysis, and application of genome-wide chromatin immunoprecipitation experiments. Genomics 83:349–360
2. Furey TS (2012) ChIP-seq and beyond: new and improved methodologies to detect and characterize protein-DNA interactions. Nat Rev Genet 13:840–852
3. Bolstad BM, Collin F, Simpson KM et al (2004) Experimental design and low-level analysis of microarray data. Int Rev Neurobiol 60:25–58
4. Bonhoure N, Bounova G, Bernasconi D et al (2014) Quantifying ChIP-seq data: a spiking method providing an internal reference for sample-to-sample normalization. Genome Res 24:1157–1168
5. Jeronimo C, Watanabe S, Kaplan CD et al (2015) The Histone Chaperones FACT and Spt6 Restrict H2A.Z from Intragenic Locations. Mol Cell 58:1113–1123

Chapter 14

High-Resolution Genome-Wide Mapping of Nucleosome Positioning and Occupancy Level Using Paired-End Sequencing Technology

Mylène Brunelle, Sébastien Rodrigue, Pierre-Étienne Jacques, and Nicolas Gévry

Abstract

Because of its profound influence on DNA accessibility for protein binding and thus on the regulation of diverse biological processes, nucleosome positioning has been studied for many years. In the past decade, high-throughput sequencing technologies have opened new perspectives in this research field by allowing the study of nucleosome positioning and occupancy on a genome-wide scale, therefore providing understanding on important aspects of chromatin packaging, as well as on various chromatin-template processes like transcription. In this chapter, we provide the protocol of MNase sequencing for the genome-wide mapping of nucleosomes using MNase to generate mononucleosomal DNA fragments and next-generation sequencing technology to identify their individual location.

Key words Histone, DNA, Nucleosome, MNase, Chromatin, Next-generation sequencing

1 Introduction

The nucleosome is a repetitive unit that forms chromatin primary structure, which refers to a "beads-on-a-string" fiber. Each nucleosome ("the beads") consists of approximately 147 bp of DNA wrapped around an octamer core of histone proteins, containing two H2A, H2B, H3, and H4, and is connected to the next nucleosome by free DNA ("the string"), called the linker region. This structure aims at packaging DNA, as well as regulating many chromatin-template processes such as transcription, replication, and DNA repair, partly by dynamically limiting the accessibility of DNA for binding by proteins. Consequently, gaining access to DNA wrapped around nucleosomes requires the eviction or the sliding of nucleosomes, or unwrapping DNA. Importantly, the timing and location of these dynamic processes, as well as the position of nucleosomes, depend on the cell type, its physiological

Benoit Guillemette and Luc Gaudreau (eds.), *Histones: Methods and Protocols*, Methods in Molecular Biology, vol. 1528, DOI 10.1007/978-1-4939-6630-1_14, © Springer Science+Business Media New York 2017

state, and various hormonal or environmental stimuli. Therefore, analyses of chromatin primary structure in different cell types or tissues and its dynamics in various conditions are an important research area. Studies are aimed at identifying key aspects of nucleosome organization in the entire genome and the biological relevance of its variation and potential deregulation in diseases.

Intense research has provided a variety of observations and a significant understanding of the mechanisms that influence nucleosome organization (reviewed by [1]). For instance, it has been recently shown that nucleosome organization around transcription start sites (TSS) and transcription-factor-binding sites (TF-BS) is highly heterogeneous and asymmetric [2], unlike previously anticipated. Moreover, the average spacing of nucleosomes varies between cell types and between different domains of posttranslational modifications of histones, and is also influenced by transcriptional activity levels [3]. Histone modifications can also affect nucleosome compaction by modifying its charge of by recruiting specific chromatin remodelers, which in turn modify nucleosome compaction, composition, or occupancy [1]. In addition, DNA sequences influence nucleosome positioning: G/C-rich sequences promote nucleosome formation, whereas A/T-rich sequences prevent it [2–8]. This has led to the recent concept of "container sites," an extensive phenomenon observed in the vicinity of TF-BS [2, 3]. Despite DNA sequences being drivers of nucleosome organization, the cellular context can override this intrinsic sequence preference and introduce or evict nucleosomes at unfavorable or favorable DNA regions [2, 3]. Moreover, the determinants of nucleosome organization influence each other, making their identification and interpretation more complex. Chromatin structure analyses benefit from the development of new technologies and tools that provide increased availability of high-quality data sets. Fundamental aspects of chromatin structure and dynamics can be achieved by mapping the exact position and occupancy level of nucleosomes in different cell types or states. Furthermore, meaningful knowledge on the regulation of biological and pathological processes can be more readily addressed by generating these data sets. In the past decade, huge progress has been made in high-throughput sequencing technologies, allowing their utilization to dissect the structure of chromatin at a reasonable cost.

The method of choice to interrogate nucleosome organization is the isolation of nucleosomes by chromatin digestion with micrococcal nuclease (MNase), an endonuclease that preferentially hydrolyzes linker DNA, releasing the nucleosomal-protected DNA (reviewed by [9]). Under appropriate experimental conditions, the size distribution of DNA when analyzed by gel electrophoresis is usually factors of 150 bp, corresponding to mono-, di-, and so on nucleosomal DNA fragments. The purification of mononucleosomal DNA fragments allows the subsequent study of nucleosome distribution along the genome of interest. Different analyses can be done to extract new information about chromatin structure. These necessarily involve occupancy maps, which refers to the fraction of cells from the population in which a particular genomic region is occupied by a nucleosome, and position maps, which refers to the localization of an individual nucleosome with respect to the genomic DNA sequence (both concepts reviewed by [10]). In this chapter, we describe an approach for genome-wide mapping of nucleosomes using MNase to generate mononucleosomes, and next-generation sequencing technology to detect their individual location (MNase-seq) (Fig. 1). The originality of this protocol comes from (1) using on-chip electrophoresis to resolve MNase-digested DNA fragments, which reduces technical variations between different chromatin preparations by allowing validation of the uniformity and reproducibility of the size distribution between replicates or experimental conditions; (2) using solid-phase paramagnetic bead technology, which facilitates sequencing library preparation and minimizes the loss of sample; and (3) using

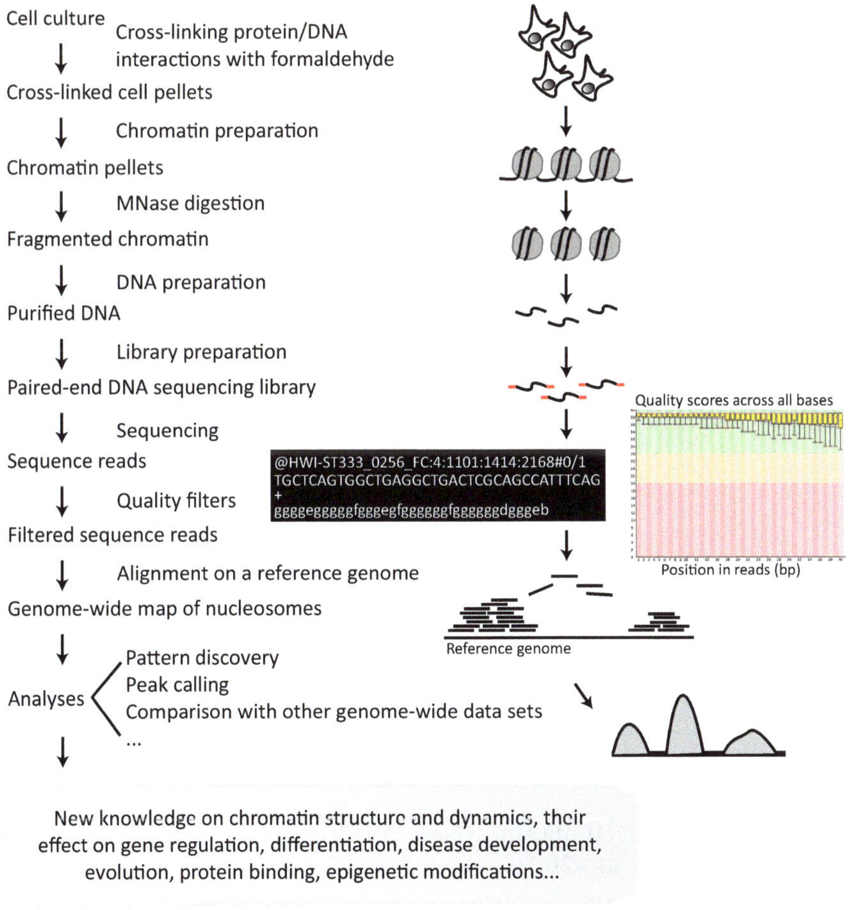

Fig. 1 General outlines of the protocol proposed for genome-wide mapping of nucleosomes

paired-end sequencing to identify each extremity of the same nucleosomal DNA fragments, therefore obtaining precise genome-wide nucleosome occupancy and position maps.

2 Materials

2.1 Cross-Link of DNA-Protein Complexes with Formaldehyde and Chromatin Preparation

1. 1× Phosphate-buffered saline (PBS, pH 7.4).
2. 37% (vol/vol) formaldehyde (molecular biology grade).
3. 1.1% (vol/vol) formaldehyde-PBS: From 37% formaldehyde and 1× PBS (pH 7.4). Make fresh and keep at room temperature until use.
4. 2.5 M Glycine.
5. 100× Protease inhibitor cocktail: 0.2 mM Pepstatin A, 72 μM leupeptin, and 26 μM aprotinin dissolved in ethanol. Keep at -20 °C.
6. 100 mM Phenylmethanesulfonyl fluoride (PMSF): Dissolve in ethanol and keep at −20 °C.
7. Buffer I: 0.25% (vol/vol) Triton X-100, 10 mM EDTA, 0.5 mM EGTA, 10 mM HEPES-KOH (pH 6.5) (*see* **Note 1**). Store at 4 °C and add 1 mM PMSF and 1× protease inhibitor cocktail to the required volume before use.
8. Buffer II: 200 mM NaCl, 1 mM EDTA, 0.5 mM EGTA, 10 mM HEPES-KOH (pH 6.5). Store at 4 °C and add 1 mM PMSF and 1× protease inhibitor cocktail to the required volume before use.

2.2 MNase Digestion

1. MNase digestion buffer: 50 mM Tris–HCl (pH 8), 1 mM $CaCl_2$, 0.2% (vol/vol) Triton X-100. Make fresh and keep at room temperature until use. Before use, add 1 mM PMSF and 1× protease inhibitor cocktail to the required volume.
2. Micrococcal nuclease, lyophilized: Follow the manufacturer's instructions for MNase preparation. It is suggested to purchase enough MNase for all planned experiments and their replicates, and to avoid repeated freeze-thaw cycles. Prepare small aliquots and store at -20 °C.
3. Stop buffer: 10 mM Tris–HCl (pH 8), 5 mM EDTA, 0.5% (wt/vol) sodium dodecyl sulfate (SDS, *see* **Note 2**). Make fresh and keep at room temperature until use. Before use, add 1 mM PMSF and 1× protease inhibitor cocktail to the required volume.

2.3 DNA Extraction and Preparation of DNA for Sequencing

1. 10 mg/ml RNase A (DNase-free): Aliquots should be stored at −20 °C.
2. 20 mg/ml Proteinase K: Aliquots should be stored at −20 °C.
3. QIAquick PCR purification kit (Qiagen) or other commercial DNA purification kit.

4. Agilent High Sensitivity DNA kit (Agilent Technologies).
5. Quant-iT PicroGreen dsDNA Assay Kit (Invitrogen).
6. Agencourt AMPure XP PCR Purification system (Beckman Coulter).
7. Paired-End DNA Sample Preparation Kits (Illumina).
8. SYBR Green I: Prepare aliquots of 10× stock solution from the concentrate and store at −20 °C.

3 Methods

This protocol is designed for mammalian cells cultured in 150 mm dishes. The amount of cells required varies depending on the cell type used, typically from 10 to 30 million cells (*see* **Note 3**). For lower cell numbers and smaller dishes, scale all volumes down.

3.1 Cross-Linking of DNA-Protein Complexes with Formaldehyde

1. Remove culture medium from dish and wash cells twice with 1× PBS at room temperature.
2. Add 20 ml of 1.1 % formaldehyde-PBS mix per dish and incubate for 10 min at room temperature (*see* **Note 4**) with gentle rocking.
3. Add 1 ml of 2.5 M glycine per dish to stop the cross-link reaction by formaldehyde quenching and incubate for 5 min at room temperature. Gently rock cell plate.
4. Remove the formaldehyde-PBS-glycine solution and wash the cells twice with 1× PBS at 4 °C.
5. Add 1 ml of 1× PBS at 4 °C and collect cells from dish using a cell scraper. Cell pellet must be kept on ice.
6. Centrifuge the cross-linked cells for 6 min at 1000 ×*g* at 4 °C and discard supernatant.
7. Quickly freeze the cross-linked cell pellet in liquid nitrogen and store at −80 °C until use.

3.2 Preparation of Chromatin

1. Thaw cells on ice.
2. Resuspend cellular pellet in 1 ml of Buffer I and incubate on ice for 10 min (*see* **Note 5**).
3. Centrifuge the cells for 6 min at 1000 ×*g* at 4 °C, and discard supernatant.
4. Resuspend cellular pellet in 1 ml of Buffer II and incubate on ice for 10 min.
5. Centrifuge cells for 6 min at 1000 ×*g* at 4 °C, and discard supernatant (remove any residual liquid).

3.3 MNase Digestion

1. Resuspend cellular pellet in 1 ml of MNase digestion buffer (*see* **Note 6**).
2. Divide each chromatin sample in 500 μL aliquot.
3. Incubate for 2 min at 37 °C (*see* **Note 7**).
4. Add 150 U of MNase per aliquot (determined by MNase titration, *see* **Notes** 7 and **8** and Fig. 2) and incubate for 5 min at 37 °C.
5. Add 500 μL of stop buffer per aliquot and put samples immediately on ice (*see* **Note 7**).
6. Sonicate samples briefly (*see* **Note 9**).
7. Centrifuge for 10 min at 15,000 ×*g* at 4 °C, and keep supernatant (*see* **Note 10**).
8. Reverse the DNA-protein cross-links by incubating samples for 4–6 h or overnight at 65 °C.

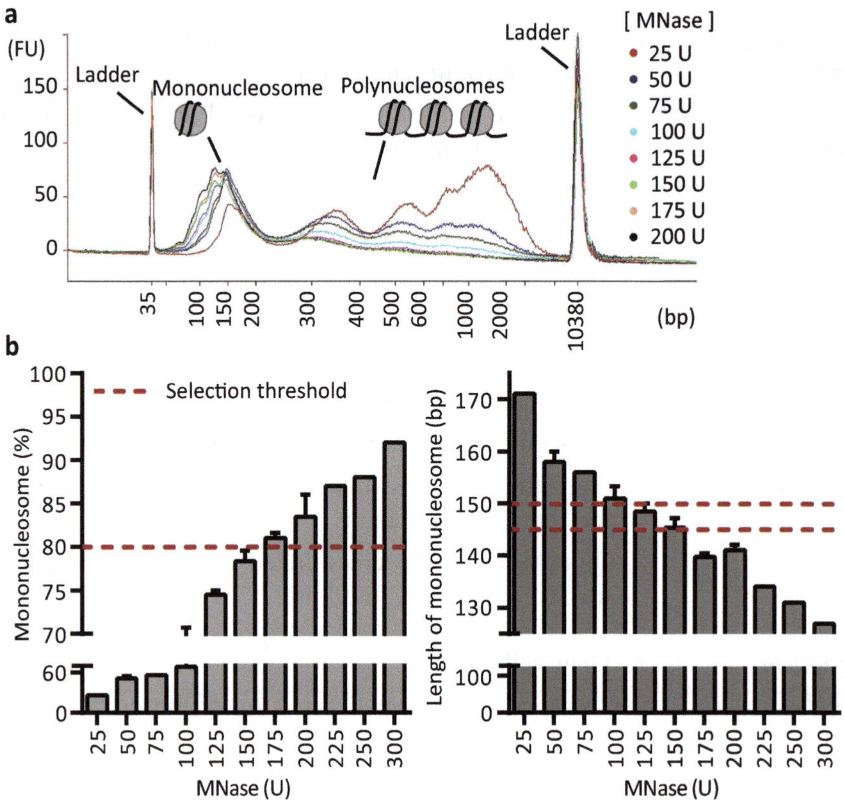

Fig. 2 Optimization of MNase digestion step. (**a**) Electrophoregram analysis of a typical MNase titration experiment showing the DNA size distribution (bp) after digestion using increasing amounts of MNase. The analysis was performed using a 2100 Bioanalyzer instrument (Agilent Technologies). (**b**) Selection of MNase concentration (results of three independent biological replicates). The proportion of mononucleosomes (*upper panel*) as well as the length of mononucleosomal DNA fragments (*lower panel*) were computed. A value of 150 U of MNase was selected for subsequent MNase ChIP-seq experiments, since it gives 80 % of mononucleosomes and an expected mononucleosome DNA length between 145 and 150 bp

3.4 DNA Preparation

1. Add 20 μg of RNase A and incubate for 1 h at 37 °C.
2. Add 100 μg of Proteinase K and incubate for 2–3 h at 37 °C.
3. Purify the DNA with the QIAquick PCR purification kit as described by the manufacturer (*see* **Notes 11** and **12**).
4. Assess DNA quality and examine the MNase digestion using the Agilent High Sensitivity DNA kit and the 2100 Bioanalyzer instrument (Agilent Technologies) according to the manufacturer's instructions (*see* **Note 13** and Fig. 3).

3.5 Preparation of DNA for Sequencing

The sequencing library is performed according to the Illumina library preparation protocol, with some minor modifications. All gel and DNA purifications are replaced by the use of solid-phase paramagnetic bead technology (AMPure XP PCR Purification system), which allows more precision, minimizes tube transfers and loss of sample, and efficiently eliminates unligated adapters and primer dimers (shorter than the library of nucleosome fragments). We recommend the construction of paired-end DNA sequencing library, which provides more accurate maps by identifying each border of the nucleosome, and thus revealing the entire nucleosome position and their length (*see* **Note 14**). Other sequencing methodologies could also be used. In that case, follow the instructions of the manufacturer for preparing library.

1. Quantify DNA using Quant-iT PicroGreen dsDNA Assay Kit according to the manufacturer's instructions and use 50 ng of each sample for the preparation of the sequencing library (*see* **Note 15**).
2. Follow the instructions of the Paired-End DNA Sample Preparation Kit to prepare sequencing library, except that:

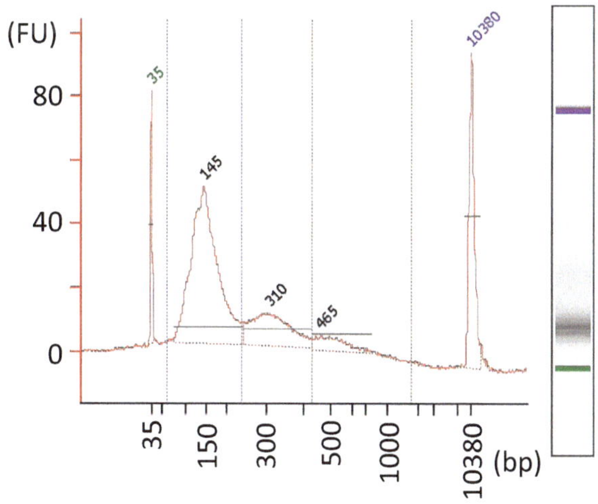

Fig. 3 Electrophoregram analysis of a purified DNA library ready for sequencing

(a) Replace all the steps that use the "QIAquick PCR Purification Kit" or the "MinElute PCR Purification Kit" by the AMPure XP PCR Purification system, which removes the excess of adapters and primers, nucleotides, salts, and enzymes (*see* **Note 16**).

(b) If possible, decrease the reaction volumes to 25–30 μl to minimize the amount of AMPure XP beads used.

(c) Omit the "Purify Ligation Products" step (*see* **Note 17**).

(d) Add SYBR Green I to the PCR master mix (final 0.2×) to monitor the "Enrich DNA fragments" step by qPCR and stop the amplification in the exponential phase to avoid PCR duplicates.

(e) Use a ratio of AMPure XP beads:DNA of 0.9 after the "Enrich DNA Fragments" step (instead of 1.8) to improve the elimination of potential primer dimers and primers in excess.

(f) Omit the "Purify Final Product" step.

3. Verify the absence of primer-dimers and the appropriate template size distribution and quantify the DNA concentration by running an aliquot of the library on the 2100 Bioanalyzer instrument (*see* Fig. 4 for examples of libraries prepared using this modified protocol). The DNA is now ready for sequencing.

3.6 Sequencing and Data Analysis

The analysis of MNase-sequencing data depends largely on the research objective and is not a trivial issue. Here, we propose non-exhaustive analysis steps and list of useful tools. However, the analysis of genome-wide MNase sequencing data evolves rapidly and new tools are published frequently.

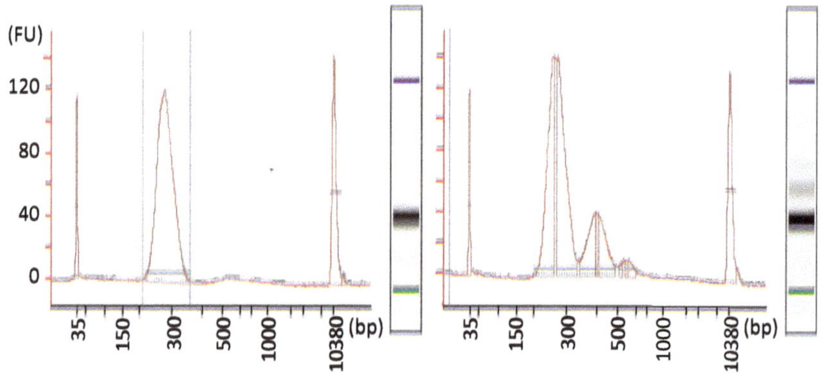

Fig. 4 Electrophoregram analysis of paired-end DNA sequencing libraries with (*left panel*) or without (*right panel*) mononucleosomal DNA fragment size selection

1. Use the DNA library for sequencing analysis using the Illumina HiSeq system following the manufacturer's protocols (*see* **Note 18**). You will obtain a file in FASTQ format containing the short sequences corresponding to either both ends of each nucleosome ("paired-end" sequencing) or to a single end ("single-end" sequencing) and its corresponding quality score.
2. Data analysis:
   (a) Before performing any analysis, verify the quality of the raw sequences (*see* **Note 19**).
   (b) Eliminate low-quality sequences (*see* **Note 20**).
   (c) Align all the resulting filtered high-quality sequences to the reference genome (e.g., hg19 human genome) and retain unique map sequences (*see* **Note 21**). You will obtain a genome-wide signal track corresponding to nucleosome borders that can be visualized using genome browser tools (*see* **Note 22**) and can be used for downstream analyses (*see* **Note 23**).
   (d) As an additional quality control, verify the length distribution of the sequenced nucleosomal DNA fragments (see Fig. 5 for typical length distributions and *see* **Note 24**).
   (e) A "peak calling" approach can be used to identify the most common/frequent (or consensus) position of nucleosomes through cell population (*see* **Note 25**).
   (f) Unsupervised pattern discovery can be applied to resolve patterns of nucleosome positioning in MNase-sequencing data [2] (*see* **Note 26**).

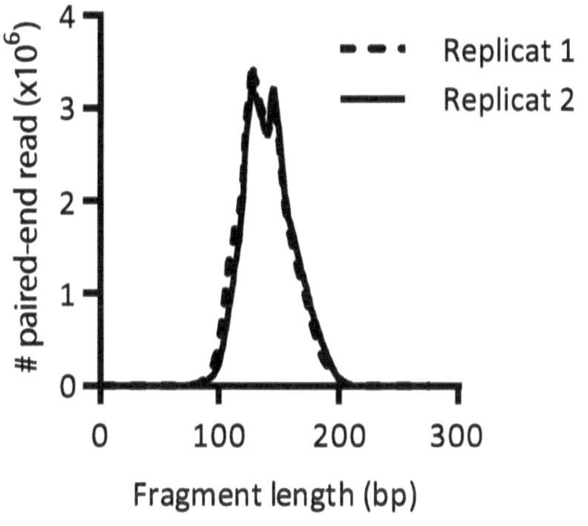

Fig. 5 Length distribution of typical mononucleosomal paired-end libraries

4 Notes

1. Triton X-100 should be added from a 10% (vol/vol) stock in H_2O.
2. The addition of SDS in the stop buffer was found to improve DNA recovery after MNase digestion. Note that SDS is not compatible with MNase digestion.
3. This protocol has been used successfully with the cell line MCF-7. The use of different cell lines may require the optimization of some parameters, in particular, the number of cells and the amount and time of the MNase digestion (*see* **Note 8** for MNase titration details).
4. The formaldehyde creates covalent bonds between DNA and proteins, thereby maintaining the in vivo status of interactions and preventing spurious reorganization during cell harvesting or chromatin preparation. Moreover, it has been shown that unstable nucleosomes containing both histone variants H3.3 and H2A.Z are easily lost during conventional chromatin preparations if they are not cross-linked, leading to misinterpretation of the nucleosomal organization in some genomic regions, mainly at regulatory elements like promoters, enhancers, and insulators [11, 12]. However, non-nucleosome structures resistant to MNase digestion are also stabilized following formaldehyde incubation, it is thus important to prevent excessive cross-linking by fixing cells no longer than 10 min.
5. To allow MNase digestion uniformity, cell pellets of the same experimental condition can be pooled at this point.
6. If different experimental conditions are to be compared, the amount of chromatin can be estimated in each sample by measuring the DNA content by spectrophotometry. Dilute 5 μl of the chromatin sample in 100 μl of 1 M NaOH and measure the absorbance at 260 nm (OD_{260}).
7. These steps have been found reproducible when working with four aliquots at a time. To insure uniformity between different samples and reproducibility within replicates, it is important to proceed quickly and meticulously at this critical step; that is, start the reaction at different time intervals to achieve the same digestion time. After the addition of MNase and stop buffer, mix well by pipetting up and down three times using a p1000 (or repeat enough times to achieve a complete homogenate). Avoid generating bubbles.
8. It is recommended to perform an MNase titration experiment in order to choose the appropriate amount of MNase that achieves 80% of mononucleosome-sized DNA fragments. Depending on the cell type used, MNase supplier, or lot, the MNase digestion can be more or less efficient. Importantly, the

extent of digestion affects the nucleosome mapping of particular genomic regions, for instance, the occupancy level in promoter regions where nucleosomal DNA are easily digested [13]. The MNase digestion is thus a critical step and must be optimized. We usually perform MNase titration in triplicate using a wide range of MNase concentrations (e.g., from 0 to 300 U/ml). We recommend the use of Agilent High Sensitivity DNA kit and the 2100 Bioanalyzer instrument to resolve aliquots of DNA fragments following MNase digestion and find which concentration works the best. The bioanalyzer instrument reaches a precision level that cannot be obtained by gel electrophoresis, separating DNA fragments with a resolution of 5 bp (*see* Fig. 2a for a typical MNase titration experiment). Using the Agilent 2100 Expert software, compute the proportion and length of the mononucleosomal-size DNA fragments and select the amount of MNase that gives 80% of mononucleosomes and an expected mononucleosomal-DNA length between 145 and 150 bp (Fig. 2b). This approach insures uniformity and repeatability of the digestion and minimizes potential artifacts caused by over- or under-trimming of DNA, hardly detectable by gel electrophoresis. The digestion reproducibility is essential between replicates and when different conditions have to be compared to avoid misinterpretations. Indeed, the over-digestion of one sample compared to the others can lead to a false interpretation of nucleosome loss [13]. The 2100 Bioanalyzer instrument allows direct comparison of samples and their appropriate matching, minimizing unwanted differences in the digestion extent between different samples.

9. A brief sonication of MNase-digested chromatin is required to improve the solubilization of chromatin fragments. The level of sonication should be optimized because it varies depending on the cell type and the apparatus. Test different sonication settings by preparing cells up to this step and monitoring the presence of a pellet as well as the level of MNase digestion on agarose gel. No visible pellet should be detected after a centrifugation at 15,000 × *g* at 4 °C for 10 min. Keep 5–10% aliquots of each test and proceed to a rapid DNA preparation. Digest samples with RNase A (20 μg) for 15 min at 37 °C, followed by heat treatment and proteinase K digestion (100 μg) for 1 h at 65 °C to reverse cross-linking. Perform standard phenol/chloroform extraction. Run on a 2% agarose gel to confirm full extraction and verify the size of DNA fragments.
10. The MNase-digested chromatin is ready to perform chromatin immunoprecipitation experiments (ChIP). Use one aliquot by ChIP. It might be necessary to adjust the immunoprecipitation buffer composition in order to avoid excessive amount of detergents that could interfere with the reaction between antigen and antibody.

11. Alternatively, phenol/chloroform extraction may be used. In that case, it is very important to remove all traces of phenol from your sample. The DNA may be stored indefinitely at −20 °C.
12. As a control of DNA quality, qPCR can be used to map the position of nucleosomes at a region of interest. Design primers amplifying ~100 bp amplicons overlapping by ~70 bp. A higher PCR amplification rate should be obtained for genomic regions showing highly positioned nucleosomes compared to regions with variable positioning across the cell population, and to regions devoid of nucleosome. Thus, peaks and valleys should be observed corresponding, respectively, to nucleosomes and linkers, or nucleosome-free regions. Sonicated DNA, instead of MNase-digested DNA, can be used as a control for amplification efficiency.
13. We recommend the use of a 2100 Bioanalyzer instrument to resolve DNA fragments rather than conventional gel electrophoresis for the reasons mentioned above (*see* **Note 8**).
14. Depending on the experimental goal, or to decrease the sequencing cost, the single-end strategy, which identifies only one end of the nucleosomal DNA molecule, could also be used. In that case, it is preferable to eliminate polynucleosomal particles (mainly dinucleosomes) before preparing sequencing libraries. Use gel electrophoresis and QIAquick Gel Extraction Kit or solid-phase paramagnetic beads [14] to select for mononucleosome-length DNA fragments. To minimize the risk of cross-contamination between libraries, it is important to use one gel and a new razor blade for each gel sample. A paired-end strategy allows the identification of the exact length of each sequencing fragment and their classification; thus the size selection of mononucleosomal DNA fragment is facultative. In fact, it could be interesting to include all MNase digestion products (sub-, mono-, and dinucleosomal DNA fragments), since they can be easily identified by their length after paired-end strategy and can give additional information on chromatin structure. Indeed, a recent study has characterized subnucleosomal DNA fragments revealing subtle features of chromatin structure and their dynamics [15].
15. This protocol has also been used successfully for the preparation of sequencing libraries of MNase-ChIP samples with as little as 10 ng of starting material.
16. Follow the instructions of the manufacturer to use the AMPure XP PCR Purification system. When a ratio of AMPure XP beads per DNA of 1.8 is used, AMPure XP beads bind DNA fragments larger or equal to 100 bp. Therefore, the DNA purification step following the "End Repair" step serves also to eliminate subnucleosomal DNA fragments.

17. The use of AMPure XP beads added at the end of the "Ligate Adapters" step efficiently removes excess adapters and adapter dimers. Moreover, the MNase digestion parameters used in this protocol already give a size range of sequencing library that is appropriate for cluster generation. Alternatively, only mononucleosomal length DNA fragments could be selected using QIAquick Gel Extraction Kit or the solid-phase paramagnetic bead technology [14].
18. In theory, 150 million sequencing reads for single end or 300 million for paired end are sufficient to obtain a 10× coverage for all the nucleosomes in the human genome (the size of the genome divided by the average size of the mononucleosomal DNA including linker). However, according to the experimental goal, extended coverage could be required. Moreover, depending on the number of sequencing reads required and sequencer performance, the samples could be multiplexed to decrease the sequencing cost.
19. Proposed tool: FastQC (http://www.bioinformatics.babraham.ac.uk/projects/fastqc/), also implemented in the Web-based platform Galaxy ("NGS: QC and manipulation") [16–18].
20. Proposed tools: Trimmomatic [19] and the Web-based platform Galaxy ("NGS: QC and manipulation") [16–18].
21. Many aligner tools may be used, including BWA [20], Bowtie [21], or those available in the Web-based platform Galaxy ("NGS: Mapping") [16–18]. If a high proportion of duplicates are observed in the FastQC report, it could be necessary to eliminate them using the subcommand "MarkDuplicates" of the Picard command-line tools (http://picard.sourceforge.net/command-line-overview.shtml) or "rmdup" of Samtools [22] or these functionalities implemented in the Web-based platform Galaxy [16–18].
22. Proposed tools for visualization: UCSC genome browser (http://genome.ucsc.edu/cgi-bin/hgGateway) and Integrative Genomics Viewer (IGV [23]). If different samples have to be compared, it is important to normalize the genome signal track, particularly for nucleosome occupancy analysis. Proposed tools for normalization: BEDTools [24], Dynamic Analysis of Nucleosome Position and Occupancy by Sequencing (DANPOS [25]), and Wiggler tool [26]).
23. If needed, use this signal file as-is (representing the 5′ ends of the nucleosomes) by extending each sequence in their 3′ direction to obtain the entire nucleosome (i.e., use the coordinates of the complete paired-end sequences as one "real" fragment or extend single-end sequences to 147 bp as one "inferred" fragment) by assuming that each sequence is obtained from

perfectly trimmed nucleosome core. Alternatively, use only the center of the nucleosome, i.e., "real" midpoint coordinates of paired-end sequences or "inferred" midpoint from the extend single-end sequences. BEDtools [24], DANPOS [25], and Wiggler tool [26], allowing different tag shifts, can be used to perform this step. The visualization tools proposed above can then be used to examine the nucleosome organization at particular genomic regions. The Versatile Aggregate Profiler (VAP) is also useful to generate profiles over groups of genomic regions of interest [27].

24. Use the "CollectInsertSizeMetrics" option of the Picard command-line tools (http://picard.sourceforge.net/command-line-overview.shtml). This step is particularly important to compare different samples, as any difference in length distribution might affect the interpretation. Using this protocol, we have obtained close agreement between biological replicates or between cells exposed to different treatments.
25. Proposed tools: Nucleosome Positioning from Sequencing (NPS [28]) and DANPOS [25]). DANPOS is powerful to map the highly probable nucleosome positions, as well as their potential dynamic through different samples, i.e., position shifts and occupancy changes. However, since the majority of nucleosomes are not at fixed positions when averaged through cell population this approach is a simplification of reality.
26. The Clustered Aggregation Tool (CAGT [2]) is an effective method to identify different patterns in the signal track of a large number of genomics data sets (e.g., MNase-seq, ChIP-seq, and DNase-seq data) around regions of interest (e.g., binding sites of a DNA-binding protein and TSS).

References

1. Hughes AL, Rando OJ (2014) Mechanisms underlying nucleosome positioning in vivo. Annu Rev Biophys 43:41–63
2. Kundaje A, Kyriazopoulou-Panagiotopoulou S, Libbrecht M, Smith CL, Raha D, Winters EE et al (2012) Ubiquitous heterogeneity and asymmetry of the chromatin environment at regulatory elements. Genome Res 22:1735–1747
3. Valouev A, Johnson SM, Boyd SD, Smith CL, Fire AZ, Sidow A (2011) Determinants of nucleosome organization in primary human cells. Nature 474:516–520
4. Kaplan N, Moore IK, Fondufe-Mittendorf Y, Gossett AJ, Tillo D, Field Y et al (2009) The DNA-encoded nucleosome organization of a eukaryotic genome. Nature 458:362–366
5. Peckham HE, Thurman RE, Fu Y, Stamatoyannopoulos JA, Noble WS, Struhl K et al (2007) Nucleosome positioning signals in genomic DNA. Genome Res 17:1170–1177
6. Segal E, Fondufe-Mittendorf Y, Chen L, Thastrom A, Field Y, Moore IK et al (2006) A genomic code for nucleosome positioning. Nature 442:772–778
7. Tillo D, Hughes TR (2009) G+C content dominates intrinsic nucleosome occupancy. BMC Bioinformatics 10:442
8. Tillo D, Kaplan N, Moore IK, Fondufe-Mittendorf Y, Gossett AJ, Field Y et al (2010) High nucleosome occupancy is encoded at human regulatory sequences. PLoS One 5, e9129
9. Telford DJ, Stewart BW (1989) Micrococcal nuclease: its specificity and use for chromatin analysis. Int J Biochem 21:127–137

10. Struhl K, Segal E (2013) Determinants of nucleosome positioning. Nat Struct Mol Biol 20:267–273
11. Jin C, Felsenfeld G (2007) Nucleosome stability mediated by histone variants H3.3 and H2A.Z. Genes Dev 21:1519–1529
12. Jin C, Zang C, Wei G, Cui K, Peng W, Zhao K et al (2009) H3.3/H2A.Z double variant-containing nucleosomes mark 'nucleosome-free regions' of active promoters and other regulatory regions. Nat Genet 41:941–945
13. Weiner A, Hughes A, Yassour M, Rando OJ, Friedman N (2010) High-resolution nucleosome mapping reveals transcription-dependent promoter packaging. Genome Res 20:90–100
14. Rodrigue S, Materna AC, Timberlake SC, Blackburn MC, Malmstrom RR, Alm EJ et al (2010) Unlocking short read sequencing for metagenomics. PLoS One 5, e11840
15. Henikoff JG, Belsky JA, Krassovsky K, MacAlpine DM, Henikoff S (2011) Epigenome characterization at single base-pair resolution. Proc Natl Acad Sci U S A 108:18318–18323
16. Blankenberg D, Von Kuster G, Coraor N, Ananda G, Lazarus R, Mangan M et al (2010) Galaxy: a web-based genome analysis tool for experimentalists. Curr Protoc Mol Biol Chapter 19:Unit 19.10.1–19.10.21
17. Giardine B, Riemer C, Hardison RC, Burhans R, Elnitski L, Shah P et al (2005) Galaxy: a platform for interactive large-scale genome analysis. Genome Res 15:1451–1455
18. Goecks J, Nekrutenko A, Taylor J, Galaxy T (2010) Galaxy: a comprehensive approach for supporting accessible, reproducible, and transparent computational research in the life sciences. Genome Biol 11:R86
19. Bolger AM, Lohse M, Usadel B (2014) Trimmomatic: a flexible trimmer for Illumina sequence data. Bioinformatics 30(15):2114–2120
20. Li H, Durbin R (2009) Fast and accurate short read alignment with Burrows-Wheeler transform. Bioinformatics 25:1754–1760
21. Langmead B, Trapnell C, Pop M, Salzberg SL (2009) Ultrafast and memory-efficient alignment of short DNA sequences to the human genome. Genome Biol 10:R25
22. Li H, Handsaker B, Wysoker A, Fennell T, Ruan J, Homer N et al (2009) The Sequence Alignment/Map format and SAMtools. Bioinformatics 25:2078–2079
23. Robinson JT, Thorvaldsdottir H, Winckler W, Guttman M, Lander ES, Getz G et al (2011) Integrative genomics viewer. Nat Biotechnol 29:24–26
24. Quinlan AR, Hall IM (2010) BEDTools: a flexible suite of utilities for comparing genomic features. Bioinformatics 26:841–842
25. Chen K, Xi Y, Pan X, Li Z, Kaestner K, Tyler J et al (2013) DANPOS: dynamic analysis of nucleosome position and occupancy by sequencing. Genome Res 23:341–351
26. Hoffman MM, Ernst J, Wilder SP, Kundaje A, Harris RS, Libbrecht M et al (2013) Integrative annotation of chromatin elements from ENCODE data. Nucleic Acids Res 41:827–841
27. Coulombe C, Poitras C, Nordell-Markovits A, Brunelle M, Lavoie MA, Robert F et al (2014) VAP: a versatile aggregate profiler for efficient genome-wide data representation and discovery. Nucleic Acids Res 42(Web Server issue):W485–W493
28. Zhang Y, Shin H, Song JS, Lei Y, Liu XS (2008) Identifying positioned nucleosomes with epigenetic marks in human from ChIP-Seq. BMC Genomics 9:537

Chapter 15

Physarum polycephalum for Studying the Function of Histone Modifications In Vivo

Vanessa Menil-Philippot and Christophe Thiriet

Abstract

Histone modifications have been widely correlated with genetic activities. However, how these posttranslational modifications affect the dynamics and the structure of chromatin is poorly understood. Here, we describe the incorporation of the exogenous histone proteins into the slime mold *Physarum polycephalum*, which has been revealed to be a valuable tool for examining different facets of the function histones in chromatin dynamics like replication-coupled chromatin assembly, histone exchange, and nucleosome turnover.

Key words Chromatin, Nucleosome, Histone modification, *Physarum polycephalum*, Assembly, Turnover, Exchange

1 Introduction

Since the pioneer works from Allfrey and colleagues that evidenced a correlation between histone modifications and genetic activities [1], the histone posttranslational modifications have been the focus of extensive research over the past decades. The bevy of reports in this field has revealed the presence of hundreds of these modifications in living cells [2]. The vast number of the histone modifications and their link with chromatin activities have led to propose the existence of the histone code [3, 4]. Although histone modifications are considered as hallmarks of chromatin activities, they required other factors for directly and indirectly processing DNA. Some factors are associated to histone modifications and are defined in three categories, the writers, the readers, and the erasers, respectively [5].

As the modifications of amino acid residues within the histones are enzymatically catalyzed, it is generally believed that the histone tail domains are preferential sites of the posttranslational modifications. Indeed, in contrast to the histone fold domains, which are involved in histone-histone and histone-DNA interaction within

Benoit Guillemette and Luc Gaudreau (eds.), *Histones: Methods and Protocols*, Methods in Molecular Biology, vol. 1528, DOI 10.1007/978-1-4939-6630-1_15, © Springer Science+Business Media New York 2017

the nucleosome, the histone tail domains extend out of the nucleosome [6]. The specific antibodies to histone modifications are valuable and reproducible tools for examining histone modifications. These modification probes have allowed the development of ChIP procedures, which have significantly improved the studies of chromatin and provided important insights into our understanding of the histone modifications. However, this technique gives a snapshot of the chromatin landscape but is rarely indicative of the dynamic nature of chromatin and the actual role of the histone modifications.

The important role of histone modifications has been primarily estimated by genetic analyses in yeast [7, 8]. It has been proposed that histone acetylation might neutralize the charge of the lysine residues [9]. Thus, to mimic lysine acetylation and unacetylated lysine within histones, lysine residues were substituted to glutamine and arginine, respectively. Using this strategy, it has been shown that the mutation of only four lysines of the histone H4 tail led to cell lethality, demonstrating the essential function of lysine acetylation [7]. However, there were caveats to the genetic approach. The genetic mutations implied that most, if not all, histones presented the mutations, which did not reflect a normal cell situation. Furthermore, the cell lethality induced by the mutations did not provide information on the cellular process that was affected by the substitutions. One can also argue that mimicking histone acetylation using amino acid residue substitution might not correspond to the natural modification. Although the substitution is not reversible unlike the lysine acetylation, biochemical experiments have shown that the genetic mimics of histone acetylation recapitulate enzymatic acetylation effects in vitro [10]. Clearly, if the genetic mimics have been very helpful for studying histone acetylation, this strategy cannot be developed for all histone modifications, such as methylation, ubiquitination, and glcNacylation.

Here, we describe a powerful and very useful method for studying histone modification function in chromatin dynamics and genetic activities. The procedure takes advantage of the unique properties of the slime mold *Physarum polycephalum*. This model organism can be cultured as a giant cell (~5–6 cm diameter) wherein several hundreds of nuclei stream within the cytoplasm, and are perfectly synchronous each other throughout the cell cycle [11]. Another characteristic is the ability for the *Physarum* cell to internalize exogenous proteins into the intracellular medium and to utilize the proteins when they are functionally relevant. This methodology has been first described by Oppenheim and Katzir for examining the biological effects of cell cycle stage-dependent cellular extracts [12]. Then, Bradbury and colleagues have incorporated calf thymus H1 kinase for demonstrating a role of the phosphorylation of the *Physarum* linker histone in mitosis [13]. A similar procedure has also been used by Allfrey and collaborators for examining alteration of the nucleosome structure in vivo [14,

15]. The incorporation of exogenous proteins into *Physarum* macroplasmodia offers the possibility to investigate the biological function of factors and constitutive proteins composed of several subunits. Although our experiments have focused on the incorporation of histone complexes, we believe that any protein complex can be internalized into *Physarum* cells. Indeed, we did observe that even proteins with a large size like IgG incorporated efficiently into the cells. Certainly, the incorporation of multimeric complexes requires biochemical preparation of the proteins, which might appear laborious. However, this unique approach also presents the undeniable advantage to examine in vivo function of the actual complex rather than only single subunit. Additionally, the perfect control of the amount of protein complex incorporated into *Physarum* enables to prevent the disturbance of the cellular metabolism by internalizing trace amount of exogenous proteins. However, for discriminating the exogenous proteins from the endogenous ones, it is required to specifically label the exogenous proteins. This can be achieved by using a tag epitope or fluorescent labeling. Clearly, if this experimental approach of incorporating exogenous proteins into *Physarum* might seem more laborious than genetic approaches, it also provides better control of critical steps. However, with the expansion of the molecular biology and the techniques of protein purification, the difficult biochemistry processes should not be a limitation. Furthermore, the necessity of preparing the proteins prior to the incorporation provides also the opportunity to carry out specific chemical reactions that can mimic posttranslational modifications for determining their cellular functions.

2 Materials

The *Physarum* cultures like any cell culture require of working in sterile conditions. Commonly, we used the strain TU 291 of *Physarum*, but the different procedures can be applied to any other strain. All buffers and media are prepared using distilled and deionized water (ddH_2O) unless indicated otherwise.

2.1 Culture Media

The different media are sterilized by autoclaving for 20 min. The semi-defined medium is the culture medium used to maintain the *Physarum* cells. The saline medium is used to induce spherules of *Physarum* that allow the conservation of the *Physarum* strains at 4 °C. The minimum medium can be used for performing different cell labeling with precursors (*see* **Note 1**).

1. Semi-defined medium: 18.425 mM Citric acid, 302 μM $FeCl_2$, 1.470 mM KH_2PO_4, 2.434 mM $MgSO_4$, 0.424 mM $MnCl_2$ $4H_2O$ (0.084 g/L), 0.118 mM $ZnSO_4$ $7H_2O$ (0.034 g/L), 4.081 mM $CaCl_2$ $2H_2O$ (0.6 g/L), 10 g/L Bacto tryptone, 55.506 mM glucose (10 g/L), 1.5 g/L yeast extract. The pH

of the culture media is then adjusted between 4.4 and 4.6 with 20% KOH. The culture media is sterilized and kept away from light to avoid the formation of precipitates.

2. Saline medium: 18.425 mM Citric acid monohydrate, 302 μM $FeCl_2$, 1.470 mM KH_2PO_4, 2.434 mM $MgSO_4$, 0.424 mM $MnCl_2$, 0.118 mM $ZnSO_4$, 4.081 mM $CaCl_2$. The pH of the culture media is then adjusted between 4.4 and 4.6 with 20% KOH. The culture media is sterilized and kept away from light to avoid the formation of precipitates.
3. Minimum medium: 18.425 mM Citric acid monohydrate, 302 μM $FeCl_2$, 1.470 mM KH_2PO_4, 2.434 mM $MgSO_4$, 0.424 mM $MnCl_2$, 0.118 mM $ZnSO_4$, 4.081 mM $CaCl_2$, 10 g/L bacto tryptone, 56.894 mM glucose, 1.5 g/L yeast extract without amino acids, 0.655 μM biotin, 1.257 μM thiamine HCl, 42.404 μM pyridoxine HCl, 0.922 μM folic acid, 1.689 mM dl-methionine, 6.048 mM glycine, 2.872 mM L-arginine HCl, 3.995 mM L-leucine, 2.653 mM L-isoleucine, 27.354 mM L-alanine. The pH of the culture media is then adjusted between 4.4 and 4.6 with 20 % KOH. The culture media is sterilized and kept away from light to avoid the formation of precipitates.
4. Hemin solution: Dissolve 50 mg of Hemin chloride in 100 ml of 0.25 M NaOH. The hemin solution is sterilized and kept at 4 °C.
5. Mounting medium for microscopic observation, glycerol/ ethanol (1:1).

2.2 Buffers for Histone Purification, Nuclei Isolation, and Chromatin Preparation

1. BL21 *E. coli* transformed with the desired histone expression vector (we typically use pET3a vector).
2. LB medium.
3. 5 mg/ml Ampicillin.
4. Homogenization buffer: 15 mM $MgCl_2$, 1 mM $CaCl_2$, 15 mM Tris–HCl pH 8.0, 5 mM EGTA, 250 mM hexylene glycol, 0.6% Surfynol, 3 mM dithioerythritol (DTE).
5. Percoll gradient buffer: 15 mM $MgCl_2$, 1 mM $CaCl_2$, 15 mM Tris–HCl, pH 8.0, 5 mM EGTA, 250 mM hexylene glycol, 0.6% Surfynol, 3 mM DTE, 25% Percoll.
6. Micrococcal nuclease buffer: 60 mM KCl, 15 mM NaCl, 10 mM Tris–HCl pH 8.0, 1 mM $CaCl_2$.
7. 15 U/μl Micrococcal nuclease in ddH_2O.
8. UNOsphere S cation exchange resin (BioRad).
9. Biorex 70 cation exchange resin (BioRad).
10. SAU_0: 7 M Urea, 20 mM sodium acetate pH 5.2, 5 mM 2-mercaptoethanol, 1 mM EDTA.

11. SAU_{200}: 7 M Urea, 20 mM sodium acetate pH 5.2, 5 mM 2-mercaptoethanol, 1 mM EDTA, 200 mM NaCl.
12. SAU_{600}: 7 M Urea, 20 mM sodium acetate pH 5.2, 5 mM 2-mercaptoethanol, 1 mM EDTA, 600 mM NaCl.
13. $SAU_{1000:}$ 7 M Urea, 20 mM sodium acetate pH 5.2, 5 mM 2-mercaptoethanol, 1 mM EDTA, 1 M NaCl.
14. Histone dialysis buffer: 2 M NaCl, 10 mM Tris–HCl, pH 8.0, 1 mM EDTA.
15. Biorex wash buffer: 0.6 M NaCl, 10 mM Tris–HCl pH 8.0, 1 mM EDTA.
16. Biorex elution buffer: 1 M NaCl, 10 mM Tris–HCl, pH 8.0, 1 mM EDTA.
17. 10 mg/ml Lysozyme.
18. 10% Triton X-100.
19. 11.6 M HCl.
20. 10 M NaOH.
21. 1 M IPTG.
22. 200 mM ATP.
23. Dialysis tubing, cutoff 6000–8000.
24. 3 mm Glass beads.

3 Methods

For the experiments of incorporation of exogenous histones into *Physarum* cells, we usually use FLAG-tagged histone complexes (Fig. 1). The utilization of FLAG-tagged complexes allows the discrimination of the exogenous (tagged) and the endogenous histones (untagged) in subsequent analyses with minimal background [16–19]. The histones H2A and H2B are typically soluble in bacteria, in contrast to H3 and H4, which are typically expressed in inclusion bodies.

3.1 Preparation of H3/H4 Histone Complexes

The different histone classes are expressed in *E. coli* as described in a previous publication [18]. For the complexes of H3 and H4, each histone is individually purified and refolded to form a complex of H3/H4. We find that preparing the H3/H4 complex by a two-step procedure leads to greater yield than purifying H3/H4 with Biorex beads [18].

1. 1 L Cultures of BL21 *E. coli* transformed with the desired histone expression vector (we typically used pET3a vector) is grown at 37 °C in LB supplemented with 5 μg/ml ampicillin up to 0.4 OD_{600}. Then the culture is induced with 1 mM IPTG for 2 h.

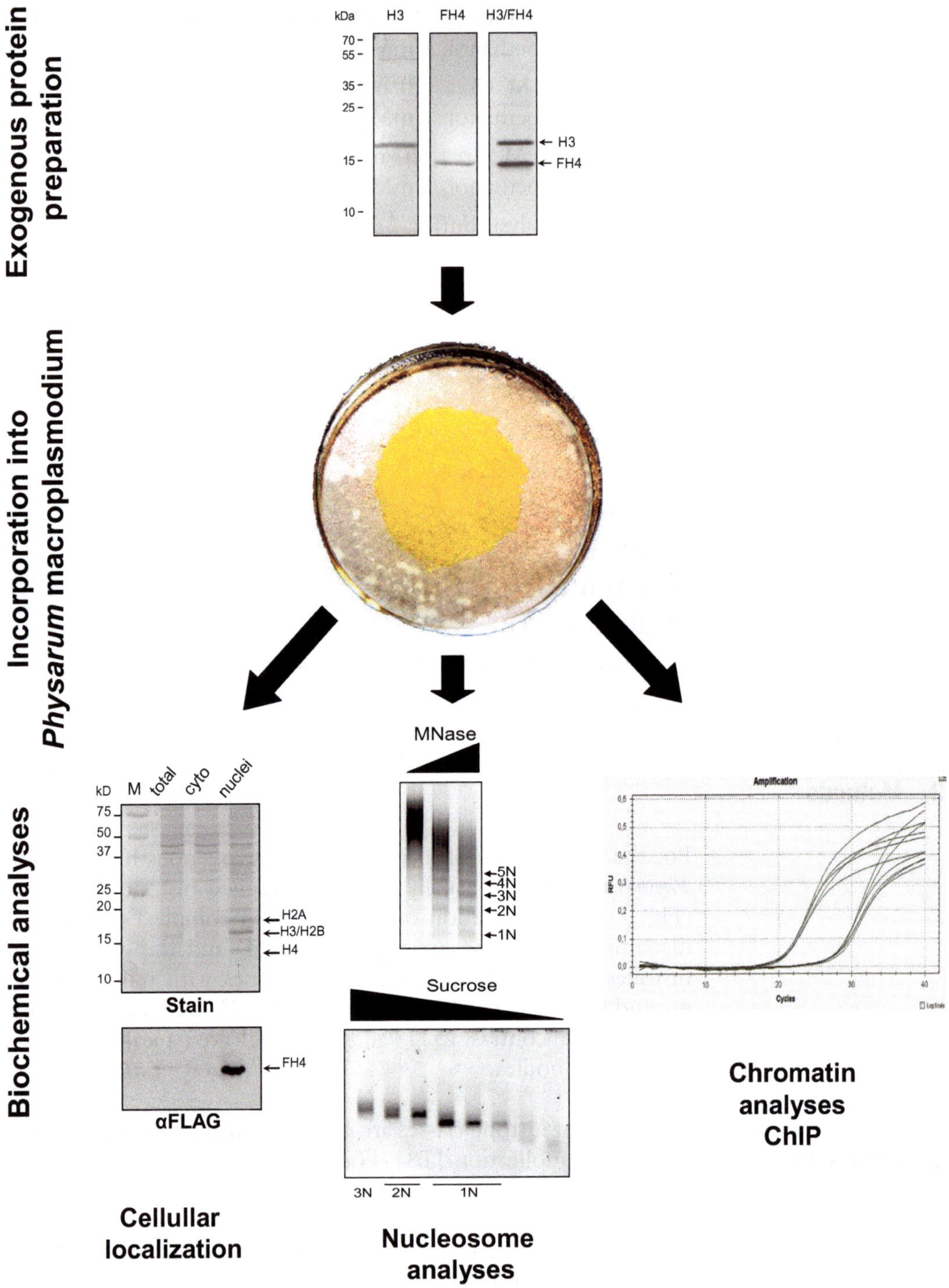

Fig. 1 Flowchart of the incorporation of exogenous proteins into *Physarum* macroplasmodium. Shown is the example of the incorporation of exogenous H3/FLAG-H4 (FH4) with the three major steps: (1) the preparation of the exogenous proteins; (2) the incorporation of the exogenous proteins, and (3) the biochemical analyses of the incorporation

2. Collect the cells by centrifugation at 3000×*g* for 10 min at 4 °C.
3. Resuspend the cells in ddH_2O (10 ml for ~200 ml cultures) and add 4 µl of 10 mg/mL lysozyme and 100 µL of 10% Triton X-100. Keep on ice for 15–30 min.
4. The cell lysate is then acidified with 345 µl of HCl (11.6 M) to reach 0.4 M final to solubilize the acid-soluble material from the bacteria. Keep on ice for 30 min (*see* **Note 2**).
5. Collect the inclusion bodies by centrifugation at 11,000×*g* for 10 min at 4 °C.
6. Solubilize the inclusion bodies in 10 ml of SAU_0 (*see* **Note 3**).
7. The solution is then acidified with 345 µl of HCl (11.6 M) and kept on ice for 30 min.
8. Clear the histone solution by centrifugation at 18,000×*g* for 30 min at 4 °C. Histones are soluble in acidic conditions; the precipitate contains nucleic acids and nonhistone proteins.
9. Transfer the histone solution in a 50 ml tube and neutralize with NaOH 10 M (*see* **Note 4**).
10. Equilibrate the column containing ~1–1.5 ml UNOsphere S beads in SAU_{200}.
11. Load the histone solution onto the column.
12. Wash the column to remove unbound proteins (*see* **Note 5**).
13. The histones are eluted with SAU_{1000} (*see* **Note 6**).
14. H3/H4 complex is formed by mixing stoichiometric amounts of H3 and H4 followed by dialysis overnight in histone dialysis buffer (*see* **Note 7**).
15. The solution is then centrifuged at 13,000×*g*, for 5 min at 4 °C, and the folded histone complex in solution is collected.

3.2 Preparation of H2A/H2B Histone Complexes

In contrast to the H3 and the H4, the histones H2A and H2B are typically soluble when expressed in bacteria, which eases the purification and preparation of the H2A/H2B dimer 20.

(**Steps 1–4** are identical to those reported above.)

1. 1 l Culture of BL21 *E. coli* transformed with the desired histone expression vector (we typically used pET3a vector) is grown at 37 °C in LB supplemented with 5 µg/ml ampicillin up to 0.4 OD_{600}. Then the culture is induced with 1 mM IPTG for 2 h.
2. Collect the cells by centrifugation at 3000×*g* for 10 min at 4 °C.
3. Resuspend the cells in ddH_2O (10 ml for ~200 ml cultures) and add 4 µl of 10 mg/mL lysozyme and 100 µL of 10% Triton X-100. Keep on ice for 15–30 min.

4. The cell lysate is then acidified with 345 μl of HCl (11.6 M) to reach 0.4 M final to solubilize the acid-soluble material from the bacteria. Keep on ice for 30 min (*see* **Note 2**).
5. The acidified cell lysate is then centrifuged at 18,000 × *g* for 30 min at 4 °C.
6. Harvest the supernatant and neutralize with 10 M NaOH.
7. H2A and H2B complexes are formed by mixing stoichiometric amounts of H2A and H2B (*see* **Note 7**). Transfer the mix onto a Biorex 70 column.
8. Wash the column with Biorex wash buffer.
9. Elute H2A/H2B complexes with Biorex elution buffer.
10. Histone complexes are then dialyzed against H_2O for subsequent incorporation into *Physarum* cells.

3.3 Culture of Physarum

Following spherule germination, *Physarum* is maintained as microplasmodia corresponding to cell suspension.

1. Plasmodia of *Physarum* are transferred into baffled flasks containing 50 ml of semi-defined media and 0.5 ml of hemin solution.
2. The maintenance of the cultures is done by successive passages of plasmodia, similar to cell lines. Typically, 1 ml of culture is transferred into 50 ml of fresh culture medium and the culture is shacked for ~3 days.
3. The preparation of *Physarum* stock is performed by induction of spherulation in saline media, and the transfer of the spherules onto sterile filter paper pieces.

3.4 Preparation of Physarum Macroplasmodia

1. Transfer *Physarum* suspension culture into a 15 ml conical tube, centrifuge the microplasmodia at 50 × *g* for 2 min, and note cellular volume of the pellet.
2. Wash the pellet of microplasmodia with sterile water by inverting the tube. This step allows the removal of slime produced by *Physarum* plasmodia. The microplasmodia are then centrifuged at 50 × *g* for 2 min. The cellular volume should be close to twice the initial volume, and adjust the volume to 2× the initial volume with sterile water.
3. Resuspend the microplasmodia by pipetting with a 2 ml pipette and drop 2 ml of the suspension onto a filter paper, which is placed at the surface of glass beads (3 mm diameter) within a Petri dish. Typically, the inoculums of *Physarum* are about 4–5 cm diameter.
4. Place the dish at 26 °C in the dark for 90 min. This step of starvation of the *Physarum* microplasmodia induces the coalescence of the plasmodia.

5. Add semi-defined medium with hemin underneath the filter paper. When using a 10 cm plastic Petri dish and 3 mm glass beads, we add 10.5 ml of culture medium. The dish is then placed at 26 °C in the dark (*see* **Notes 8–10**).

3.5 Determination of the Cell Cycle Stage

The perfect synchrony of a large population of nuclei within a single macroplasmodium of *Physarum* is a very useful tool for examining cell cycle-dependent features. However, this requires a perfect knowledge of the cell cycle stage at which experiments are carried out. The cell cycle in Physarum lasts ~9 h, and is composed of a 3-h S phase, a 6-h G2 phase, and ~15-min mitosis. Note that the plasmodial stage of the *Physarum* life cycle lacks a G1 phase. We use mitosis as a reference stage for determining the cell stage because pre-mitotic nuclear features leading to the establishment of mitosis are microscopically observable (*see* ref. 11).

1. Tiny explants of *Physarum* macroplasmodia are harvested and smeared on microscope slides.
2. The slide is then mounted with a solution of glycerol/ethanol (1:1) and observed under phase-contrast microscope.
3. Typically, we use the objective 100× to observe the nuclei.

3.6 Incorporation of Exogenous Proteins into Physarum

1. At the desired cell cycle stage, the filter paper supporting *Physarum* macroplasmodium is cut in equal parts and the filter paper exceeding the cellular surface is removed (*see* **Notes 11** and **12**).
2. The macroplasmodium fragments are carefully deposited onto glass beads of a dish with culture media.
3. The proteins of interest are deposited in small drops (20–40 μl) at the surface of the macroplasmodium fragments (*see* **Note 13**).
4. The drops are then uniformly spread over the cellular surface using a glass Pasteur pipette while being very careful not to damage the cell surface (*see* **Note 14**).

3.7 Isolation of Nuclei

The nuclei from *Physarum* macroplasmodium fragments are isolated using a method of Percoll gradient.

1. Wash the macroplasmodium fragment in 5 mM EDTA to remove unincorporated protein from the cellular surface.
2. Scrape the macroplasmodium with a microscope slide and transfer the cell fragment into a Dounce homogenizer.
3. Add 5 ml of homogenization buffer and homogenize the cell with the loose pestle (20 strokes).
4. Transfer the solution into a 15 ml conical tube and wash the homogenizer with 5 ml of homogenization buffer.
5. Centrifuge the cellular material at 700 × *g* for 5 min at 4 °C.

6. Remove the supernatant. Add 5 ml Percoll buffer and vortex to resuspend the pellet.
7. Transfer the solution into an ultracentrifuge tube and fill the rest of the tube with Percoll buffer.
8. Centrifuge at 48,000 ×*g* for 40 min at 4 °C in a Beckman ultracentrifuge with a Ti90 rotor (or equivalent) (*see* **Note 15**).
9. Transfer the nuclei which form a white-yellowish disc at the bottom of the tube into a 1.5 ml tube.
10. Wash the nuclei with homogenization buffer to remove any trace of Percoll. The nuclei are then pelleted by centrifugation at 700 ×*g* for 5 min at 4 °C. At this time, the pellet of nuclei can be resuspended in any buffer depending upon the subsequent experiments.
11. To ensure the efficiency of the isolation of nuclei, we typically quantify DNA by measuring the OD260. Aliquots of nucleus suspension are solubilized in 2 M NaCl and 5 M urea and the absorbance is measured in a spectrophotometer. Typically, nucleus concentration is adjusted to 800–1000 μg/ml.

3.8 Micrococcal Nuclease Digestion

1. Micrococcal nuclease digestion is carried out at room temperature in MNase buffer supplemented with 0.075 units of enzyme for different durations for determining the appropriated digestion. Typically, the digestion kinetics from 1 to 15 min is carried out using a 50 μl nucleus suspension aliquot and analyzed in agarose gel.
2. Digestion is stopped by adding 1/10 V of a 100 mM EDTA stock and the nuclei are incubated for 10 min on ice.
3. Soluble chromatin is recovered in the supernatant by centrifugation at 10,000 ×*g* for 5 min.
4. Subsequently, soluble chromatin can be analyzed by different techniques (electrophoresis, analytical ultracentrifugation, etc.).

4 Notes

1. We found that the *Physarum* growth in minimum medium is slower than with the semi-defined medium. Importantly, the hemin solution is absolutely required for *Physarum* growth.
2. Although histones are acid-soluble proteins, the inclusion bodies prevent their solubilization at this time.
3. It is possible to use a sonicator (Branson 50% pulse, 20% amplitude) to achieve the solubilization.
4. For subsequent purification using UNOsphere S (BioRad) the H3 solution is diluted twofold with SAU_0 reaching ~0.2 M NaCl, while the H4 solution can be used directly.

5. For H3 histone the washing buffer is composed of 20% SAU_{600} + 80% SAU_{200} (280 mM NaCl final). For H4 histone the washing buffer is composed of 40% SAU_{600} + 60% SAU_{200} (360 mM NaCl final).
6. OD_{280} is then measured and fractions above 0.1 are pooled.
7. Stoichiometry of individual histones is evaluated by OD_{280} and Coomassie-stained SDS-PAGE. If cysteine residues are present in the primary structure, add 5 mM 2-mercaptoethanol into the refolding buffer.
8. Typically, the second synchronous mitosis takes place after about 16–17 h and the third synchronous mitosis after about 24 h.
9. Commonly, we carry out the experiments between the second and the third mitosis. However, the population of nuclei is still perfectly synchronous between the third and the fourth mitosis. Afterwards, the culture medium begins to be exhausted and asynchrony within the nucleus population is observed.
10. The different fragments of *Physarum* resulting of the same macroplasmodium remain synchronous each other for more than one cell cycle.
11. There is no limitation in the number of pieces resulting of a single macroplasmodium. We usually cut one macroplasmodium between 2 and 8 fragments depending upon the number of different samples we want to examine with a unique cell.
12. The volume of the protein solution depends upon the macroplasmodium surface.
13. The solution of protein is buffered with 10 mM phosphate buffer pH 7.4 and ATP is added to 10 mM final concentration. For histones, we find that ATP significantly improves the incorporation. However, it is unclear whether ATP is utilized as an energy source or if the phospho groups of the ATP neutralize the histone charges.
14. We find that histones are rapidly uptaken by the cell and recovered into nuclei. Indeed, as early as 5 min after the deposition of the protein solution, the exogenous proteins are recovered into nuclei.
15. Alternatively if no ultracentrifuge is available, it is possible to carry out the procedure in an Eppendorf centrifuge at 25,000 × *g* for 40 min using 2 ml Eppendorf tubes.

Acknowledgments

This work is supported by grants of "La Ligue contre le Cancer" région grand ouest.

References

1. Allfrey VG, Faulkner R, Mirsky AE (1964) Acetylation and methylation of histones and their possible role in the regulation of RNA synthesis. Proc Natl Acad Sci U S A 51: 786–794
2. Huang H, Sabari BR, Garcia BA, Allis CD, Zhao Y (2014) SnapShot: histone modifications. Cell 159:458–458.e1
3. Jenuwein T, Allis CD (2001) Translating the histone code. Science 293:1074–1080
4. Strahl BD, Allis CD (2000) The language of covalent histone modifications. Nature 403: 41–45
5. Gardner KE, Allis CD, Strahl BD (2011) Operating on chromatin, a colorful language where context matters. J Mol Biol 409:36–46
6. Wolffe AP, Hayes JJ (1999) Chromatin disruption and modification. Nucleic Acids Res 27:711–720
7. Megee PC, Morgan BA, Mittman BA, Smith MM (1990) Genetic analysis of histone H4: essential role of lysines subject to reversible acetylation. Science 247:841–845
8. Megee PC, Morgan BA, Smith MM (1995) Histone H4 and the maintenance of genome integrity. Genes Dev 9:1716–1727
9. Hong L, Schroth GP, Matthews HR, Yau P, Bradbury EM (1993) Studies of the DNA binding properties of histone H4 amino terminus. Thermal denaturation studies reveal that acetylation markedly reduces the binding constant of the H4 "tail" to DNA. J Biol Chem 268:305–314
10. Wang X, Hayes JJ (2008) Acetylation mimics within individual core histone tail domains indicate distinct roles in regulating the stability of higher-order chromatin structure. Mol Cell Biol 28:227–236
11. Thiriet C, Hayes JJ (1999) Histone proteins in vivo: cell-cycle-dependent physiological effects of exogenous linker histones incorporated into Physarum polycephalum. Methods 17:140–150
12. Oppenheim A, Katzir N (1971) Advancing the onset of mitosis by cell free preparations of Physarum polycephalum. Exp Cell Res 68:224–226
13. Bradbury EM, Inglis RJ, Matthews HR, Langan TA (1974) Molecular basis of control of mitotic cell division in eukaryotes. Nature 249:553–556
14. Prior CP, Cantor CR, Johnson EM, Allfrey VG (1980) Incorporation of exogenous pyrene-labeled histone into Physarum chromatin: a system for studying changes in nucleosomes assembled in vivo. Cell 20:597–608
15. Prior CP, Cantor CR, Johnson EM, Littau VC, Allfrey VG (1983) Reversible changes in nucleosome structure and histone H3 accessibility in transcriptionally active and inactive states of rDNA chromatin. Cell 34:1033–1042
16. Ejlassi-Lassallette A, Mocquard E, Arnaud MC, Thiriet C (2011) H4 replication-dependent diacetylation and Hat1 promote S-phase chromatin assembly in vivo. Mol Biol Cell 22:245–255
17. Elliott OG, Murphy K, Hayes JJ, Thiriet C (2013) Replication-independent nucleosome exchange is enhanced by local and specific acetylation of histone H4. Nucleic Acids Res 4:2228–2238
18. Thiriet C (2004) Analysis of chromatin assembled in vivo using exogenous histones in Physarum polycephalum. Methods 33:86–92
19. Thiriet C, Hayes JJ (2005) Replication-independent core histone dynamics at transcriptionally active loci in vivo. Genes Dev 19:677–682
20. Hayes JJ, Lee KM (1997) In vitro reconstitution and analysis of mononucleosomes containing defined DNAs and proteins. Methods 12:2–9

Chapter 16

A Method for Large-Scale Screening of Random Sequence Libraries to Determine the Function of Unstructured Regions from Essential Proteins

Jean-François Millau*, Benoit Guillemette*, and Luc Gaudreau

Abstract

In this chapter we present a method allowing the screening of random sequences to discover essential aspects of unstructured protein regions in yeast. The approach can be applied to any protein with unstructured peptide sequences for which functions are difficult to decipher, for example the N-terminal tails of histones. The protocol first describes the building and preparation of a large library of random peptides in fusion with a protein of interest. Recent technical advances in oligonucleotide synthesis allow the construction of long random sequences up to 35 residues long. The protocol details the screening of the library in yeast for sequences that can functionally replace an unstructured domain in an essential protein in vivo. Our method typically identifies sequences that, while being totally different from the wild type, retain essential features allowing yeast to live. This collection of proteins with functional synthetic sequences can subsequently be used in phenotypic tests or genetic screens in order to discover genetic interaction.

Key words Histone tails, Synthetic biology, H3, Chromatin, Nucleosome, Genetic screen, Histone modification, Yeast

1 Introduction

Deciphering the function of proteins and unstructured protein domains often proves to be difficult as we are studying biological systems that have evolved over billions of years. Essential functions often show redundancy among proteins or protein domains as a fail-safe mechanism. A perfect example of this is the N-terminal tail of the histone H3. This sequence of roughly 40 amino acids is the target of a plethora of posttranslational modifications involved in important cellular function such as gene transcription, replication, and DNA repair [1]. However, despite all the experimentation done thus far, several observations challenge our understanding of the H3 tail function. First, although the conservation of the peptide

*These authors contributed equally to this work.

Benoit Guillemette and Luc Gaudreau (eds.), *Histones: Methods and Protocols*, Methods in Molecular Biology, vol. 1528,
DOI 10.1007/978-1-4939-6630-1_16, © Springer Science+Business Media New York 2017

sequence in the H3 tail is virtually perfect from yeast to mammals, several histone modifications observed in mammals are not present in yeast species. Second, no single modification seems to be essential and there is redundancy in these modifications not only within the same histone tail but also between different histones [2, 3], especially when lysine acetylation is involved. Because of this, our ability to decipher the minimal essential features required for histone tails to be functional has been severely hampered.

In this chapter, we describe a new method allowing the screening of synthetic unstructured protein regions in yeast and use the H3 protein as an example. Synthetic H3 tails, while having totally different sequences from the wild-type tail, retain essential function allowing yeast to live, providing precious tools to study the biology of the H3 tail in a simplified context.

Essentially, we built a vector library containing hundreds of thousands of H3 in which random peptides replace the N-terminal tail (residues 1–32) in a centromeric *TRP1* plasmid. The screen then relies on a modified plasmid shuffle strategy based on pioneering work by the Grunstein lab [4]. A yeast strain, for which endogenous H3 and H4 genes have been deleted and are expressed from an inducible/repressible GAL1-10 promoter on a *URA3* centromeric plasmid, is transformed with the library. Transformed yeasts are then plated on a selective medium (without tryptophan) with glucose. In these specific conditions, the presence of glucose switches off the promoter driving the wild-type copies of H3 and H4 and the absence of tryptophan forces the selection of the vector carrying the copies of H3 with random tails. Only yeasts having a functional synthetic H3 tail selected from the random peptide collection will form colonies on this medium. Each colony is then isolated and patched on medium containing 5-FOA. This drug counter selects vectors containing *URA3* and the wild-type copies of H3 and H4. Only yeasts with a functional random H3 tail and no copy of wild-type H3 and H4 histones will pass this last selection step and can be stored in glycerol stock for subsequent sequencing of the synthetic tail and phenotypic assays (Fig. 1).

2 Materials

2.1 Enzymes

1. VeraSeq 2.0 High-Fidelity Polymerase (Enzymatics).
2. 20,000 U/ml AgeI-HF (NEB).
3. 10,000 U/ml ClaI (NEB).
4. Instant Ligase Sticky-end mastermix (NEB).

2.2 Solutions

All solutions are prepared in ultrapure filtered distilled water.

1. 10 mM dNTP mix.
2. VeraSeq Buffer 5× (Enzymatics).

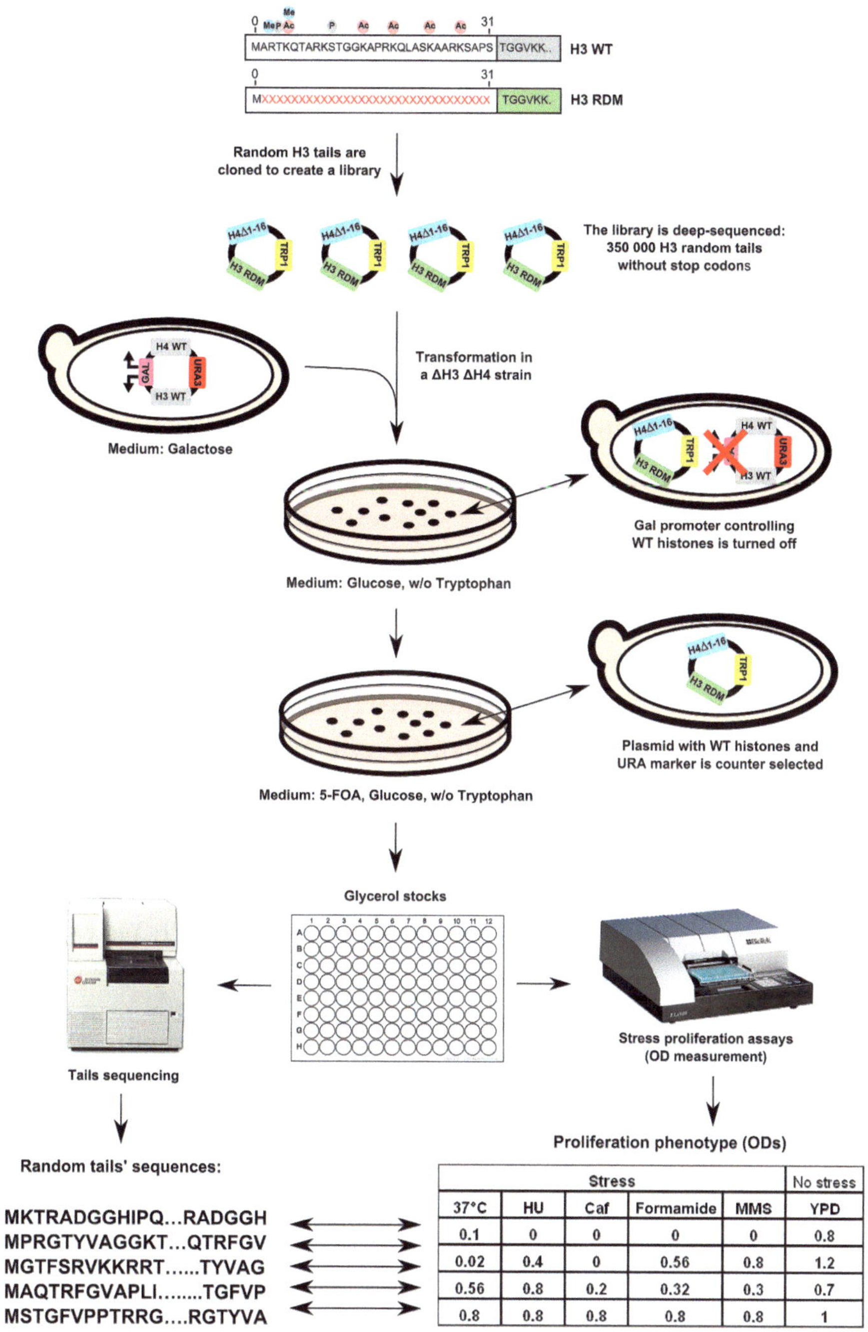

Fig. 1 H3 synthetic tail screen overview

3. CutSmart 10× buffer (NEB).
4. Buffer QG (Qiagen).
5. Isopropanol.
6. 70% Ethanol.
7. 6× DNA loading buffer: 20 mM Tris–HCl ph 8.0, 15% Ficoll, 0.9% Orange G.

8. Super optimal broth with catabolite repression (S.O.C.) medium: 2% Tryptone-peptone, 0.5% yeast extract, 10 mM NaCl, 2.5 mM KCl, 10 mM $MgCl_2$, 10 mM $MgSO_4$, 0.36% glucose (*see* **Note 1**).
9. LB: 25 g/L Lysogeny broth culture medium, autoclave.
10. LB agar ampicillin plates: 25 g/L LB, 2% agar, autoclave and add 100 μg/ml ampicillin when media is cooled down to approximately 55 °C. Pour in 100 mm or 150 mm plastic petri dishes.
11. Ampicillin 100 mg/ml filter sterilized.
12. 20% Galactose, filter sterilized.
13. 40% Glucose, autoclaved.
14. YPD: 1% Yeast extract, 2% peptone, 2% D-glucose (*see* **Note 1**).
15. YPD agar: 1% Yeast extract, 2% peptone, 2% D-glucose, 2% agar (*see* **Note 1**).
16. YPGal medium: 1% Yeast extract, 2% peptone, 4% galactose (*see* **Note 1**).
17. YPGal agar medium plates: 1% Yeast extract, 2% peptone, 2% agar, 4% galactose (*see* **Note 1**).
18. SD-trp medium: 6.7 g/L Yeast nitrogen base, 1 g/L amino acid dropout mix without tryptophan, 2% glucose (*see* **Note 1**).
19. SD-trp agar medium plates: 6.7 g/L Yeast nitrogen base, 1 g/L amino acid dropout mix without tryptophan, 2% agar, 2% glucose (*see* **Notes 1** and **2**).
20. SDGal-trp agar medium plates: 6.7 g/L Yeast nitrogen base, 1 g/L amino acid dropout mix without tryptophan, 2% agar, 4% galactose (*see* **Notes 1** and **2**).
21. SD-trp + FoA agar plates: 6.7 g/L Yeast nitrogen base, 1 g/L amino acid dropout mix without tryptophan, 2% agar, 2% glucose, 1 g/L 5-fluoro-orotic acid (added after autoclaving at 60 °C) (*see* **Notes 1** and **2**).
22. 50% PEG 3350: 50 g Polyethylene glycol 3350 in distilled water, 100 ml total volume. Autoclave or filter sterilize through a 0.2 μM filter with a syringe.
23. 1 M LiAc: Autoclave or filter sterilize through a 0.2 μM filter with a syringe.
24. 2 mg/ml Salmon sperm DNA (ssDNA) (*see* **Note 3**).

2.3 Bacteria and Yeast

1. Electrocompetent *E. coli* MM294 bacteria (*see* **Note 4**).
2. *Saccharomyces cerevisiae* strain BGY539: *MATa trp1-1 ura3-1 his3-11,15 leu2-3,112 ade2-1 can1-100 hht1-hhf1::LEU2 hht2-hhf2::kanMX3 [p211-URA3-GAL1-10-HHT2-HHF2]* (*see* **Note 5**).

2.4 DNA Purification Kits

1. QIAquick PCR Purification Kit (Qiagen).
2. QIAGEN Plasmid Maxi Kit (Qiagen).
3. Agencourt AMPure XP DNA purification system (Beckman Coulter).

2.5 Apparati and Consumables

1. Thermocycler.
2. Nanodrop spectrophotometer.
3. Magnetic separation rack for 1.5 ml tubes.
4. Electroporator (Bio-Rad Gene Pulser Xcell).
5. Electroporation cuvettes, 2 mm gap.
6. Microplate reader/spectrophotometer (BioTek).
7. Sterile round-bottom 96-well plates.
8. Sterile flat-bottom 96-well plates.
9. Parafilm.
10. Cell scraper.
11. 6.3 mm Glass beads for plating, autoclaved.
12. 150 mm Plastic petri dishes.

2.6 Custom Oligonucleotides

Order custom oligonucleotides using an appropriate service (*see* **Note 6** before ordering). All oligonucleotides are noted 5′ end to 3′ end.

1. H3 random tail oligo (*see* **Notes** 7 and **8**):
   CTGTGGCTCTGATTGGCTTTATCGATATGN
   NBNNBNNBNNBNNBNNBNNBNBNNBNNBNN
   BNNBNNBNNBNNBNNBNNBNNBNNBNNBNN
   BNNBNNBNNBNNBNNBNNBNNBNNBACCG
   GTGAGGAAGGGGATGGTAGGAG
   (N = A, T, G, C and B = T, G, C)
2. H3 random forward primer:
   CTGTGGCTCTGATTGGCTTT
3. H3 random reverse primer:
   CTCCTACCATCCCCTTCCTC
4. H3 deep Seq forward primer:
   AATGATACGGCGACCACCGAGATCTACACTCT
   TTCCCTACACGACGCTCTTCCGATCTtccaagc
   aaacactccacaa
5. H3 deep Seq reverse primer:
   CAAGCAGAAGACGGCATACGAGATGTGACTGGAG
   TTCAGACGTGTGCTCTTCCGATCtctgtgtggcttcttaacacca
   DNA sequences in capital correspond to the Illumina adaptors and in lower case to the pJD62-specific primers amplifying the H3 random tail library.

6. H3 sequencing forward primer (for Sanger sequencing): GGACCACTGTTTTGTGACTTC
7. H3 tail forward primer: CGCACAATCACGGCTATGG
8. H3 tail reverse primer: AGGCCTCAACAGATTCTTGC

3 Methods

3.1 Converting Single-Strand H3 Tail Random Oligo to Double-Stranded DNA

1. Assemble the following PCR mix in a 1.5 ml tube: 50 μl of VeraSeq 2.0 Buffer 5×, 5 μl of 10 mM dNTPs, 53.8 μl of 5 μM H3 random forward primer, 52.8 μl of 5 μM H3 random reverse primer, 3.3 μl of 5 μM H3 random tail oligo, 2.5 μl of VeraSeq 2.0, and 83.6 μl of H_2O.
2. Vortex, quick spin (5 s in a microcentrifuge), and divide the mix in 50 μl aliquots in 0.2 ml PCR tubes.
3. Incubate on a PCR machine using the following program: 98 °C for 2 min, followed by 30 cycles of the following three steps: 10 s at 98 °C, 30 s at 60 °C, and 30 s at 72 °C, followed by 5 min at 72 °C and a rest at 4 °C.
4. Verify that the reaction worked by running 5 μl of the PCR on a 2.5 % agarose gel; you should have a single band at 150 bp.

3.2 Purification of Double-Stranded H3 Random Tail DNA

1. Use the QIAquick PCR Purification Kit from QIAGEN to purify the double-stranded H3 random tail DNA.
2. Pool the five PCR reactions from Subheading 3.1, **step 3**, into one 1.5 ml tube.
3. Add 1250 μl of PB buffer, then mix, and quick spin in a microcentrifuge.
4. Load 650 μl onto a column and spin for 1 min at full speed in a microcentrifuge.
5. Discard flow-through and add the remaining 650 μl to the column.
6. Spin for 1 min at full speed in a microcentrifuge and then discard the flow-through.
7. Add 750 μl of PE wash buffer to the column and then spin at full speed for 30 s in a microcentrifuge.
8. Discard the flow-through and spin for an additional 2 min in a microcentrifuge to dry the column.
9. Transfer the column to a fresh 1.5 ml collection tube, add 43 μl of water, and spin 1 min at full speed in a microcentrifuge. Approximately 41 μl can be recovered.

3.3 Digestion and Purification of H3 Random Tail DNA (See Note 9)

1. To the 41 μl of double-stranded H3 random tail DNA from the previous section, add 5 μl of NEB CutSmart 10× buffer, 2 μl of AgeI-HF, and 2 μl of ClaI.
2. Mix and incubate for 4 h at 37 °C.
3. For DNA purification, use the QIAquick PCR Purification Kit from QIAGEN.
4. Add 250 μl of PB buffer to the reaction, then mix, and quick spin in a microcentrifuge.
5. Load the 300 μl onto a column and spin for 1 min at full speed in a microcentrifuge.
6. Discard flow-through, add 750 μl of PE wash buffer to the column, and then spin at full speed for 30 s in a microcentrifuge.
7. Discard the flow-through and spin for an additional 2 min in a microcentrifuge to dry the column.
8. Transfer the column to a 1.5 ml collection tube, add 50 μl of EB buffer, and spin for 1 min at full speed in a microcentrifuge.
9. Determine the DNA concentration using a NanoDrop. For the subsequent ligation carried out in Subheading 3.6, you will need 168 ng of ClaI-AgeI-HF-digested H3 random tail DNA. If you do not have this amount of DNA, redo the experiments described in Subheadings 3.1–3.3 until you get enough.

3.4 Digestion of the pJD62 Vector for Cloning H3 Random Tails (See Note 10)

1. Start the digestion of pJD62 with ClaI. Assemble the following reaction mix: 30 μg of pJD62 (30 μl at 1 μg/μl), 25 μl of NEB4 10× buffer, 25 μl of 10× BSA, 5 μl of ClaI, and 165 μl of H_2O.
2. Mix and incubate for 4 h at 37 °C.
3. Add 5 μl of ClaI, mix, and incubate for 4 h at 37 °C.
4. Purify ClaI-digested pJD62 using the QIAquick PCR Purification Kit from QIAGEN: first add 1250 μl of buffer PB to the pJD62 digestion, mix, and quick spin in a microcentrifuge.
5. Use 3 columns and load 500 μl per column.
6. Spin for 1 min at full speed in a microcentrifuge.
7. Discard the flow-through and add to each column 750 μl of PE buffer.
8. Spin for 1 min at full speed in a microcentrifuge.
9. Discard the flow-through and spin the columns for 2 min at full speed in a microcentrifuge in order to completely dry them.

10. Transfer the columns to 1.5 ml collection tubes, add 53 μl of H_2O, and spin for 1 min at full speed in a microcentrifuge.
11. Pool the eluates altogether for a final volume of 150 μl.
12. For the AgeI-HF digestion of pJD62, assemble the following mix: 150 μl of ClaI-digested pJD62, 25 μl of CutSmart 10× buffer, 5 μl of AgeI-HF, and 70 μl of H_2O.
13. Incubate overnight at 37 °C.
14. Add 5 μl of AgeI-HF, mix, and incubate for 4 h at 37 °C.

3.5 Purification of ClaI AgeI-HF-Digested pJD62

1. Prepare a 0.6% agarose gel with at least 7 wells able to accommodate 60 μl.
2. Add 50 μl of 6× DNA loading buffer to the ClaI AgeI-HF-digested pJD62 vector from Subheading 3.4 **step 14**.
3. Load 5 × 60 μl of digested vector on the 0.6% agarose gel, and migrate the digested pJD62 vector. As a control also run undigested pJD62 vector.
4. Cut the gel to isolate the bands containing the digested pJD62.
5. Transfer each band in a 1.5 ml tube, add 450 μl of QG buffer, incubate for 10 min at 50 °C, and vortex every 2 min.
6. Add 150 μl of isopropanol to each tube, mix, and quick spin at full speed in a microcentrifuge.
7. Transfer the content of each tube into QIAquick columns and spin for 1 min at full speed in a microcentrifuge.
8. Discard the flow-through, add 750 μl of PE wash buffer, and incubate for 5 min at room temperature.
9. Spin for 30 s at full speed in a microcentrifuge and discard the flow-through.
10. Add 750 μl of PE wash buffer and incubate for 5 min at room temperature.
11. Spin for 30 s at full speed in a microcentrifuge and discard flow-through.
12. Spin for 2 min at full speed in a microcentrifuge to dry the columns.
13. Transfer the columns to 1.5 ml collection tubes, open the column caps, and let the membrane dry for 5 min at room temperature.
14. Proceed to sequential elution of the columns: add 100 μl of EB buffer to one column. Spin for 1 min at full speed in a microcentrifuge and use the flow-through to sequentially elute the other columns.
15. Determine the DNA concentration using a NanoDrop. If the 260/230 ratio is below 2.0 proceed to an additional purification step (*see* **Note 11**). For the subsequent ligation carried in

Subheading 3.6, you will need 2100 ng of ClaI-AgeI-HF-digested pJD62 vector. If you do not have this amount of digested vector, redo the experiment described in Subheadings 3.4 and 3.5 until you get enough DNA (*see* **Note 12**).

3.6 H3 Random Tail Library Ligation

1. Assemble the following mix: 2100 ng of ClaI-AgeI-HF-digested pJD62 vector, 168 μl of ClaI AgeI-HF-digested H3 random tail DNA, complete volume to 210 μl with H_2O.
2. Divide in 5 μl aliquots in 0.2 ml PCR tubes (42 aliquots total).
3. Add 5 μl of Instant Ligase Sticky-End mix to each 5 μl of DNA and mix with 10 up-and-downs.
4. Incubate for 30 s at room temperature and then place on ice for 5 min.
5. Pool the 42 reactions together for a total volume of 420 μl and divide in 70 μl aliquots in 1.5 ml tubes (6 aliquots total).

3.7 H3 Random Tail Library Purification

1. Add 105 μl of Agencourt AMPure XP beads previously warmed up at room temperature to each tube containing 70 μl of ligation mix from the previous section. Care should be taken to resuspend the beads homogenously before pipetting.
2. Mix well by pipetting and incubate for 5 min at room temperature.
3. Incubate for 5 min on a magnetic rack.
4. While the tubes are still on the magnetic rack, carefully remove the liquid and add 800 μl of ethanol 70%.
5. Remove the ethanol and add 800 μl of ethanol 70%.
6. Remove all the ethanol and let the Agencourt AMPure XP beads dry for 8 min at room temperature.
7. Remove the tubes from the magnetic rack, resuspend the Agencourt AMPure XP beads into 75 μl of H_2O, and incubate for 5 min at room temperature.
8. Place the tubes on the magnetic rack and incubate for 5 min.
9. Collect 70 μl from each tube and pool all the samples altogether for a final volume of 420 μl.

3.8 Transformation of the H3 Random Tail Library

1. Before doing the transformation, make sure that you already have prepared 150 mm LB agar ampicillin plates (at least 70).
2. Prepare 35 electroporation cuvettes filled with 350 μl of electrocompetent MM294 cells, and keep on ice (*see* **Note 4**).
3. Add 12 μl of ligated library from the previous section to each cuvette and gently mix by pipetting using a 1 ml pipet.
4. Electroporate at 2.5 kV on a Bio-Rad Gene Pulser Xcell.
5. Add 700 μl of SOC to each cuvette.

6. Pool all the electroporated bacteria in a 100 ml sterile flask, then add 35 ml of LB for a final volume of 70 ml, and mix.
7. Plate 1 ml of electroporated bacteria per 150 mm LB-ampicillin plate.
8. Incubate the plates overnight at 37 °C.
9. The next day count the number of colonies per plates. You should have 10,000 colonies per plate and thus a total of 700,000 H3 random tail clones.

3.9 H3 Random Tail Vector Library Maxiprep

1. Prepare 110 ml of fresh LB + medium 100 μg/ml ampicillin.
2. Add 1.5 ml of LB medium + 100 μg/ml ampicillin per plate from Subheading 3.8 **step 9** and gently scrape the colonies using a flexible cell scraper.
3. Collect the cell suspension and pool in a 250 ml centrifugation bottle.
4. Centrifuge for 15 min at 6000 × *g* at 4 °C.
5. Discard supernatant, weigh the bacteria pellet, and resuspend the cells in P1 buffer from the QIAGEN Plasmid Maxi Kit accordingly: add 10 ml of P1 per 1.5 g of bacteria.
6. Perform the maxiprep as indicated in the kit protocol but adjust P2 and P3 volumes accordingly to the P1 volume. At the end of the kit protocol, dissolve the DNA pellet in 500 μl of H_2O and dose using a NanoDrop. Adjust the library concentration to 1 μg/μl.

3.10 Deep Sequencing of the H3 Random Tail Library

1. Assemble the following reaction mix: 20 μl VeraSeq 2.0 5× Buffer, 2 μl of dNTP 10 mM, 4 μl of 5 mM H3 deep Seq forward primer, 4 μl of 5 μM H3 deep Seq reverse primer, 1 μl of VeraSeq 2.0 polymerase, 200 ng of H3 random tail vector library, complete to 100 μl with H_2O.
2. Vortex, quick spin, and incubate on a PCR machine using the following program: 2 min at 98 °C, 11 cycles of the following steps: 10 s at 98 °C, 30 s at 60 °C, and 30 s at 72 °C, followed by 5 min at 72 °C and a rest at 4 °C.
3. Run the PCR product on a 2.5% agarose gel; a single band around 270 bp should be obtained (*see* **Note 13**).
4. Purify the PCR product using the QIAGEN QIAquick Gel Extraction Kit:
5. Cut the gel and isolate the band in a 1.5 ml tube.
6. Add 450 μl of QG buffer. Incubate at 50 °C for 10 min and vortex every 2 min.
7. Add 150 μl of isopropanol, mix, and quick spin in a microcentrifuge.

8. Load onto a QIAquick column and spin for 1 min at full speed in a microcentrifuge.
9. Discard the flow-through, add 750 μl of PE wash buffer, and incubate for 5 min at room temperature.
10. Spin for 30 s at full speed in a microcentrifuge and discard the flow-through.
11. Add 750 μl of PE wash buffer and spin for 30 s at full speed in a microcentrifuge.
12. Discard the flow-through and spin the column for 2 min at full speed in a microcentrifuge.
13. Transfer the column to a 1.5 ml collection tube, elute the PCR product by adding 30 μl of EB elution buffer, and spin the column for 1 min at full speed in a microcentrifuge.
14. Quantify the concentration of the DNA using a NanoDrop. If the ratio 260/230 is below 2, repeat **steps 6–14** and elute with 30 μl of EB elution buffer.
15. The H3 random tail library can then be directly sequenced on any Illumina sequencing platform compatible with the paired-end TruSeq chemistry. Only few nanograms of DNA are required for the sequencing. We recommend at least 3.5 million reads for sequencing the library, which ensures a 5× coverage of the 700,000 random tails.

3.11 Bioinformatic Analysis

1. The bioinformatic analysis is carried out as follows. First, because the H3 random tail library is sequenced paired-end, the R2 reads (complementary to R1) are reverse complemented.
2. Then R1 and R2 reads are pooled together. Subsequently, the H3 random tail sequences are isolated by recovering the sequence comprised between TCCAAGCAAACA CTCCACAATCGATATG and ACCGGTGGTGTTA AGAAGCCACACAGA.
3. Next to ensure that the sequenced H3 random tails do not contain errors, only tails for which identical sequences were found more than one time in the pool are kept. This means that these sequences were at least sequenced twice and each time the same sequence was obtained.
4. The remaining H3 random tail sequences are then converted into amino acid sequences; typically we obtained approximately 525,000 unique sequences from the 700,000 clones.
5. Finally, the sequences that do not contain stop codons are isolated and will be subsequently used as a reference population against which screened and functional H3 random tails will be compared.

3.12 Transformation of the Library in Saccharomyces cerevisiae

Our protocol for the transformation of the library is adapted from the high-efficiency protocol by Gietz and Schiestl [5]. We use a yeast strain (BGY539 *MATa trp1-1 ura3-1 his3-11,15 leu2-3,112 ade2-1 can1-100 hht1-hhf1::LEU2 hht2-hhf2::kanMX3 [p211-URA3-GAL1-10-HHT2-HHF2]*) in which both copies of endogenous genes encoding H3 and H4 are deleted. This strain survives with a plasmid bearing the H3- and H4-encoding genes (*HHT2* and *HHF2*, respectively) under control of the divergent *GAL1-10* promoter, and a *URA3* selection marker. This strain is only viable when it is grown in the absence of glucose and in the presence of galactose.

1. Streak a YPGal agar plate from a glycerol stock of BGY539. Incubate at 30 °C for 3–5 days, until colonies reach 1–2 mm in diameter. Make sure that there are isolated colonies on your plate.
2. Pick five to six isolated colonies from a freshly streaked plate and inoculate a starter culture of 50 ml of YPGal medium in a 250 ml flask. Incubate overnight (16–20 h) at 30 °C with shaking at 200–250 rpm.
3. From the starter culture, inoculate 600 ml of YPGal medium in order to reach an OD_{600nm} of 1.0–1.5 (*see* **Note 14**).
4. Split the culture in 300 ml bottles and centrifuge at 4000 × *g* for 5 min in a JA-10 rotor at room temperature.
5. Discard supernatant and add 250 ml of sterile distilled water at room temperature. Resuspend the pellet by shaking the bottle vigorously.
6. Centrifuge at 4000 × *g* for 5 min at room temperature.
7. Discard supernatant, resuspend in 15 ml of sterile distilled water, and pool both samples together in a single 50 ml conical tube.
8. Centrifuge at 4000 × *g* for 5 min in a tabletop centrifuge at room temperature.
9. Discard supernatant and resuspend in a volume of sterile distilled water depending on the density of cells harvested. Follow the following equation: $V_{H2O} = 10 \times OD_{600nm} \times V_C$, where V_{H2O} is the volume of water to add to the pellet in μl, OD_{600nm} is the optical density of the culture at the time of harvest, and V_C is the volume of culture at the time of harvest. For example, the volume for a culture of 600 ml harvested at OD_{600nm} of 1.0 is 6 ml.
10. Prepare the following master mix: (for one reaction) 240 μl of 50% PEG 3350, 36 μl of 1 M LiAc, 50 μl of 2 mg/ml of ssDNA (*see* **Note 14**), 2 μg (1–34 μl) of H3 random tail DNA library, 100 μl of yeast suspension from **step 9**, and H_2O (1–33 μl) for a total volume of 460 μl. We typically prepare a master mix for 70–90 reactions (between 140 and 180 μg of plasmid library) for a 600 ml culture.

11. Divide the master mix in 500 μl aliquots in sterile 1.5 ml tubes.
12. Place tubes in a 42 °C water bath, and incubate for 45 min.
13. Spin at 5000 × *g* (6000 rpm) in a microcentrifuge for 3 min at room temperature.
14. Discard supernatant and resuspend in 200 μl of sterile distilled water. Let the aliquots rest at room temp for 10 min; this allows the yeast pellet to soften before performing up/down with the 1 ml pipette.
15. Spread each aliquot on an SD-trp agar plate using 12–15 sterile glass beads and shaking sideways back and forth to spread evenly. Optionally, let the plates dry open with the beads in a laminar flow hood until no trace of moisture is evident. If no flow hood is available, simply let the plates dry with the cover opened 1/10th on a clean bench. Once all the liquid has been absorbed in the agar, discard the beads by inverting the plate with a swift movement of the wrist.
16. To measure the efficiency of transformation, we use one of the aliquots and plate on SDGal-trp agar. This is a selective media (without tryptophan) which allows the expression of the WT copies of H3 and H4 from the original plasmid in the strain. From the aliquot, perform a 1/5 and a 1/10 dilution in 250 μl of water in a separate tube and spread on SDGal-trp agar.
17. Incubate at 30 °C for 5–10 days. Wrap plates in parafilm to avoid drying of the agar.
18. Count colonies on the SDGal-trp and convert to colony-forming units per μg (CFU/μg) of plasmid library. We typically obtained 10,000–15,000 CFU/μg with our library and strain.

3.13 Selection and Isolation of Clones Carrying Synthetic H3 Tails

Plates are incubated until colonies are at least 1 mm in diameter. Because our screen generated strains with variable fitness, certain colonies grew much more rapidly than others. We harvested colonies as soon as they reached approximately 1 mm in diameter and placed the plates back to incubate at 30 °C for other colonies to grow.

1. Pick a single colony using a sterile flat toothpick. Spread in a 1 cm × 1 cm square "patch" on an SD-trp + FoA agar plate. This causes a counter selection of the *URA3*-bearing plasmid and ensures that the survival of the clone is independent of the original plasmid.
2. Let the patches grow for 3–5 days at 30 °C. At this point, certain colonies might not grow into patches; these are simply discarded.
3. From the patches, inoculate 5 ml of SD-trp medium and incubate at 30 °C with shaking. Inoculate at an initial OD_{600} of approximately 0.2. At this point, only strains that grow back to an OD_{600} of at least 1.0 are kept for further analyses. This may take up to 7 days in certain cases.

4. On the day of harvest, prepare a sterile flat-bottom 96-well plate containing 200 μl of sterile 50% glycerol in each well. This will serve as storage and backup for the new yeast clones. Also prepare a fresh block of dry-ice snow in a cooler for snap-freezing the plate.
5. When liquid cultures reach an OD_{600} of at least 1.0, centrifuge the whole 5 ml at 4000 × *g* at room temperature for 5 min.
6. Discard the supernatant by carefully inverting the tube over a waste bucket (always in sterile conditions). This usually leaves a few drops of media in the tube (approximately 200–400 μl).
7. Using a clean filtered tip, resuspend the yeast pellet in the media remaining in the tube and transfer 200 μl to a specific well in the 96-well plate. Mix by up-and-down with the 50% glycerol already in the well.
8. In parallel, add 50 μl of the yeast suspension to the wells of a 96-well PCR plates that will serve for colony PCR. This plate is then kept at −80 °C.
9. Repeat for every strain that has grown in **step 5**. We typically fill the first 90 wells of the plate with different strains, leaving six empty wells (H7–H12) that will serve for control strains during the growth assays (*see* **Note 15**).
10. To flash-freeze the plate, put the 96-well plate on a bed of snowy dry-ice and leave for 5 min, and then store the plate at −80 °C (*see* **Note 16**).

3.14 Colony PCR for Sequencing of the Synthetic H3 Tail Clones

For our initial screen, we sequenced 900 clones from ten plates by standard Sanger sequencing (via a commercial sequencing service) of the PCR products obtained by resuspending yeast directly in the PCR reaction mix (colony PCR).

1. For each PCR reaction, prepare the following mix (50 μl total): 10 μl of VeraSeq 2.0 5× Buffer, 1 μl of dNTP 10 mM, 0.5 μl of 100 μM H3 tail forward primer, 0.5 μl of 100 μM H3 tail reverse primer, 0.25 μl of VeraSeq 2.0 polymerase, and 32.75 μl H_2O. Prepare a master mix for the total amount of reactions plus three. Split the master mix in 45 μl aliquots in a PCR reaction plate. We then use a multichannel pipette to transfer 5 μl of yeast suspension from the plate prepared in Subheading 3.13 **step 8**. Mix with several up-and-downs, and do not centrifuge.
2. Incubate in a PCR machine using the following program: 95 °C for 5 min, followed by 35 cycles of the following three steps: 30 s at 95 °C, 30 s at 60 °C, and 1 min at 72 °C, followed by 5 min at 72 °C and a rest at 4 °C.
3. Verify that the reaction worked by running 5 μl of the PCR on a 2.5% agarose gel; you should have a single band of 513 bp.

4. At this point, centrifuge at 2000 × *g* and transfer 20 μl of the PCR reaction to a fresh plate. We sent the PCR samples for standard DNA purification and Sanger sequencing with the H3-seq primer.
5. To decrease the false-positive discovery rate, we recommend keeping sequences that have been selected at least twice.

3.15 Phenotypic Analyses of the Strains Carrying Synthetic H3 Tails

Large-scale phenotypic analyses can be performed by replicating the strains into 96-well plates containing fresh media combined with spectrophotometry using a plate reader. Monitoring cell density at fixed time points is a convenient way to measure phenotypes simultaneously over a wide range of conditions (temperature stress, osmotic stress, DNA damage, etc.). Because many synthetic-H3 mutant strains tend to flocculate, we devise a method that avoids the need to resuspend the yeast cultures in the plate by simply letting them settle in a round-bottom well (i.e., without shaking). The problem is that even strong shaking is often not sufficient to resuspend a flocculating strain homogeneously, resulting in large variations in density even from repeated measurements and variable access to oxygen and nutrients, especially when comparing strains with different levels of flocculation. By letting all cultures settle (flocculating or not), there is no bias in measurement and oxygen or nutrient supply. The results obtained by this method are semiquantitative at best, as differences in growth rate are lost when strains reach saturation.

1. Prepare a round-bottom 96-well plate for pre-cultures that will be used for inoculating experimental stress plates. Add 200 μl of YPD to the bottom of the plate.
2. Slowly thaw the glycerol stock plate prepared in Subheading 3.13 by leaving it on ice (*see* **Note 17**).
3. Using a multichannel pipette, transfer 10 μl of glycerol stock into the pre-culture well. For simplicity, maintain well numbers between plates as much as possible. You can also inoculate control cultures (wild-type or selected mutants), but always leave at least one media-containing well as a blank.
4. Wrap the plates with parafilm to avoid evaporation and leave the plates to grow at 30 °C for several days, until all strains reach stationary phase.
5. For each experimental growth plate, prepare 20 ml of growth media in a sterile 50 ml conical tube (*see* **Note 18**).
6. Transfer 200 μl of media to a sterile round-bottom 96-well plate.
7. Using a multichannel pipette, transfer 5 μl of pre-culture into the experimental plate well. Mix with several up-and-downs.
8. Wrap the plates with parafilm to avoid evaporation and incubate at 30 °C (or any indicated stressful temperature).

9. Monitor growth by reading the optical density at 600 nm on a plate reader at fixed time points. We typically read at 24, 48, and 72 h.
10. For analyses, normalize density at different time points in experimental media to growth in YPD. A sensitivity threshold should be chosen according to the distribution of the data.

4 Notes

1. Growth media containing any type of sugar (glucose or galactose) must not be autoclaved because of the risk of producing Maillard reactions between amino acids, peptides, and sugars. These media can either be filter sterilized or prepared as follows: prepare separately the sugar solution and the rest of the media. For the media, mix all the components except the sugar in a reduced volume of water, leaving a volume space for the sugar solution. Autoclave the sugar and media separately. Add the sugar solution to the media just before use.
2. Never autoclave the minimal media with the agar; this can result in agar breakdown and lead to "soft" plates. Instead, prepare a solution of 4% agar and a solution of 2.5× SD-trp (keeping a volume space for the sugar solution), autoclave separately, and mix before pouring into plates.
3. Salmon sperm ssDNA 2 mg/ml (ssDNA): Resuspend 1 g of salmon sperm DNA in 50 ml of H_2O. Place the solution on ice and sonicate at 50% power for 5 min. Sonicate until there is a uniform and obvious decrease in viscosity, but do not allow the solution to heat up. If heating occurs, rest the solution for several minutes before sonicating again. After sonication, dilute the DNA with H_2O to a final concentration of 2 mg/ml. Prepare 500 μl aliquots in 1.5 ml tubes. We usually prepare one hundred 500 μl aliquots and freeze the rest of the solution in 50 ml aliquots in conical tubes. When required, 500 μl aliquots are boiled for 5 min and rapidly chilled on ice before use for transformation. For optimal transformation efficiency, use 500 μl ssDNA aliquots only once.
4. Protocol to prepare electrocompetent MM294 cells:
   (a) Inoculate 100 ml of LB with a single colony isolated from a freshly streaked plate of MM293. Incubate overnight at 37 °C.
   (b) Dilute the pre-culture in 4000 ml of LB and grow the culture until exponential log phase (OD_{600} between 0.4 and 0.6).
   (c) Transfer the bacterial culture into 8 × 500 ml centrifuge bottles.
   (d) Incubate on ice for 30 min.
   (e) Centrifuge at 3500 × *g* for 15 min at 4 °C.

(f) Remove the supernatant and resuspend the bacterial pellet from each bottle in 500 ml of cold sterile water.

(g) Centrifuge at 3500 × *g* for 15 min at 4 °C.

(h) Remove the supernatant (carefully as the pellet is fragile) and resuspend the bacterial pellet from each bottle in 350 ml of cold sterile water.

(i) Centrifuge at 3500 × *g* for 15 min at 4 °C.

(j) Remove the supernatant and resuspend the pellet from each bottle in 60 ml cold sterile 10% glycerol solution. Pool together and divide in 10 × 48 ml into 50 ml Falcon tube.

(k) Centrifuge at 2800 × *g* for 15 min at 4 °C.

(l) Remove the supernatant, then resuspend, and pool all the bacterial pellets in 8 ml of cold sterile 10% glycerol solution. You should have a total volume of 16 ml.

(m) Aliquot the electrocompetent cells by 300 μl in 1.5 ml tubes and immediately snap-freeze in liquid nitrogen.

(n) We highly recommend testing the bacteria competency before preparing the bank by electroporating 50 μl of bacteria with 1 μg of pUC19. A minimum of 1×10^9 colonies per μg should be obtained.

5. This specific strain is used to screen for viable synthetic H3 from a collection of random N-terminal tails. Our protocol can be adapted to screen any part of any protein or DNA as long as it is essential for growth on a given plate media in your strain of interest.

6. The synthesis of the H3 random tail oligo is custom made in order to minimize biases during the synthesis of the random part of the sequence. One wants the random sequence to be perfectly random with no bases overrepresented. During the synthesis you need to ensure the following:

   (a) The dNTPs are pooled together in the same bottle in order for them to be injected exactly at the same time in the reaction chamber. If the dNTPs are in four different bottles a bias occurs during the injection step, as each dNTP will not reach the reaction chamber at the same time. This can cause a base to be overrepresented within the random sequence.

   (b) When the dNTPs are pooled together, the ratio of each dNTP needs to be adjusted as each dNTP does not react with the same efficiency during the synthesis chemical reaction. This also may result in a base to be overrepresented within the random sequence.

   Most companies specialized into DNA synthesis will be able to do this type of synthesis and have the know-how to deal with these biases if asked.

7. The protocol depicted in this chapter is designed to replace 30 amino acids from positions 2 to 31 of the wild-type H3 tail and consequently preserve the original length of the tail. However, it is possible to screen for shorter tails as our screen results showed that sequences as short as six amino acids, obtained serendipitously, can be functional.
8. We use NNB instead of NNN in the random part of the sequence to prevent the occurrence of the TAA and TGA stop codons in the open reading frame 1. When using NNB, statistically 52% of 31-amino acid-long sequences will not have a stop codon instead of 22.5% when using NNN.
9. During the digestion of the H3 random tail DNA it is important to note that certain random sequences will have AgeI and ClaI site. These few sequences will be cleaved and will generate shorter random tails.
10. Although the digestion of the pJD62 vector could be done in one step (AgeI-HF and ClaI are compatible), we prefer to do the digestion in two steps. We noted that it gives better digestion efficiency. This is very important as any vector not totally digested will generate false-positive clone during the preparation of the library.
11. For the library ligation, the purity of the vector is very important in order to have the maximum efficiency. If the 260/230 ratio for your plasmid preparation is below 2, it is highly recommended to proceed to the following purification step using the QIAquick PCR purification kit:
    (a) Add 500 μl of PB buffer to the 100 μl of digested vector from Subheading 3.5, **step 15**, mix, and quick spin.
    (b) Load on a QIAquick column and spin for 1 min at full speed.
    (c) Discard flow-through.
    (d) Add 750 μl of PE wash buffer and incubate for 5 min.
    (e) Spin for 30 s at full speed and discard the flow-through.
    (f) Spin for 2 min at full speed to dry the column.
    (g) Elute with 50 μl of EB elution buffer.
    (h) Dose using a NanoDrop. Your vector preparation 260/230 ratio should be superior to 2.
12. Before performing the library ligation, we highly recommend to test if the vector digestion was optimal by performing the experiments of Subheadings 3.6–3.8 in the absence of the H3 random tail DNA insert. The number of colonies obtained in Subheading 3.8 **step 9**, should be minimal (<100).
13. The PCR product band is "fuzzy" in an agarose gel because part of the amplicon is random and does not hybridize prop-

erly during the annealing step. This forms a "bubble" in the amplicon population that alters its migration within the gel.

14. You can estimate the initial volume of starter culture needed to obtain a target density by using the following formula: $V_I = (V_F \times OD_F)/(2n \times OD_I)$, where V_I is the volume of the starter culture to be inoculated, V_F is the desired final volume of culture, OD_F is the desired optical density of the exponential culture, n is the doubling time, and OD_I is the OD_{600nm} of the starter culture.
15. Care should be taken in identifying strains in the 96-well plates. Not only each plate and each well should be identified with a unique serial number, but details on the strains can also be noted in a spreadsheet, such as the date of selection, the size of the colony, and the time it took to obtain high density.
16. If you wish to add new colonies to an incomplete plate which has already been frozen, the following method can be used to avoid thawing the whole plate: resuspend the yeast pellet in a small volume of remaining media; pipet 200 μl of the suspension in a fresh and sterile 1.5 ml tube; add 200 μl of 50% glycerol and mix well; add the mix to the plate over snowy dry ice.
17. It is not advisable to repeatedly thaw/refreeze glycerol stocks, as this will decrease viability of the stock. Unfortunately, careful thawing was the only reliable way to obtain pre-cultures from all strains in a plate. For this reason, we advise to carefully plan ahead your phenotypic experiments and test as many stress conditions as possible using the same pre-culture.
18. We typically prepared a wide range of stress conditions to monitor strain phenotype. Many types of chemicals can be added directly to YPD at different concentrations, such as caffeine (0.5–8 mM), hydroxyurea (50–200 mM), camptothecin (1–10 μM), methyl-methane sulfonate (MMS, 0.001–0.05%), and formamide (0.5–3%). Other chemicals need specific media, such as various carbon sources or 6-azauracil (100 μg/ml in complete synthetic media). Other stresses can be imposed using YPD as the growth media: heat stress, cold stress, irradiation, etc.

References

1. Kouzarides T (2007) Chromatin modifications and their function. Cell 128:693–705
2. Kim JA, Hsu JY, Smith MM, Allis CD (2012) Mutagenesis of pairwise combinations of histone amino-terminal tails reveals functional redundancy in budding yeast. Proc Natl Acad Sci U S A 109:5779–5784
3. Zhang W, Bone JR, Edmondson DG, Turner BM, Roth SY (1998) Essential and redundant functions of histone acetylation revealed by mutation of target lysines and loss of the Gcn5p acetyltransferase. EMBO J 17:3155–3167
4. Ma XJ, Lu Q, Grunstein M (1996) A search for proteins that interact genetically with histone H3 and H4 amino termini uncovers novel regulators of the Swe1 kinase in Saccharomyces cerevisiae. Genes Dev 10:1327–1340
5. Gietz RD, Schiestl RH (2007) High-efficiency yeast transformation using the LiAc/SS carrier DNA/PEG method. Nat Protoc 2:31–34

Index

Benoit Guillemette and Luc Gaudreau (eds.), *Histones: Methods and Protocols*, Methods in Molecular Biology, vol. 1528, DOI 10.1007/978-1-4939-6630-1, © Springer Science+Business Media New York 2017

MIX
Papier aus verantwortungsvollen Quellen
Paper from responsible sources
FSC® C105338

If you have any concerns about our products,
you can contact us on
ProductSafety@springernature.com

In case Publisher is established outside the EU,
the EU authorized representative is:
Springer Nature Customer Service Center GmbH
Europaplatz 3, 69115 Heidelberg, Germany

Printed by Libri Plureos GmbH
in Hamburg, Germany